AF358742

Current Research in Pulsed Laser Deposition

Current Research in Pulsed Laser Deposition

Editors

Liviu Duta
Andrei C. Popescu

MDPI • Basel • Beijing • Wuhan • Barcelona • Belgrade • Manchester • Tokyo • Cluj • Tianjin

Editors
Liviu Duta
National Institute for Lasers, Plasma and Radiation Physics (NILPRP)
Romania
Andrei C. Popescu
National Institute for Lasers, Plasma and Radiation Physics (NILPRP)
Romania

Editorial Office
MDPI
St. Alban-Anlage 66
4052 Basel, Switzerland

This is a reprint of articles from the Special Issue published online in the open access journal *Coatings* (ISSN 2079-6412) (available at: https://www.mdpi.com/journal/coatings/special_issues/pulse_laser_depos).

For citation purposes, cite each article independently as indicated on the article page online and as indicated below:

LastName, A.A.; LastName, B.B.; LastName, C.C. Article Title. *Journal Name* **Year**, *Volume Number*, Page Range.

ISBN 978-3-0365-1044-6 (Hbk)
ISBN 978-3-0365-1045-3 (PDF)

Cover image courtesy of AC Popescu, L Duta, et al.
Journal Of Applied Physics 110, 064321 (2011).

© 2021 by the authors. Articles in this book are Open Access and distributed under the Creative Commons Attribution (CC BY) license, which allows users to download, copy and build upon published articles, as long as the author and publisher are properly credited, which ensures maximum dissemination and a wider impact of our publications.
The book as a whole is distributed by MDPI under the terms and conditions of the Creative Commons license CC BY-NC-ND.

Contents

About the Editors

Liviu Duta is a II-nd degree Senior Scientific Researcher at the National Institute for Lasers, Plasma and Radiation Physics, Lasers Department, "Lasers–Surface–Plasma" Interactions (LSPI) laboratory (http://lspi.inflpr.ro/People/CV%20Liviu%20Duta.html). His main activity fields and current research interests are related to laser interactions, lasers and plasma physics, nanostructured thin film deposition (by Pulsed laser deposition, PLD, and Matrix Assisted Pulsed Laser Evaporation), biomaterials and protective coatings, thin film characterization methods, surface physics and engineering, natural origin calcium phosphates as sustainable biofunctional coatings for medical applications, biomimetic metallic implants. Dr. Liviu Duta is one of the pioneers of simple and doped biological-derived calcium phosphate coatings fabricated by the PLD technique. He has written 4 book chapters and authored and co-authored more than 60 scientific publications, including +50 ISI papers, his work being cited more than 850 times in the literature (H index = 18). He is a member of the Editorial board and co-guest editor for two Special Issues of the journal Coatings. He has two Patents, he is an active reviewer for important international journals, evaluator of national projects and between 2009 and 2019 he was responsible for ten specialization stages of foreign students.

Andrei C. Popescu is the head of Center for Advanced Laser Technologies (CETAL), the newest research department of the National Institute for Lasers, Plasma and Radiation Physics. He worked for more than 10 years on laser thin film deposition techniques, dealing with organic and inorganic bioactive materials, superhydrophobic coatings, gas sensors layers or antimicrobial barriers. Currently, his interests lie in 3D printing of metallic and composite materials. Besides scientific activities, he and his team work with companies in Romania to help them implement additive manufacturing technologies in the production line. He is trying to stay close to his roots by combining the 3D printing of biomaterials with laser coating techniques to enhance the surface properties of additive manufactured parts. Andrei Popescu conducted research stages in prestigious European laboratories, and his research took the form of in 70 ISI papers, 6 book chapters, 6 national patent requests and 1 new technology internally approved in his institute. He is member of the editorial board of the journal Metals of MDPI and regularly presents his work in invited and oral presentations at international laser themed conferences

Editorial

Current Research in Pulsed Laser Deposition

Liviu Duta [1],* and **Andrei C. Popescu [2]**

[1] Lasers Department, National Institute for Lasers, Plasma, and Radiation Physics, 409 Atomistilor Street, 077125 Magurele, Romania
[2] Center for Advanced Laser Technologies (CETAL), National Institute for Lasers, Plasma and Radiation Physics, 077125 Magurele, Romania; andrei.popescu@inflpr.ro
* Correspondence: liviu.duta@inflpr.ro

check for
updates

Citation: Duta, L.; Popescu, A.C. Current Research in Pulsed Laser Deposition. *Coatings* **2021**, *11*, 274. https://doi.org/10.3390/coatings 11030274

Received: 19 February 2021
Accepted: 22 February 2021
Published: 26 February 2021

Publisher's Note: MDPI stays neutral with regard to jurisdictional claims in published maps and institutional affiliations.

Copyright: © 2021 by the authors. Licensee MDPI, Basel, Switzerland. This article is an open access article distributed under the terms and conditions of the Creative Commons Attribution (CC BY) license (https:// creativecommons.org/licenses/by/ 4.0/).

In industry, thin films proved invaluable for protection of tools withstanding high frictions and elevated temperatures, but also found successful applications as sensors, solar cells, bioactive coatings for implants, photocatalysis and in lithography. Other envisaged applications are currently researched for energy-storage devices, drug delivery or in situ microstructuring for enhancing the surface properties.

When compared to other plasma-assisted deposition methods, such as plasma spraying or plasma enhanced chemical vapor deposition, the Pulsed Laser Deposition (PLD) technique stands as a simple, versatile, rapid, and cost-effective approach for the fabrication of high-quality structures from a wide range of materials [1,2]. PLD as a thin film synthesis method, although limited in terms of surface covered area, still gathers interest among researchers due to some important advantages such as stoichiometric transfer of the target composition to the synthesized coatings, high adherence of the deposited structures to the substrate, improved uniformity in terms of morphology and chemical composition, controlled degree of phase, crystallinity, and thickness of deposited coatings, lower porosity, decreased tendency of the deposited structures to crack or delaminate, and relatively simple experimental set-up.

Over the last years, an increasing research interest was observed in the field of PLD synthesis of oxides and nitrides. From the first class, transparent conductive oxides, with indium tin oxide (ITO) as one of the most representative materials, were intensively investigated due to their applications in many technological areas, including sensors [3,4]. By combining UV nanoimprint lithography and PLD, the fabrication of nanopatterned ITO films was demonstrated to be appropriate for applications in the field of organic photovoltaic cells or organic light emitting devices. In addition, iron oxide (Fe_2O_3) nanostructures have been also studied due to their unique electrical and magnetic properties, corroborated with the preservation of their chemical compatibility with biological tissues for various medical applications. Thus, the formation of iron oxide nanostructures was demonstrated using combined infrared pulsed laser ablation and carbothermal heat treatment of Fe microspheres [5]. It was shown that iron oxide nanowires and nanoflakes could be synthesized without the necessity of a catalyst. A possible growth mechanism of iron oxide nanowires was introduced considering both the Fe_3O_4/Fe_2O_3 interface and the chemical driving force for the diffusion of Fe species through the microsphere, which are kinetically and thermodynamically controlled by the amount of Fe and carbon gaseous species. An important aspect that should be mentioned is related to the targets' preparation for PLD synthesis of iron-doped indium oxide films. Therefore, at low oxygen pressures, it was shown that, besides the existence of iron ions in the Fe^{2+} state, rather than the Fe^{3+} one, there was little metallic iron. All these characteristics were followed by an important magnetization. When increasing the oxygen amount, the band gap increased also, whilst the number of defect phases and the saturation magnetization decreased. It was demonstrated that the oxygen amount in the target due to the precursor was an important factor

1

but not a defining one for the overall quality of the films. It was also indicated that FeO, in comparison to Fe_2O_3, could be a better choice due to a higher magnetization.

It should be emphasized that thin though hard coatings demonstrated to be of key-importance for the fabrication of mechanical parts or tools due to their hardness and wear-resistance characteristics [6]. Therefore, nitride-based films were recently investigated as protective coatings, due to their physical–chemical, electronic, thermal, or mechanical properties [7–10]. Particularly, aluminum nitride coatings own such characteristics, which qualify them as relevant candidates for a wide range of applications, including photodetectors, light-emitting and laser diodes, acoustic devices, insulating and buffer layers, or designs of self-sustainable opto- and micro-electronic devices [11–13].

Solid-electrolyte thin-films are mainly fabricated by thermal vacuum evaporation and radiofrequency-sputtering techniques. It is important to note that their synthesis by the PLD method generated improvements in their intrinsic characteristics. Due to an increasing demand for smaller power sources, the current interest in rechargeable microbatteries has expanded. Therefore, the recent progress in the field of lithium microbatteries is quite remarkable, and this is mainly due to the PLD growth of high-quality, pinhole-free, solid-state electrolyte thin films (i.e., $Li_{6.1}V_{0.61}Si_{0.39}O_{5.36}$). In this respect, using the sequential PLD technique, rechargeable thin-film lithium-ion batteries were designed [14].

Aiming to expand the field of hard tissue engineering and regeneration, many biomaterials were designed and exploited in different strategies, either for controlled drug delivery [15], or to address various bone-related traumas [16]. For this aim, Matrix-Assisted Pulsed Laser Evaporation (MAPLE), which is a variation of the PLD technique, was recently used to deposit thin films of shellac for targeted drug release [17]. Thus, it was demonstrated that the laser fluence is a key parameter in preserving the chemical structure of shellac. In the case of low fluences, all the characteristic absorption bands were shown and the best concordance to the pristine shellac was indicated. One innovative approach to boost the efficiency of a biomaterial could be to coat it with a natural protein or a biological-derived compound. To the difference of injecting a bioactive compound, which has the disadvantage of a rapid removal from the living body, the deposition of a protein onto the surface of the implanted material can determine an increase in its concentration, which further results in a controlled local release. Post-surgical infections in bone implants could be therefore prevented [18], and a local induction of osteogenic differentiation can be also favored [15]. In this respect, lactoferrin (Lf) is a protein of interest, due to its antimicrobial and anticarcinogenic activities, corroborated with an anti-inflammatory function and an osteogenic role [19,20]. Thus, it was demonstrated that the controlled release of the components by polymeric coating could better accommodate the combination of the properties of both biological compounds, (i.e., the bone mechanical stability and osteogenic capacity of hydroxyapatite, HA) with the anti-inflammatory and osteogenic effect of the Lf component, which make the composition an excellent support for further bone regeneration.

Calcium phosphates (CaP) are the main inorganic component of bone tissues [21]. Due to their excellent biocompatibility, high adherence to the substrate, and osseointegration and osteoconduction characteristics, CaP-based bioceramic materials are widely used in the domain of bone regeneration, both in orthopedics and dentistry [22–25], mainly as coatings for metallic implants [26]. In the last few decades, research focus was put on HA, which represents the most frequently used CaP due to some important properties, like its role as a scaffold for osteogenic differentiation [27] and its capacity to stimulate and accelerate the formation of new bone around implants [28,29]. The advantages of using animal-origin HA (BioHA) coatings, fabricated by the PLD technique, as viable alternatives to synthetic HA ones for various medical applications, were recently emphasized [16]. In this respect, the biocompatibility of BioHA materials was reported to be superior to that of synthetic HA, with respect to cells' morphology, proliferation, and bioactivity [16]. Moreover, when doped with various agents (i.e., Ti or Li_2CO_3), the effect was an increase in cell proliferation [30,31]. Therefore, animal-origin HA could represent a reliable, promising, safe, and, most importantly, low-cost resource for coatings that aim for biomimetism. For

PLD, this domain is still in his incipient stage of development, offering endless possibilities for expansion in terms of natural-CaP new resources, various concentrations of doping elements, or controlled degree of the obtained coatings' morphology and structure. It was also demonstrated that, to the difference of synthetic HA, BioHA coatings could exhibit antimicrobial properties, besides an increased bioactivity. One could therefore conclude that, the future of biomimetic coatings might belong mainly to the natural-origin CaPs source materials, and their stoichiometric transfer and delicate tailoring in terms of thickness, crystallinity, and functional groups can be performed by PLD as an optimal deposition method. In addition, two decades of achievements focused on the in vivo performances of CaP-based coatings fabricated by the PLD technique were also reported [32]. One should note that, in the dedicated literature, there are studies on the in vivo testing of CaP-based coatings (especially HA) synthesized by numerous physical vapor deposition methods, but only few of them addressed the PLD technique. It was demonstrated that the values inferred from the mechanical tests (concerning bone attachment) were all superior for the functionalized Ti implants as compared to simple Ti ones. Both the PLD surface functionalization of metallic implants and a longer implantation time period were shown to have a key role on the overall bone bonding strength characteristics of the investigated medical devices. Despite the interesting progress made in vitro and in vivo in the field of CaP-coated metallic implants by PLD, not enough comparable clinical results were delivered so far for an easier assessment. This was mainly because of the lack of standardization of the coating properties and in vivo models. As a consequence, additional tests are still needed to be performed, both to advance a certain "recipe" for optimum in vivo results, and to evidence the relative influence of implant design, surgical procedure, and coating characteristics (thickness and surface structure and morphology), on the short- and long-term characteristics of the CaP-based synthesized structures.

Taking into consideration all these facts, one can conclude that a leap forward for PLD, which is mainly a laboratory technique, could be its recognition as an industrial coating technique in the near future. This is now more achievable than ever due to progress in robotics, which can help with automatic and precise translation of targets and substrates, but also to advances in vacuum techniques, allowing for generation of a high vacuum in large enclosed spaces.

Author Contributions: Writing—original draft preparation, L.D.; writing—review and editing, L.D. and A.C.P. All authors have read and agreed to the published version of the manuscript.

Funding: L.D. acknowledges the support from two grants of Ministry of Research and Innovation, CNCS-UEFISCDI, PN-III-P1-1.1-PD-2016-1568 (PD 6/2018) and PN-III-P1-1.1-TE2019-1449 (TE189/2021), and the Romanian Ministry of Education and Research, under Romanian National Nucleus Program LAPLAS VI—Contract 16N/2019.

Acknowledgments: The Guest Editors, L.D. and A.C.P., thank to all the authors for their contributions to this Special Issue "Current Research in Pulsed Laser Deposition" and to the editorial staff of the journal Coatings.

Conflicts of Interest: The authors declare no conflict of interest.

References

1. Chrisey, D.B.; Hubler, G.K. *Pulsed Laser Deposition of Thin Films*, 1st ed.; John Wiley & Sons: Hoboken, NJ, USA, 1994.
2. Eason, R. *Pulsed Laser Deposition of Thin Films-Applications-Led Growth of Functional Materials*; Wiley-Interscience: Hoboken, NJ, USA, 2006.
3. Socol, M.; Preda, N.; Rasoga, O.; Costas, A.; Stanculescu, A.; Breazu, C.; Gherendi, F.; Socol, G. Pulsed Laser Deposition of Indium Tin Oxide Thin Films on Nanopatterned Glass Substrates. *Coatings* **2019**, *9*, 19. [CrossRef]
4. Albargi, H.B.; Alshammari, M.S.; Museery, K.Y.; Heald, S.M.; Jiang, F.X.; Saeedi, A.M.A.; Fox, A.M.; Gehring, G.A. Relevance of the Preparation of the Target for PLD on the Magnetic Properties of Films of Iron-Doped Indium Oxide. *Coatings* **2019**, *9*, 381. [CrossRef]

5. De Vero, J.C.; Jasmin, A.C.; Dasallas, L.L.; Garcia, W.O.; Sarmago, R.V. Synthesis of Iron Oxide Nanostructures via Carbothermal Reaction of Fe Microspheres Generated by Infrared Pulsed Laser Ablation. *Coatings* **2019**, *9*, 179. [CrossRef]
6. Kolaklieva, L.; Chitanov, V.; Szekeres, A.; Antonova, K.; Terziyska, P.; Fogarassy, Z.; Petrik, P.; Mihailescu, I.N.; Duta, L. Pulsed Laser Deposition of Aluminum Nitride Films: Correlation between Mechanical, Optical, and Structural Properties. *Coatings* **2019**, *9*, 195. [CrossRef]
7. Jianxin, D.; Aihua, L. Dry sliding wear behavior of PVD TiN, Ti55Al45N, and Ti35Al65N coatings at temperatures up to 600 °C. *Int. J. Refract. Met. Hard Mater.* **2013**, *41*, 241–249. [CrossRef]
8. Jianxin, D.; Fengfang, W.; Yunsong, L.; Youqiang, X.; Shipeng, L. Erosion wear of CrN, TiN, CrAlN, and TiAlN PVD nitride coatings. *Int. J. Refract. Met. Hard Mater.* **2012**, *35*, 10–16. [CrossRef]
9. Liang, C.L.; Cheng, G.A.; Zheng, R.T.; Liu, H.P. Fabrication and performance of TiN/TiAlN nanometer modulated coatings. *Thin Solid Films* **2011**, *520*, 813–817. [CrossRef]
10. Cecchini, R.; Fabrizi, A.; Cabibbo, M.; Paternoster, C.; Mavrin, B.N.; Denisov, V.N.; Novikova, N.N.; Haïdopoulo, M. Mechanical, microstructural and oxidation properties of reactively sputtered thin CrN coatings on steel. *Thin Solid Films* **2011**, *519*, 6515–6521. [CrossRef]
11. Duta, L.; Stan, G.E.; Stroescu, H.; Gartner, M.; Anastasescu, M.; Fogarassy, Z.; Mihailescu, N.; Szekeres, A.; Bakalova, S.; Mihailescu, I.N. Multi-stage pulsed laser deposition of aluminum nitride at different Temperatures. *Appl. Surf. Sci.* **2016**, *374*, 143–150. [CrossRef]
12. Antonova, K.; Duta, L.; Szekeres, A.; Stan, G.E.; Mihailescu, I.N.; Anastasescu, M.; Stroescu, H.; Gartner, M. Influence of laser pulse frequency on the microstructure of aluminum nitride thin films synthesized by pulsed laser deposition. *Appl. Surf. Sci.* **2017**, *394*, 197–204. [CrossRef]
13. Fogarassy, Z.; Petrik, P.; Duta, L.; Mihailescu, I.N.; Anastasescu, M.; Gartner, M.; Antonova, K.; Szekeres, A. TEM and AFM studies of aluminium nitride films synthesized by pulsed laser deposition. *Appl. Phys. A* **2017**, *123*, 756. [CrossRef]
14. Julien, M.C.; Mauger, A. Pulsed Laser Deposited Films for Microbatteries. *Coatings* **2019**, *9*, 386. [CrossRef]
15. Icriverzi, M.; Rusen, L.; Brajnicov, S.; Bonciu, A.; Dinescu, M.; Cimpean, A.; Evans, R.W.; Dinca, V.; Roseanu, A. Macrophage in vitro Response on Hybrid Coatings Obtained by Matrix Assisted Pulsed Laser Evaporation. *Coatings* **2019**, *9*, 236. [CrossRef]
16. Duta, L.; Popescu, A.C. Current Status on Pulsed Laser Deposition of Coatings from Animal-Origin Calcium Phosphate Sources. *Coatings* **2019**, *9*, 335. [CrossRef]
17. Brajnicov, S.; Bercea, A.; Marascu, V.; Matei, A.; Mitu, B. Shellac Thin Films Obtained by Matrix-Assisted Pulsed Laser Evaporation (MAPLE). *Coatings* **2018**, *8*, 275. [CrossRef]
18. Nagano-Takebe, F.; Miyakawa, H.; Nakazawa, F.; Endo, K. Inhibition of initial bacterial adhesion on titanium surfaces by lactoferrin coating. *Biointerphases* **2014**, *9*, 029006. [CrossRef] [PubMed]
19. Moreno-Exposito, L.; Illescas-Montes, R.; Melguizo-Rodriguez, L.; Ruiz, C.; Ramos-Torrecillas, J.; de Luna-Bertos, E. Multifunctional capacity and therapeutic potential of lactoferrin. *Life Sci.* **2018**, *195*, 61–64. [CrossRef] [PubMed]
20. Garcia-Montoya, I.A.; Cendon, T.S.; Arevalo-Gallegos, S.; Rascon-Cruz, Q. Lactoferrin a multiple bioactive protein: An overview. *BBA Gen. Subj.* **2012**, *1820*, 226–236. [CrossRef] [PubMed]
21. Eliaz, N.; Metoki, N. Calcium Phosphate Bioceramics: A Review of Their History, Structure, Properties, Coating Technologies and Biomedical Applications. *Materials* **2017**, *10*, 334. [CrossRef]
22. Oladele, I.O.; Agbabiaka, O.; Olasunkanmi, O.G.; Balogun, A.O.; Popoola, M.O. Non-synthetic sources for the development of hydroxyapatite. *J. Appl. Biotechnol. Bioeng.* **2018**, *5*, 88–95. [CrossRef]
23. Akram, M.; Ahmed, R.; Shakir, I.; Ibrahim, W.A.W.; Hussain, R. Extracting hydroxyapatite and its precursors from natural resources. *J. Mater. Sci.* **2014**, *49*, 1461–1475. [CrossRef]
24. Šupová, M. Substituted hydroxyapatites for biomedical applications: A review. *Ceram. Int.* **2015**, *41*, 9203–9231. [CrossRef]
25. Graziani, G.; Boi, M.; Bianchi, M. A Review on Ionic Substitutions in Hydroxyapatite Thin Films: Towards Complete Biomimetism. *Coatings* **2018**, *8*, 269. [CrossRef]
26. Tite, T.; Popa, A.C.; Balescu, L.M.; Bogdan, I.M.; Pasuk, I.; Ferreira, J.M.F.; Stan, G.E. Cationic substitutions in hydroxyapatite: Current status of the derived biofunctional effects and their *in vitro* interrogation methods. *Materials* **2018**, *11*, 2081. [CrossRef] [PubMed]
27. Ballini, A.; Mastrangelo, F.; Gastaldi, G.; Tettamanti, L.; Bukvic, N.; Cantore, S.; Cocco, T.; Saini, R.; Desiate, A.; Gherlone, E.; et al. Osteogenic differentiation and gene expression of dental pulp stem cells under low-level laser irradiation: A good promise for tissue engineering. *J. Biol. Regul. Homeost. Agents* **2015**, *29*, 813–822. [CrossRef] [PubMed]
28. Surmenev, R.A.; Surmeneva, M.A. A critical review of decades of research on calcium phosphate–based coatings: How far are we from their widespread clinical application? *Curr. Opin. Biomed. Eng.* **2019**, *10*, 35–44. [CrossRef]
29. Dorozhkin, S.V. Biphasic, triphasic, and multiphasic calcium orthophosphates. In *Advanced Ceramic Materials*; Tiwari, A., Gerhardt, R.A., Szutkowska, M., Eds.; Wiley, Scrivener Publishing: Austin, TX, USA, 2016; pp. 33–95.
30. Popescu, A.C.; Florian, P.E.; Stan, G.E.; Popescu-Pelin, G.; Zgura, I.; Enculescu, M.; Oktar, F.N.; Trusca, R.; Sima, L.E.; Roseanu, A.; et al. Physical-chemical characterization and biological assessment of simple and lithium-doped biological-derived hydroxyapatite thin films for a new generation of metallic implants. *Appl. Surf. Sci.* **2018**, *439*, 724–735. [CrossRef]

31. Duta, L.; Mihailescu, N.; Popescu, A.C.; Luculescu, C.R.; Mihailescu, I.N.; Cetin, G.; Gunduz, O.; Oktar, F.N.; Popa, A.C.; Kuncser, A.; et al. Comparative physical, chemical and biological assessment of simple and titanium-doped ovine dentine-derived hydroxyapatite coatings fabricated by pulsed laser deposition. *Appl. Surf. Sci.* **2017**, *413*, 129–139. [CrossRef]
32. Duta, L. In Vivo Assessment of Synthetic and Biological-Derived Calcium Phosphate-Based Coatings Fabricated by Pulsed Laser Deposition: A Review. *Coatings* **2021**, *11*, 99. [CrossRef]

Article

Pulsed Laser Deposition of Indium Tin Oxide Thin Films on Nanopatterned Glass Substrates

Marcela Socol [1],*, Nicoleta Preda [1],*, Oana Rasoga [1], Andreea Costas [1], Anca Stanculescu [1], Carmen Breazu [1], Florin Gherendi [2] and Gabriel Socol [2]

[1] National Institute of Material Physics, 405A Atomistilor Street, 077125 Bucharest-Magurele, Romania; oana@infim.ro (O.R.); andreea.costas@infim.ro (A.C.); sanca@infim.ro (A.S.); carmen.breazu@infim.ro (C.B.)

[2] National Institute for Lasers, Plasma and Radiation Physics, 409 Atomistilor Street, 077125 Bucharest-Magurele, Romania; florin.gherendi@infim.ro (F.G.); gabriel.socol@inflpr.ro (G.S.)

* Correspondence: marcela.socol@infim.ro (M.S.); nicol@infim.ro (N.P.); Tel.: +40-21-241-8160

Received: 5 December 2018; Accepted: 25 December 2018; Published: 29 December 2018

Abstract: Indium tin oxide (ITO) thin films were grown on nanopatterned glass substrates by the pulsed laser deposition (PLD) technique. The deposition was carried out at 1.2 J/cm^2 laser fluence, low oxygen pressure (1.5 Pa) and on unheated substrate. Arrays of periodic pillars with widths of ~350 nm, heights of ~250 nm, and separation pitches of ~1100 nm were fabricated on glass substrates using UV nanoimprint lithography (UV-NIL), a simple, cost-effective, and high throughput technique used to fabricate nanopatterns on large areas. In order to emphasize the influence of the periodic patterns on the properties of the nanostructured ITO films, this transparent conductive oxide (TCO) was also grown on flat glass substrates. Therefore, the structural, compositional, morphological, optical, and electrical properties of both non-patterned and patterned ITO films were investigated in a comparative manner. The energy dispersive X-ray analysis (EDX) confirms that the ITO films preserve the In$_2$O$_3$:SnO$_2$ weight ratio from the solid ITO target. The SEM and atomic force microscopy (AFM) images prove that the deposited ITO films retain the pattern of the glass substrates. The optical investigations reveal that patterned ITO films present a good optical transmittance. The electrical measurements show that both the non-patterned and patterned ITO films are characterized by a low electrical resistivity (<2.8 × 10^{-4}). However, an improvement in the Hall mobility was achieved in the case of the nanopatterned ITO films, evidencing the potential applications of such nanopatterned TCO films obtained by PLD in photovoltaic and light emitting devices.

Keywords: PLD; ITO; nanoimprint lithography; coatings

---

1. Introduction

Transparent conductive oxides (TCO) have been intensively studied in recent years due to their applications in many technological areas, such as optoelectronics devices, automobile and aircraft windows, and antireflection coatings [1–4]. Among various metal oxides, ZnO, SnO$_2$, In$_2$O$_3$, and CdO have been extensively used in organic optoelectronic devices such as organic photovoltaics (OPVs) and organic light-emitting devices (OLEDs) [1,5–7]. Due to its unique optical (high transparency) and electrical (low electrical resistivity) features, indium tin oxide (ITO) remains the most investigated TCO [2,3]. Other ITO properties are linked to surface roughness, work function, mechanical characteristics, and environmental stability. The attributes of ITO layers are strongly influenced by the deposition techniques involved in TCO fabrication [5], the most used methods being magnetron sputtering [8,9], oxygen ion beam assisted deposition [10,11], chemical vapor deposition (CVD) [12,13], and pulsed laser deposition (PLD) [14,15]. The PLD method results in TCO coatings with appropriate electrical and optical properties [1], with its major advantage being the preservation of the target chemical composition in the transferred films. Furthermore, PLD allows the deposition of the films at low substrate temperature, thus making the technique suitable for the fabrication of TCO films even on plastic substrates [14]. In 2000, Ohta published a study regarding the PLD of an ITO layer characterized by a low electrical resistivity (7.7×10^{-5} $\Omega \cdot$cm) on an yttria-stabilized zirconia substrate [16]. In 2001, Suzuki reported the growth by PLD of an ITO film characterized by a lower electrical resistivity (7.2×10^{-5} $\Omega \cdot$cm), a transmittance in the visible part of the solar spectrum greater than 90%, and a very smooth surface (0.61 nm average surface roughness) [17]. For the preparation of ITO films with adequate optical and electrical properties, the most important parameters of PLD are the oxygen gas pressure and the substrate temperature. Thus, a decrease of the oxygen pressure (P_{O_2}) leads to an increase of the oxygen vacancy concentration in the ITO film, which has the consequence of increasing the carrier concentration and the conductivity of the deposited layer [18]. Because apparently the optical transmittance increases with the increase of the oxygen pressure [1], an optimum between optical and electrical properties must be achieved. A higher substrate temperature leads to the increase of grain size in the grown films, and, in this way, the grain boundary scattering is reduced and results in an increase of the conductivity of the ITO films [19].

In the last few years, some papers in the field of solar cells and OLEDs have focused on the improvement of device parameters using light out-coupling methods, approaches which imply the modification of substrates by patterning techniques [20–23]. For patterns fabrication, e-beam lithography [24], nanoimprint lithography (NIL) [25], and focused ion beam (FIB) [26] are the most frequently used techniques. NIL has been proven to be a powerful tool for the mass production of micro/nanostructures with controlled shape and density on different substrates like glass, Si wafers, and polymer films [27,28]. Moreover, the NIL technique provides an ideal solution for patterning large areas with good resolutions at low cost and high throughput [29]. This special peculiarity of NIL opens the way for many economically feasible applications regarding the deposition of ITO films by different routes on patterning substrates obtained by NIL. Despite this aspect of NIL, few research papers have been published on this subject [30–32]. Hence, ITO films were grown by PLD on thermal nanoimprinted glass plates using self-organized nanopattern molds of oxides (NiO, α-Al_2O_3) [30,31] and on patterned glass substrates heated at 200 °C [32].

In this context, the present work is focused on the PLD (unheated substrate and using a lower laser fluency) of ITO films on nanopatterned glass substrates fabricated by NIL. ITO layers were also deposited by PLD on flat glass substrates for evidencing the influence of the patterns on the properties of the nanostructured ITO films. The structural, compositional, morphological, optical, and electrical properties of both non-patterned and patterned ITO films were investigated and analyzed in a comparative manner. The information from this research can be helpful for developing potential applications of such patterned TCO films in organic optoelectronic devices.

2. Experimental Section

The patterns were fabricated on glass substrate with the UV nanoimprint lithography (UV-NIL) method using an EVG 620 mask aligner (EV Group, Sankt Florian, Austria). The following steps were involved in the patterning process: (i) the glass substrate was preheated at 150 °C for 2 min; (ii) the primer was spin-coated (Brewer Science Cee 200X Spin Coater, Brewer Science, Inc., Rolla, MO, USA) to improve the adhesion of the polymer photoresist layer; (iii) the UV-resist layer was deposited by spin-coating and further annealed at 120 °C for 30 s; (iv) the soft stamp (mold) with the desired pattern was placed on the photoresist layer and pressed at a uniform contact pressure of 100 mbar; (v) the photoresist layer was exposed to UV light for 90 s, allowing the cross-linking process to lead to its solidification; and (vi) the soft mold was released, which left the UV-resist pattern [25]

The ITO films were deposited on flat and nanopatterned glass substrates by the PLD technique using an excimer laser source with krypton fluoride, KrF (CompexPro 205, 248 nm wavelength, τ_{FWHM} ~25 ns, Coherent Inc., Santa Clara, CA, USA) operating at 10 Hz repetition rate. A solid ITO target (SCI Engineered Materials Inc., Columbus, OH, USA, In_2O_3:SnO_2 = 90%:10% weight) was ablated with 7000 laser pulses. The laser beam was directed on the target surface at 45° incidence angle with a MgF_2 lens (300 mm focal length) placed outside of the deposition chamber. The distance between target and substrate was fixed at 5 cm, and the target was rotated during the deposition to prevent its local damage. The deposition chamber—having 10^{-4} Pa residual pressure—was filled with oxygen 6.0 up to 1.5 Pa pressure. The P_{O_2} was chosen by taking into account that ITO films with low resistivity value can be grown at room temperature (RT) using this pressure value [19]. The deterioration of the patterned nanostructures was avoided by using low laser fluency (1.2 J/cm^2) and maintaining the substrates at RT. More details about the PLD experimental set-up are given in our previous papers [33,34]. The investigated samples were labeled as follows: glass (flat glass substrate), ITO/glass (flat glass substrate coated with ITO), NP-glass (nanopatterned glass substrate), and ITO/NP-glass (nanopatterned glass substrate coated with ITO). In addition to the samples already mentioned, a second specimen based on nanopatterned glass substrate coated with ITO (ITO/NP-glass_bis) was also analyzed to emphasize the reproducibility of this preparation procedure combining UV-NIL and PLD techniques.

The thickness of the deposited ITO/glass films (as average median between the values obtained from three different measurements) was evaluated at ~340 nm using an XP 100 profilometer (Ambios Technology, Inc., Santa Cruz, CA, USA).

The prepared samples were studied from structural, morphological, optical, and electrical points of view. The crystalline structure was analyzed using a D8 Advance (Bruker AXS, Karlsruhe, Germany) instrument operating with Bragg-Brentano geometry with a monochromatized Cu Kα1 radiation (λ = 1.4506 Å). The diffractograms were recorded in the 20°–80° range using 0.02° step size and 2 s/step. The morphology and the elemental composition were observed using a Zeiss Merlin Compact field emission scanning electron microscope and a Gemini SEM 500 (Zeiss, Oberkochen, Germany) field emission scanning electron microscope equipped with energy dispersive X-ray analysis Quantax XFlash detector 610 M (Bruker, Billerica, MA, USA)as an accessory. The roughness was evaluated using a MultiView 4000 atomic force microscope (phase feedback, Nanonics Imaging Ltd., Jerusalem, Israel). The optical properties were investigated by UV-VIS spectroscopy and photoluminescence (PL) using a Carry 5000 Spectrophotometer (Varian, Inc., Palo Alto, CA, USA) and a FL 920 spectrometer (Edinburgh Instruments Ltd., Livingston, UK) with a 450 W Xe lamp excitation and double monochromators on both excitation and emission, respectively. The optical measurements were made at room temperature. The transmission spectra were recorded in the 200–1100 nm domain, and the PL spectra were acquired at λ_{exc} = 335 nm (350–650 nm range) and at λ_{exc} = 435 nm (450–850 nm range). The parameters for recording were as follows: slit size = 3 mm, step size = 1 nm, and dwell time = 0.25 s. The carrier transport properties were evaluated from the measurements carried out at room temperature using a H50 Hall effect/Van der Pauw system (version, MMR Technologies Inc., San Jose, CA, USA).

3. Results and Discussions

The elemental composition of the TCO films grown by PLD was evaluated by energy dispersive X-ray analysis (EDX), which is presented in the Figure 1. The EDX spectra of ITO/glass and ITO/NP-glass samples evidenced the presence of In, Sn, and O components. In both cases, the atomic percentage for Sn:O:In was found to be about 3.7:63.5:32.7, confirming that the PLD target composition was maintained in the deposited ITO film. The results are in agreement to those reported for ITO films obtained by PLD [35], as the technique allows a stoichiometric transfer of the material from the target to the deposited film. The other peaks observed in the EDX spectra originated from the glass used as deposition substrate.

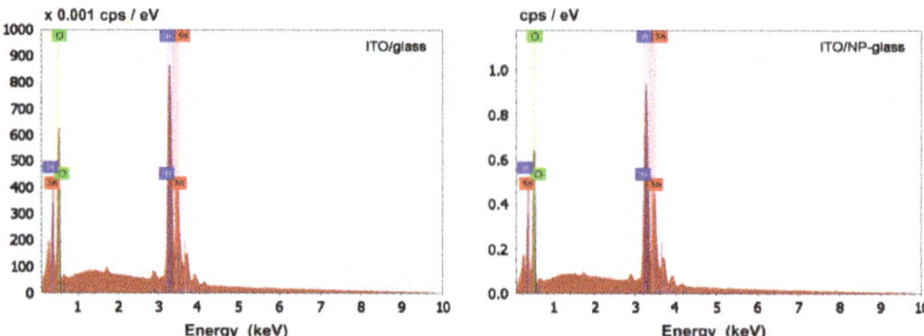

Figure 1. Energy dispersive X-ray analysis (EDX) spectra of indium tin oxide ITO/glass and ITO/nanopatterned (NP)-glass samples.

As mentioned above, substrate temperature and oxygen pressure have determinant impact on the crystalline structure of the TCO films grown by PLD. Thus, the ITO/glass and ITO/NP-glass samples were investigated by X-ray diffraction (XRD). In Figure 2, the XRD patterns of the ITO films deposited on unheated substrates exhibit only a broad peak at ~31° which can be related to a microcrystalline structure with a preferred (222) orientation [36]. The result is consistent with other studies carried out on ITO films prepared by PLD at RT and low P_{O_2}, though with higher laser fluency (4 J/cm^2) [37]. It was noticed that the patterning process has no influence on the ITO crystallinity.

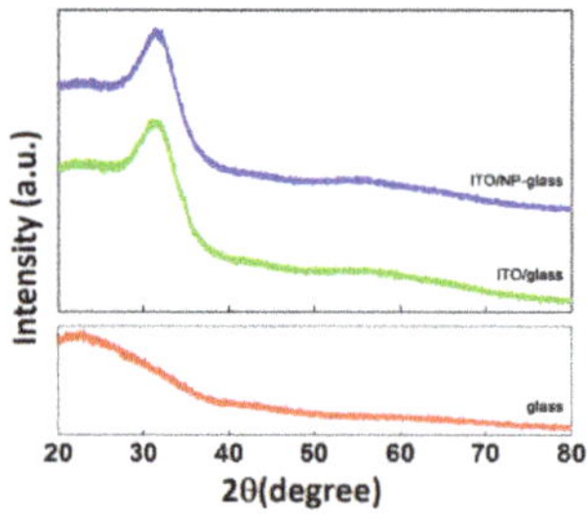

Figure 2. X-ray diffraction (XRD) patterns of glass, ITO/glass and ITO/NP-glass samples.

Also, the morphology and the roughness of the TCO thin films are key features which influence their optical and electrical properties. Ergo, in the next step, all prepared samples were analyzed by field emission scanning electron microscopy (FESEM) and atomic force microscopy (AFM). These types of analyses for ITO/glass, NP-glass, ITO/NP-glass, and ITO/NP-glass_bis can be seen in Figures 3–6, respectively.

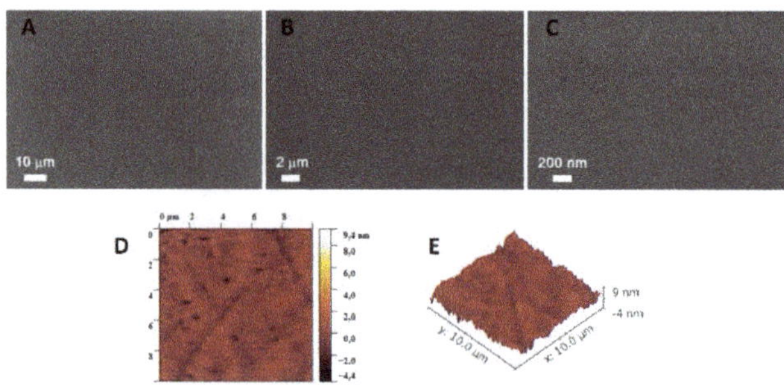

Figure 3. Field emission scanning electron microscopy (FESEM) images at different magnifications (**A,B,C**) and atomic force microscopy (AFM) images (**D,E**) in 2D and 3D format of the ITO/glass sample.

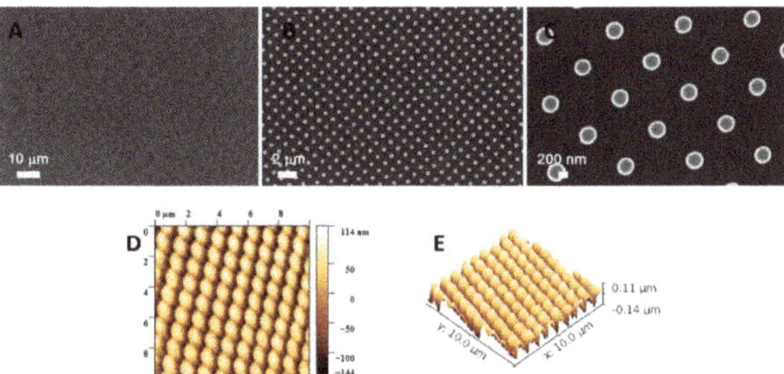

Figure 4. FESEM images at different magnifications (**A,B,C**) and AFM images (**D,E**) in 2D and 3D format of the NP-glass sample.

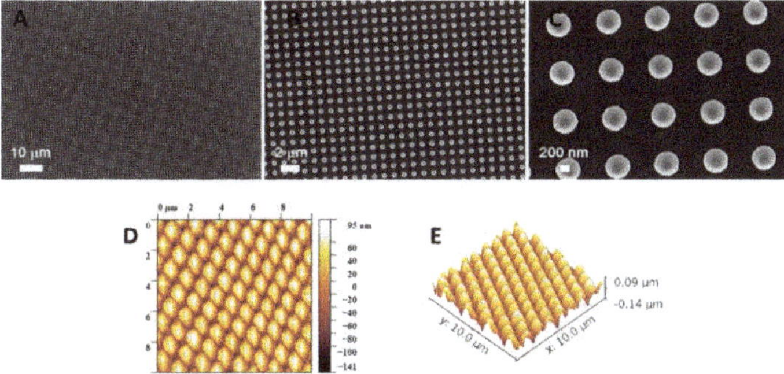

Figure 5. FESEM images at different magnifications (**A,B,C**) and AFM images (**D,E**) in 2D and 3D format of the ITO/NP-glass sample.

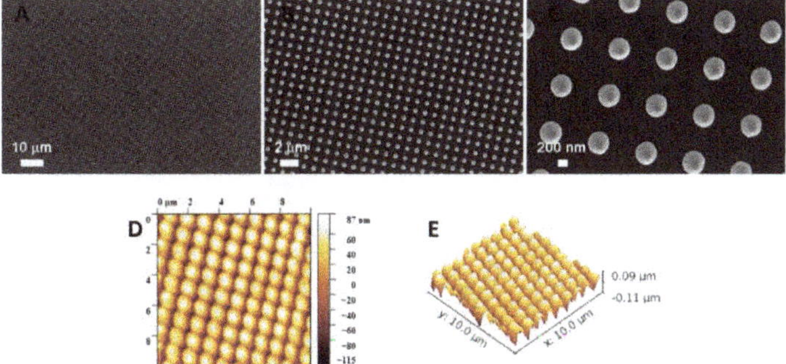

Figure 6. FESEM images at different magnifications (**A,B,C**) and AFM images (**D,E**) in 2D and 3D format of the ITO/NP-glass_bis sample.

The FESEM images of the ITO films deposited on flat glass substrate (Figure 3) disclose a smooth surface in accordance with observations reported in the case of ITO films prepared by solution-based fabrication methods [38].

In the case of the structures fabricated by UV–NIL on glass substrates, the FESEM images (Figure 4) show arrays of periodic pillars with diameters of ~350 nm and a distance between the pillars of ~1100 nm. As can be seen in Figures 5 and 6, the deposition of the ITO by PLD resulted in the increase of the pillars' width and the decrease of the separation step—the deposited films were trying to fill the space between pillars and to cover them.

Using the AFM investigations, the root mean square (RMS) parameter was evaluated as being: 1 nm for ITO/glass (Figure 3), 59 nm for NS-glass (Figure 4), 47 nm for ITO/NP-glass (Figure 5), and 44 nm for ITO/NP-glass_bis (Figure 6). The presence of the pillars lead, as was expected, to a roughness increase. In the case of the ITO/glass sample, the RMS value is lower in comparison with other values calculated for ITO films prepared by other methods [8], but it is comparable with those reported for ITO films grown by PLD, as this technique offers the possibility to obtain TCO films with a smooth surface when the deposition is made at low temperature and with a low laser fluency [19,39].

Additionally, the surface kurtosis (R_{ku}) parameter, a peakedness indicator—also determined by AFM—has the following values: 5.2 for ITO/glass, 2.1 for NP-glass, 2.3 for ITO/NP-glass, and 2.4 for ITO/NP-glass_bis. Based on the R_{ku} values, it can be mentioned that the ITO/glass film presents more peaks than valleys, while the ITO/NP-glass films present more valleys than peaks.

The EDX mapping images of the ITO films deposited on both flat (Figure 7A,B) and nanopatterned glass (Figure 7C,D) substrate illustrate a uniform distribution of the chemical elements Sn, O, and In contained in the PLD target. Over the fabricated pillars, no preferential arrangement of the elements as clusters is evidenced.

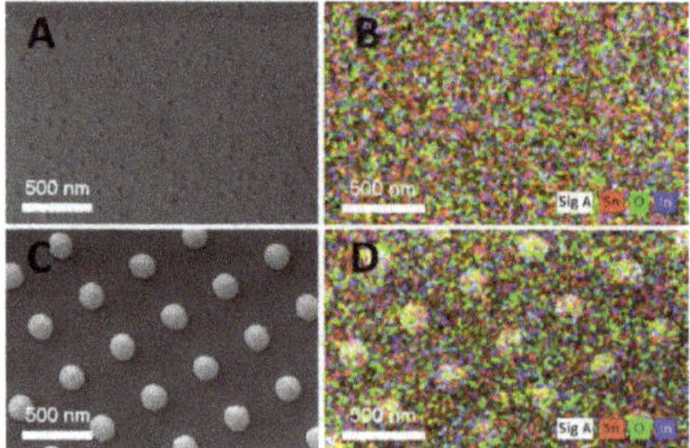

Figure 7. FESEM (**A,C**) and EDX mapping (**B,D**) images of ITO/glass (**A,B**) and ITO/NP-glass (**C,D**) samples.

The height of the pillars obtained by UV-NIL was estimated at ~250 nm from the cross-sectional FESEM images shown in Figure 8A,B, as the height was reduced after their coating with ITO films Figure 8C,D. Furthermore, these FESEM images emphasize an interesting aspect: the shape of the pillars is changed after the deposition of the TCO towards a pyramid trunk-like shape.

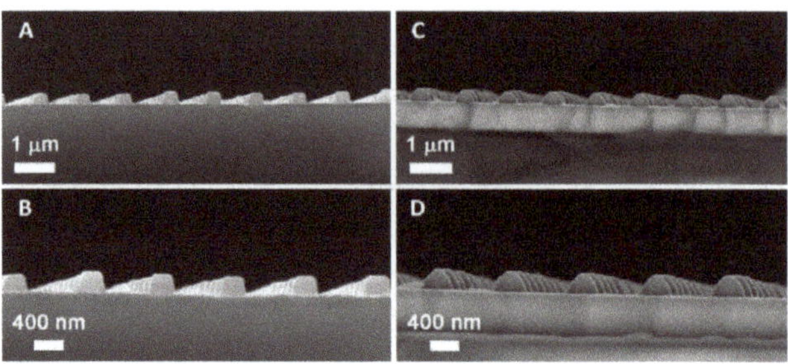

Figure 8. Cross-sectional FESEM images at two magnifications of NP-glass (**A,B**) and ITO/NP-glass (**C,D**) samples.

The optical properties are presented in Figures 9 and 10. The ITO film deposited on flat glass substrate presents a transmittance over 80% between 600 and 1100 nm, with the interference maxima and minima exhibited in the visible range being typical for the films with similar thickness (~340 nm) deposited by PLD or by magnetron sputtering [8,19]. The interference characteristics are also observed in the UV-VIS spectra of the ITO film deposited on nanopatterned glass substrate. Using these maxima and minima and the Swaneopel method [40], the refractive index (*n*) value was evaluated at ~1.9 for ITO/glass—a characteristic value for the ITO films deposited by PLD—and at ~2.4 for ITO/NP-glass. Some studies reported that the n value depended on the substrate temperature, since higher values were obtained for the ITO films prepared at a low temperature [19]. In addition, by plotting $(\alpha h v)^2$ versus photon energy, the band gap values for both ITO/glass and ITO/NP-glass samples were estimated at ~3.65 eV and ~3.21 eV, respectively. The band gap of ITO deposited on flat glass substrate is thickness-dependent and is in accordance with other reported films prepared by PLD with different SnO_2 content (5% or 10%) [37,39]. It is well known that ITO is a semiconductor with a direct wide band gap greater than 3 eV [3]. The increase of this value is usually attributed to the increasing of the carrier density, a process known as Burstein-Moss effect [19]. Regarding the lowering of the band gap value of ITO film deposited on the nanopatterned glass substrate, a similar effect, related to oxygen deficiency, was observed for the ITO deposited by spin-coating on self-assembled polymer nanopattern [41].

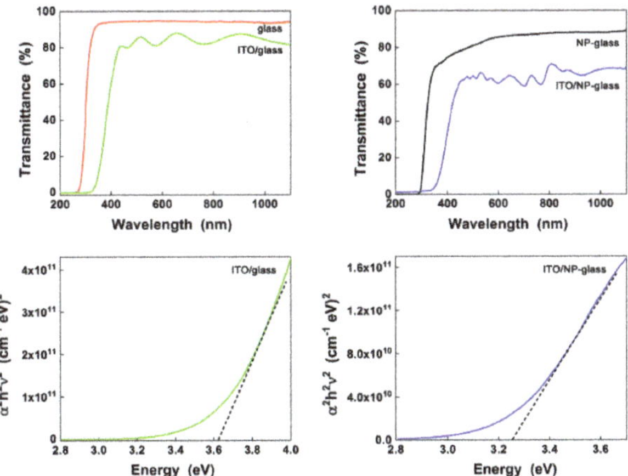

Figure 9. UV-VIS spectra and $(\alpha h v)^2$ vs. E dependency of ITO/glass and ITO/NP-glass samples.

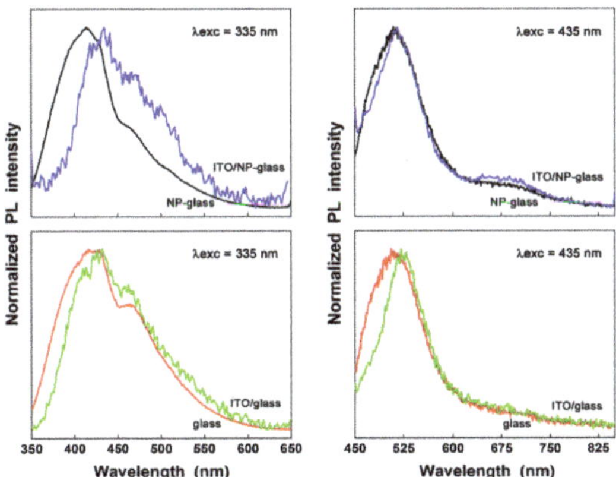

Figure 10. Photoluminescence (PL) spectra at two excitation wavelengths of glass, NP-glass, ITO/glass and ITO/NP-glass samples.

The PL spectra of all investigated samples—recorded at two excitation wavelength (335 and 435 nm)—are shown in Figure 10, as such information is useful for the TCO applications in OPV or OLED devices [42]. Under UV excitation (335 nm), the ITO/glass sample revealed a structured emission band with peaks at about 410, 430, and 460 nm which can be attributed to the ITO [25,43]. Generally, these emissions are characteristic of ITO, appearing due to the In_2O_3 oxygen deficiencies [44]. However, it is difficult to make an assignment, since the glass substrate presents also a similar emission. The ITO/NP-glass sample disclosed a shift of the structured emission band compared with that of the NP/glass sample. Furthermore, the shoulder at about 500 nm can also be associated with the ITO emission [45]. Under visible excitation (435 nm), an emission band peaked at about 550 nm is evidenced by the ITO/glass sample being narrower in comparison with that of the glass substrate [46]. In the case of the patterned samples, the PL spectra near the emission band at 550 nm show another weak broad emission band between 600 and 800 nm, which can be as also related the deposited ITO film [46].

The electrical properties of the ITO films deposited on flat and nanopatterned glass substrate by PLD were evaluated from the Hall measurements (Table 1), which revealed that the electrons are the charge carriers.

Table 1. Electrical parameters of the ITO deposited by PLD estimated from the Hall measurements.

Sample	Resistivity (Ωcm)	Mobility (cm^2/Vs)	Carrier Concentration (cm^{-3})	Sheet Resistance (Ω/sq)
ITO/glass	1.8×10^{-4}	10.6	3.3×10^{21}	5.3
ITO/NP-glass	2.8×10^{-4}	15.1	1.5×10^{21}	8
ITO/NP-glass_bis	2.7×10^{-4}	14.6	1.6×10^{21}	7.9

The calculated electrical resistivity is lower for the ITO/glass sample in comparison with the values for ITO/NP/glass and ITO/NP-glass_bis samples. Therefore, ITO samples prepared by PLD on substrates maintained at RT have resistivity values close to those reported for the ITO films deposited by PLD using a heated substrate (300 °C)—a target with 5% SnO_2, with the TCO film having 30 nm in thickness [17]. Our ITO samples are also similar to ITO films deposited using a 1 Pa O_2 atmosphere, a target having 10% SnO_2 content and different temperature substrate (beginning from 100 °C) [39,47]. In addition, the resistivity of the ITO/glass sample has a better value (1.8×10^{-4} Ω·cm) relative to others obtained in similar PLD experimental parameters (substrate kept at RT, SnO_2 content, oxygen pressure) [48]. As mentioned above, the oxygen pressure is a key parameter in the PLD, as it influences the resistivity value of the TCO films—an influence owing to the number of formed oxygen vacancies in the laser transferred ITO films [25]. Thus, the electrical resistivity decreases with the increase of the SnO_2 content over 5%. The increasing of the Sn amount leads to a higher concentration of the electron traps [19]. In our case, the ITO/glass sample has the highest carrier concentration (3.3×10^{21} cm^{-3}), the values of the ITO/NP-glass and ITO/NP-glass_bis samples being similar. The results are in consensus with other results obtained for ITO samples prepared on heated substrate [19,48]. Also, in these studies, the carrier density is related to the Sn quantity, an increase in the carrier density being reported as effect of the Sn donor electrons for the samples grown from the target containing a higher Sn concentration [19]. The carrier density is connected to the band gap value: ITO/glass samples that featured the highest carrier density had E_g = 3.65 eV, whereas ITO/NP-glass characterized by a lower carrier density has E_g = 3.21 eV. Furthermore, an increase in the Hall mobility was evidenced for the ITO/NP-glass and ITO/NP-glass_bis samples—15.1 and 14.6 cm^2/Vs, respectively—compared to the ITO/glass sample—10.6 cm^2/Vs—which has smaller values than other reported values [25] due to the carrier-carrier scattering as an effect of high carrier density [1]. Taking into consideration that the only difference between the samples investigated in this study was the use of different substrate type, the increase in the Hall mobility for the ITO films deposited by PLD on nanopatterned glass is clearly owed to this patterned substrate.

4. Conclusions

PLD is an adequate technique to deposit ITO films on both flat and patterned glass substrates at room temperature, 1.2 J/cm^2 laser fluency, and low oxygen pressure (1.5 Pa). The patterned substrates containing periodic pillars with ~350 nm in width, ~250 nm in height, and ~1100 nm separation step were fabricated by a UV-NIL process. The EDX analysis confirms that the laser-deposited ITO films preserve the stoichiometry of the solid ITO target (In$_2$O$_3$:SnO$_2$ = 90%:10% weight), and the XRD data show a (222) preferential orientation for the grown film. The SEM and AFM images prove that the deposited ITO films retain the pattern of the glass substrates, while the nanopatterned TCO films featured a good optical transmittance. A low RMS value (1 nm) was estimated for the ITO film deposited on flat glass. The optical band gap and the refractive index were evaluated from the UV-VIS spectra, and the obtained values were in agreement with other results reported for ITO films deposited by PLD. The electrical measurements revealed that a n-type semiconductor was obtained, as the TCO films were characterized by a low electrical resistivity ($<2.8 \times 10^{-4}$) regardless the substrate type. However, an increase in the Hall mobility was observed in the case of the ITO films deposited on nanostructured glass. By combining UV-NIL—a cost-effective technique to fabricate nanopatterns on large areas—and PLD, a method for growing TCO films at low substrate temperature can be obtained for nanopatterned ITO films. This could have applications in organic photovoltaic cells or organic light emitting devices.

Author Contributions: Conceptualization, M.S.; Methodology, M.S. and N.P.; Validation, M.S., N.P. and G.S.; Formal Analysis, A.C., F.G. and G.S.; Investigation, A.C., N.P., G.S., F.G., O.R. and C.B.; Resources, G.S. and A.S.; Writing—Original Draft Preparation, M.S.; Writing—Review and Editing, N.P. and F.G.; Supervision, G.S. and A.S.

Funding: This research was funded by the Romanian Ministry of Research and Innovation through National Core Program from PN18-110101 and LAPLAS V (3N/2018) contracts and from ROSA STAR 179/2017 contract.

Conflicts of Interest: The authors declare no conflict of interest.

References

1. Eason, R. *Pulsed Laser Deposition of Thin Films: Applications-Led Growth of Functional Materials*; John Wiley & Sons, Inc.: Hoboken, NJ, USA, 2007; pp. 240–255.
2. Bright, C.I. Review of transparent conductive oxides (TCO). In *50 Years of Vacuum Coating Technology and the Growth of the Society of Vacuum Coaters*; Society of Vacuum Coaters: Materials Park, OH, USA, 2007.
3. Afre, R.A.; Sharma, N.; Sharon, M.; Sharon, M. Transparent conducting oxide films for various applications: A review. *Rev. Adv. Mater. Sci.* **2018**, *53*, 79–89.
4. Hosono, H.; Ueda, K. Transparent conductive oxides. In *Springer Handbook of Electronic and Photonic Materials*; Kasap, S., Capper, P., Eds.; Springer: Berlin, Germany, 2017; pp. 1391–1404.
5. Cao, W.; Li, J.; Chen, H.; Xue, J. Transparent electrodes for organic optoelectronic devices: A review. *J. Photonics Energy* **2014**, *4*, 040990. [CrossRef]
6. Hosono, H.; Kim, J.; Toda, Y.; Kamiya, T.; Watanabe, S. Transparent amorphous oxide semiconductors for organic electronics: Application to inverted OLEDs. *Pro. Nati. Acad. Sci.* **2017**, *114*, 233–238. [CrossRef] [PubMed]
7. Mbule, P.; Wang, D.; Grieseler, R.; Schaaf, P.; Muhsin, B.; Hoppe, H.; Mothudi, B.; Dhlamini, M. Aluminum-doped ZnO thin films deposited on flat and nanostructured glass substrates: Quality and performance for applications in organic solar cells. *Sol. Energy* **2018**, *172*, 219–224. [CrossRef]
8. Prepelita, P.; Filipescu, M.; Stavarache, I.; Garoi, F.; Craciun, D. Transparent thin films of indium tin oxide: Morphology–optical investigations, inter dependence analyzes. *Appl. Surf. Sci.* **2017**, *424*, 368–373. [CrossRef]
9. Starkov, I.A.; Nyapshaev, I.A.; Starkov, A.S.; Abolmasov, S.N.; Abramov, A.S.; Levitskii, V.S.; Terukov, E.I. Influence of substrate movement on the ITO film thickness distribution during magnetron sputtering. *J. Vac. Sci. Technol. A Vac. Surf. Films* **2017**, *35*, 061301. [CrossRef]
10. Meng, L.J.; Gao, J.; Silva, R.A.; Song, S. Effect of the oxygen flow on the properties of ITO thin films deposited by ion beam assisted deposition (IBAD). *Thin Solid Films* **2008**, *516*, 5454–5459. [CrossRef]

11. Kim, S.J.; Choi, S.Y.; Choi, K. Preparation and characterization of ITO thin films deposition by ion beam assisted deposition. *J. Korean Inst. Met. Mater.* **2014**, *52*, 475–484. [CrossRef]
12. Atabaev, I.G.; Hajiev, M.U.; Pak, V.A. Growth of ITO films by modified chemical vapor deposition method. *Int. J. Thin Films Sci. Technol.* **2016**, *5*, 13–16.
13. Nishinaka, H.; Yoshimoto, M. Mist chemical vapor deposition of single-phase metastable rhombohedral indium tin oxide epitaxial thin films with high electrical conductivity and transparency on various α-Al$_2$O$_3$ substrates. *Cryst. Growth Des.* **2018**, *18*, 4022–4028. [CrossRef]
14. Socol, G.; Socol, M.; Stefan, N.; Axente, E.; Popescu-Pelin, G.; Craciun, D.; Duta, L.; Mihailescu, C.N.; Mihailescu, I.N.; Stanculescu, A.; et al. Pulsed laser deposition of transparent conductive oxide thin films on flexible substrates. *Appl. Surf. Sci.* **2012**, *260*, 42–46. [CrossRef]
15. Craciun, V.; Chiritescu, C.; Kelly, F.; Singh, R.K. Low temperature growth of smooth indium tin oxide films by ultraviolet assisted pulsed laser deposition. *J. Optoelectron. Adv. Mater.* **2002**, *4*, 21–25.
16. Ohta, H.; Orita, M.; Hirano, M.; Tanji, H.; Kawazoe, H.; Hosono, H. Highly electrically conductive indium-tin-oxide thin films epitaxially grown on yttria-stabilized zirconia (100) by pulsed-laser deposition. *Appl. Phys. Lett.* **2000**, *76*, 2740–2742. [CrossRef]
17. Suzuki, A.; Mastsushita, T.; Aoki, T.; Yoneyama, Y.; Okuda, M. Pulsed laser deposition of transparent conducting indium tin oxide films in magnetic field perpendicular to plume. *Jpn. J. Appl. Phys.* **2001**, *40*, L401. [CrossRef]
18. Kim, H.; Horwitz, J.S.; Pique, A.; Gilmore, C.M.; Chrisey, D.B. Electrical and optical properties of indium tin oxide thin films grown by pulsed laser deposition. *Appl. Phys. A* **1999**, *69*, S447–S450. [CrossRef]
19. Kim, H.; Gilmore, C.M.; Pique´, A.; Horwitz, J.S.; Mattoussi, H.; Murata, H.; Kafafi, Z.H.; Chrisey, D.B. Electrical, optical, and structural properties of indium-tin-oxide thin films for organic light-emitting devices. *J. Appl. Phys.* **1999**, *86*, 6451–6461. [CrossRef]
20. Bi, Y.-G.; Feng, J.; Ji, J.-H.; Yi, F.-S.; Li, Y.-F.; Liu, Y.-F.; Zhang, X.-L.; Sun, H.-B. Nanostructures induced light harvesting enhancement in organic photovoltaics. *Nanophotonics* **2018**, *7*, 371–391. [CrossRef]
21. Ferry, V.E.; Verschuuren, M.A.; Lare, M.C.V.; Schropp, R.E.; Atwater, H.A.; Polman, A. Optimized spatial correlations for broadband light trapping nanopatterns in high efficiency ultrathin film a-Si:H solar cells. *Nano Lett.* **2011**, *11*, 4239–4245. [CrossRef]
22. Choo, S.; Choi, J.; Choi, H.-J.; Huh, D.; Son, S.; Kim, Y.D.; Lee, H. Enhancement of light extraction efficiency for GaN-based light emitting diodes using ZrO$_2$ high-aspect-ratio pattern as scattering layer. *Ceram. Int.* **2017**, *43*, S609–S612. [CrossRef]
23. Saxena, K.; Jain, V.K.; Mehta, D.S. A review on the light extraction techniques in organic electroluminescent devices. *Opt. Mater.* **2009**, *32*, 221–233. [CrossRef]
24. Tseng, A.A.; Chen, K.; Chen, C.D.; Ma, K.J. Electron beam lithography in nanoscale fabrication: Recent development. *IEEE Trans. Electron. Packag. Manuf.* **2003**, *26*, 141–149. [CrossRef]
25. Breazu, C.; Preda, N.; Socol, M.; Stanculescu, F.; Matei, E.; Stavarache, I.; Iordache, G.; Girtan, M.; Rasoga, O.; Stanculescu, A.I. Investigations on the properties of a two-dimensional nanopatterned metallic electrode. *Dig. J. Nanomater. Biostructures* **2016**, *11*, 1213–1229.
26. Matsui, S. Three-dimensional nanostructure fabrication by focused-ion-beam chemical-vapor-deposition. *Microsc. Microanal.* **2006**, *12*, 130–131. [CrossRef]
27. Guo, L.J. Recent progress in nanoimprint technology and its applications. *J. Phys. D Appl. Phys.* **2004**, *37*, R123. [CrossRef]
28. Lee, H.; Hong, S.; Yang, K.; Choi, K. Fabrication of nano-sized resist patterns on flexible plastic film using thermal curing nano-imprint lithography. *Microelectron. Eng.* **2006**, *83*, 323–327. [CrossRef]
29. Austin, M.D.; Ge, H.; Wu, W.; Li, M.; Yu, Z.; Wasserman, D.; Lyon, S.A.; Chou, S.Y. Fabrication of 5 nm linewidth and 14 nm pitch features by nanoimprint lithography. *Appl. Phys. Lett.* **2004**, *84*, 5299–5301. [CrossRef]
30. Akita, Y.; Sugimoto, Y.; Kobayashi, K.; Suzuki, T.; Oi, H.; Mita, M.; Yoshimoto, M. Crystal growth control of functional oxide thin films on nanopatterned substrate surfaces. *J. Laser Micro/Nanoeng.* **2009**, *4*, 202–206. [CrossRef]
31. Akita, Y.; Miyake, Y.; Nakai, H.; Oi, H.; Mita, M.; Kaneko, S.; Mitsuhashi, M.; Yoshimoto, M. Evolution of atomically stepped surface of indium tin oxide thin films grown on nanoimprinted glass substrates. *Appl. Phys. Express* **2011**, *4*, 035201. [CrossRef]

32. Liu, Y.; Kirsch, C.; Gadisa, A.; Aryal, M.; Mitran, S.; Samulski, E.T.; Lopez, R. Effects of nano-patterned versus simple flat active layers in upright organic photovoltaic devices. *J. Phys. D Appl. Phys.* **2013**, *46*, 024008. [CrossRef]

33. Stanculescu, A.; Socol, M.; Socol, G.; Mihailescu, I.N.; Girtan, M.; Preda, N.; Albu, A.-M.; Stanculescu, F. Effect of maleic anhydride-aniline derivative buffer layer on the properties of flexible substrate heterostructures: Indium tin oxide/nucleic acid base/metal. *Thin Solid Films* **2011**, *520*, 1251–1258. [CrossRef]

34. Stanculescu, A.; Rasoga, O.; Preda, N.; Socol, M.; Stanculescu, F.; Ionita, I.; Albu, A.-M.; Socol, G. Preparation and characterization of polar aniline functionalized copolymers thin films for optical non-linear applications. *Ferroelectrics* **2009**, *389*, 159–173. [CrossRef]

35. Tseng, K.-S.; Lo, Y.-L. Effects of cumulative ion bombardment on ITO films deposited on PET and Si substrates by DC magnetron sputtering. *Opt. Mater. Express* **2014**, *4*, 764–775. [CrossRef]

36. Ohshima, T.; Matsunaga, T.; Kawasaki, H.; Suda, Y.; Yagyu, Y. Preparation of ITO thin films by pulsed laser deposition for use as transparent electrodes in electrochromic display devices. *Trans. Mater. Res. Soc. Jpn.* **2010**, *35*, 583–587. [CrossRef]

37. Schou, J. Physical aspects of the pulsed laser deposition technique: The stoichiometric transfer of material from target to film. *Appl. Surf. Sci.* **2009**, *255*, 5191–5198. [CrossRef]

38. Chen, Z.; Li, W.; Li, R.; Zhang, Y.; Xu, G.; Cheng, H. Fabrication of highly transparent and conductive indium-tin oxide thin films with a high figure of merit via solution processing. *Langmuir* **2013**, *29*, 13836–13842. [CrossRef] [PubMed]

39. Kim, S.H.; Park, N.M.; Kim, T.; Sung, G.Y. Electrical and optical characteristics of ITO films by pulsed laser deposition using a 10 wt.% SnO_2-doped In_2O_3 ceramic target. *Thin Solid Films* **2005**, *475*, 262–266. [CrossRef]

40. Swanepoel, R. Determination of surface roughness and optical constants of inhomogeneous amorphous silicon films. *J. Phys. E Sci. Instrum.* **1984**, *17*, 896–903. [CrossRef]

41. Xia, G.; Wang, S. Solution patterning of ultrafine ITO and $ZnRh_2O_4$ nanowire array below 20 nm without etching process. *Nanoscale* **2011**, *3*, 3598–3600. [CrossRef]

42. Babu, S.H.; Rao, N.M.; Kaleemulla, S.; Amarendra, G.; Krishnamoorthi, C. Room-temperature ferromagnetic and photoluminescence properties of indium–tin-oxide nanoparticles synthesized by solid-state reaction. *Bull. Mater. Sci.* **2017**, *40*, 17–23. [CrossRef]

43. Luo, S.; Kohiki, S.; Okada, K.; Shoji, F.; Shishido, T. Hydrogen effects on crystallinity, photoluminescence, and magnetization of indium tin oxide thin films sputter-deposited on glass substrate without heat treatment. *Phys. Status Solidi A* **2010**, *207*, 386–390. [CrossRef]

44. Lee, M.S.; Choi, W.C.; Kim, E.K.; Kim, C.K.; Min, S.K. Characterization of the oxidized indium thin films with thermal oxidation. *Thin Solid Films* **1996**, *279*, 1–3. [CrossRef]

45. Thirumoorthi, M.; Thomas Joseph Prakash, J. Structure, optical and electrical properties of indium tin oxide ultrathin films prepared by jet nebulizer spray pyrolysis technique. *J. Asian Ceram. Soc.* **2016**, *4*, 124–132. [CrossRef]

46. Venkatesh, P.S.; Ramakrishnan, V.; Jeganathan, K. Vertically aligned indium doped zinc oxide nanorods for the application of nanostructured anodes by radio frequency magnetron sputtering. *CrystEngComm* **2012**, *14*, 3907–3914. [CrossRef]

47. Fang, X.; Mak, C.L.; Zhang, S.; Wang, Z.; Yuan, W.; Ye, H. Pulsed laser deposited indium tin oxides as alternatives to noble metals in the near-infrared region. *J. Phys. Condens. Matter* **2016**, *28*, 224009. [CrossRef] [PubMed]

48. Wu, Y.; Maree, C.H.M.; Haglund, R.F., Jr.; Hamilton, J.D.; Morales Paliza, M.A.; Huang, M.B.; Feldman, L.C.; Weller, R.A. Resistivity and oxygen content of indium tin oxide films deposited at room temperature by pulsed-laser ablation. *J. Appl. Phys.* **1999**, *86*, 991–994. [CrossRef]

© 2018 by the authors. Licensee MDPI, Basel, Switzerland. This article is an open access article distributed under the terms and conditions of the Creative Commons Attribution (CC BY) license (http://creativecommons.org/licenses/by/4.0/).

Article

Relevance of the Preparation of the Target for PLD on the Magnetic Properties of Films of Iron-Doped Indium Oxide

Hasan B. Albargi [1,2], Marzook S. Alshammari [3,*], Kadi Y. Museery [3], Steve M. Heald [4],
Feng-Xian Jiang [5], Ahmad M. A. Saeedi [1], A. Mark Fox [1] and Gillian A. Gehring [1,*]

[1] Department of Physics and Astronomy, Hicks Building, The University of Sheffield, Sheffield S3 7RH, UK;
     albargih@yahoo.com (H.B.A.); asaeedi1@sheffield.ac.uk (A.M.A.S.); mark.fox@sheffield.ac.uk (A.M.F.)
[2] Department of Physics, Najran University, P.O. Box 1988, Najran 11001, Saudi Arabia
[3] The National Centre for Laser and Optoelectronics, King Abdulaziz City for Science and Technology, KACST
     P.O. Box 6086, Riyadh 11442, Saudi Arabia; kmuseery@kacst.edu.sa
[4] Advanced Photon Source, Argonne National Laboratory, Argonne, IL 60439, USA; heald@aps.anl.gov
[5] School of Chemistry and Materials Science of Shanxi Normal University & Key Laboratory of Magnetic
     Molecules and Magnetic Information Materials of the Ministry of Education, Linfen 041004, China;
     jfx9902@163.com
*   Correspondence: alshammari@kacst.edu.sa (M.S.A.); g.gehring@sheffield.ac.uk (G.A.G.)

Received: 2 May 2019; Accepted: 10 June 2019; Published: 13 June 2019

Abstract: This paper concerns the importance of the preparation of the targets that may be used for pulsed laser deposition of iron-doped indium oxide films. Targets with a fixed concentration of iron are fabricated from indium oxide and iron metal or one of the oxides of iron, FeO, Fe_3O_4 and Fe_2O_3. Films from each target were ablated onto sapphire substrates at the same temperature under different oxygen pressures such that the thickness of the films was kept approximately constant. The films were studied using X-ray diffraction, X-ray absorption (both XANES and EXAFS), optical absorption and magnetic circular dichroism. The magnetic properties were measured with a SQUID magnetometer. At the lowest oxygen pressure, there was evidence that some of the iron ions in the films were in the state Fe^{2+}, rather than Fe^{3+}, and there was also a little metallic iron; these properties were accompanied by a substantial magnetisation. As the amount of the oxygen was increased, the number of defect phases and the saturation magnetisation was reduced and the band gap increased. In each case, we found that the amount of the oxygen that had been included in the target from the precursor added to the effect of adding oxygen in the deposition chamber. It was concluded that the amount of oxygen in the target due to the precursor was an important consideration but not a defining factor in the quality of the films.

Keywords: PLD; target preparation; room temperature ferromagnetism; dilute magnetic semiconductor; Indium oxide; $(InFe)_2O_3$

---

1. Introduction to the Growth of Oxide Films Using Pulsed Laser Deposition

There is a great interest in the magnetic properties of thin oxide films for use in sensors. Pulsed laser deposition (PLD) is one of the most commonly used growth techniques [1,2]. Particular examples are pure and doped In_2O_3, ZnO and cuprate superconductors. A common feature of these oxides is that their magnetic and electrical properties depend strongly on the amount of oxygen that is incorporated into the film and the grain size [3–5]. PLD is a particularly versatile technique because the oxygen stoichiometry can be controlled by depositing the film in a chamber that contains some oxygen gas, and the grain size and quality of the films depend on the substrate temperature [5–7]. Almost all groups have used targets made using conventional solid-state reaction techniques to fabricate their targets.

An exception to this showed that good quality PLD films of oxides of indium could be made using a metallic target and using the oxygen pressure in the chamber to obtain an oxide film. This was done using targets that contained both metallic indium and tin that were ablated in an oxygen pressure of 7.5 Torr and a silicon substrate heated to 500 °C to make an NO gas sensor [8]. Films of CdO doped with indium where the target was $Cd_{1-x}In_x$ with $x = 0.049$ were ablated in an oxygen atmosphere of 75 mTorr onto a quartz glass substrate held at 300 °C to fabricate CdO which is a good transparent conductor [9].

Growth using an oxide target involves grinding and sintering powders and then pressing them into a target and finally sintering again. There are several variables here, including how the grinding was performed, how many times the powders were ground and sintered and the highest temperature used to sinter the target. In many publications, these details are not given. An interesting comparison was made between the properties of PLD-ablated films of $(InFe)_2O_3$ formed when the grinding was done mechanically compared with using hand grinding using a pestle and mortar. The Fe ions that were in the films that had been ablated from a target that had been formed using mechanical grinding were almost entirely present as a secondary phase of Fe_3O_4 [10]. Essentially all the Fe ions were on In sites in films made using the exact same protocol of grinding and sintering but using hand grinding with a pestle and mortar [5].

Another relevant factor is the maximum temperature used to anneal the target. Good films of pure and calcium-doped yttrium iron garnet (YIG) were deposited using a target that had been made from Fe_2O_3, Y_2O_3 and, where appropriate, CaO. In this case, the target had to be annealed at a high temperature, 1200 °C, for 15 h because targets sintered at lower temperatures were brittle and were destroyed during laser ablation [11,12]. Good quality films were grown on a gadolinium gallium garnet (GGG) using a substrate temperature of only 500 °C after the films had subsequently been annealed in air at 1000 °C [11,12].

The annealing temperature for targets of pure Fe_2O_3 was important for a different reason. In this case, the target retained its orange colour when it was annealed at 500 °C and then could be used to grow films of maghemite by PLD at 100 mTorr. However, if the target had been annealed at 1200 °C in air it changed its colour to black and then could be used to get films of Fe_3O_4 and FeO [12].

Another situation where a target changed colour after annealing to a high temperature occurred with ZnCoO. A target was made from metallic cobalt and ZnO and was ground and sintered repeatedly. It retained its light grey colour when the maximum temperature used for annealing was 1000 °C but the colour changed to dark green if it was annealed at a higher temperature ~1150 °C. The ordered compound of $Zn_{1-x}Co_xO$ is green (Rinman's Green) hence it was clear that in this case high temperatures are required to complete the solid-state reaction. Films that contained 10% cobalt and were of similar thickness were ablated from the targets annealed at 1000 and 1150 °C and were compared. It was found that the film made from the target with the 1000 °C anneal had a significant content of metallic cobalt present as nanoparticles that caused blocking behaviour at 30 K. The film made with the target that had been annealed at 1150 °C had a much larger saturation magnetisation and any nanoparticles of cobalt were too small to show blocking behaviour above 5 K [13].

Tuning the oxygen content in the films of pure and doped ZnO by changing the oxygen pressure in the chamber has been performed very widely. However, the oxygen content of the target can also be controlled by the precursor. To investigate the relevance of this, a study was made of films of ZnCoO using three different precursors in the targets: Metallic cobalt, CoO and Co_3O_4 [14]. Most previous work had used Co_3O_4. This study demonstrated that in this case, these different precursors produced different films even though there was no trace of the precursors in the ablated films. It also showed that using metallic cobalt as a precursor had effects beyond the concentration of oxygen and that the subtle chemistry of PLD was also affected [14].

In_2O_3 is an *n*-type transparent semiconductor material with a wide band gap of 3.75 eV that is in the ultraviolet (UV) region of the spectrum [15,16]. This material is an insulator in its stoichiometric form, while in its oxygen deficient form it has *n*-type doping levels that are induced by oxygen vacancies. The stoichiometry is an important factor in determining electrical properties [17,18]. In_2O_3 can grow in three different structures; however, all thin films that were grown by PLD grow in the cubic bixbyite structure as is seen here. In this structure, each cubic unit cell of In_2O_3 contains 16 formula units (80 atoms) and has a lattice constant of 10.118 Å [19,20].

In_2O_3 has been doped with transition metals, in particular, iron, to form $(In_{1-x}Fe_x)_2O_3$ and in this case the lattice constant decreases monotonically with increasing Fe concentration until $x = 0.2$, indicating that the maximum solubility limit of Fe ions in In_2O_3 lattice is approximately 20%. The saturation magnetisation, M_s, has also been found to increase proportionally with increasing Fe concentration, for x between 0.05 and 0.2 [15,21–25].

It is generally found that oxygen vacancies induce defects that are responsible for the ferromagnetism and that magnetism occurs when the Fe ions are reduced from Fe^{3+} to Fe^{2+} [15,26]. In addition, the value of M_s in $(In_{0.95}Fe_{0.05})O_3$ thin films has been found to be affected by the grain size, where the highest magnetisation saturation corresponds to the largest grain size implying that grain boundary magnetism is not important for these films [5].

However, there are also reports of Fe-doped In_2O_3 thin films that contain magnetic nanoparticles of Fe_2O_3 or Fe_3O_4 [20,25,27]. The presence of the Fe_3O_4 nanoparticles has been reported to enhance the room temperature magnetisation, magnetoresistance and a larger value of the coercive of ~400 Oe [10].

Previous work has considered Fe_2O_3 to be the obvious precursor to use with In_2O_3 to generate a target for the PLD because it should generate a stoichiometric target [5,10,15,19,24,26]. It is well known that if a film is ablated from a target in a high vacuum, the film will contain less oxygen than the target because some oxygen is lost in the PLD process. Hence, films are grown in different oxygen pressures so as to control the density of oxygen vacancies. In this work, we describe the effects of controlling the density of oxygen vacancies by changing the precursor used to fabricate the target as well as controlling the amount of oxygen in the growth chamber. We have made PLD films from targets that contain 5% iron using metallic iron, FeO, Fe_3O_4 and Fe_2O_3 together with In_2O_3 in the targets. All other conditions were kept constant.

The films were studied using X-ray diffraction to measure the change in the lattice constant with the changing density of oxygen vacancies and also X-ray absorption, X-ray absorption near edge structure (XANES) and extended X-ray absorption fine structure (EXAFS) techniques to measure the state of ionisation of the Fe and its environment. The hysteresis loops of the films were measured at room temperature and at 5 K. Optical measurements of the absorption and the magnetic circular dichroism (MCD) were also studied. This enables us to investigate the effects on the films of adding oxygen to the target using a different precursor with that of adding oxygen to the PLD chamber. It will also indicate if there are extra chemical effects of using different targets that occur in the PLD process beyond the effects of having a different concentration of oxygen vacancies. Such effects were found in PLD films made of ZnCoO using different compounds of Co in the fabrication of the target [14].

2. Fabrication of the Targets and Growth of the Films

The targets were made using a solid-state reaction method that was performed using the following protocol which we have found to be an effective method to produce targets that could be used to grow good quality films [14]. Appropriate weights of one of the precursors, FeO, Fe_2O_3 or Fe_3O_4 and In_2O_3 chosen so as to give a ratio of 0.05:0.95 of Fe to In, were mixed together; the amounts used and the necessary information required to obtain these values are given in Table 1. The powders were purchased from Alfa Aesar (Karlsruhe, Germany) and had purities of 99.999% for In_2O_3, 99.995% for FeO, 99.998% for Fe_3O_4 and 99.999% for Fe_2O_3. The powders were hand ground for 30 min in a ceramic pestle and mortar and calcined in air at 300 °C for 12 h. They were then ground again for a further 30 min and sintered in air at 600 °C for 12 h. The procedure was repeated with the sintering temperature raised to 900 °C. After the final anneal, the mixture was placed in a *Specac* (Specac Ltd., Kent, England) die, which was evacuated with a roughing-pump and, using a manual hydraulic press, compressed to 25000 kPa. This produced a relatively dense, cylindrical pellet of diameter 25 mm and thickness between 2 and 5 mm, depending upon the amount of the initial powders used. The pellet was then given a final anneal at a maximum temperature of 1000 °C.

Table 1. The weight of each material required to make targets weighing ~11 g.

Precursor	Molecular Weight	Element	Number Per Gram $\times 10^{21}$	Number of Grams for a Target of $(In_{1-x}Fe_x)_2O_3$
In_2O_3	277.64	In	4.3381	10
Fe	55.845	Fe	10.784	0.212
FeO	71.844	Fe	8.3822	0.272
Fe_3O_4	231.533	Fe	7.8030	0.292
Fe_2O_3	159.69	Fe	7.5422	0.303

Thin films of thickness of approximately 200 nm were deposited on double-side polished sapphire c-cut Al_2O_3 (0001) substrates that were held at 450 °C. The deposition used a Lambda Physik LEXTRA 200 XeCl excimer laser (Lambda Physik Lasertechnik, Goettingen, Germany) with a maximum power of 400 mJ per pulse, an operating wavelength of 308 nm, and a 10 Hz repetition rate of 28 ns pulses. The target was rotated at 60 rpm and was placed 5 cm from the substrate. We had previously checked that there was almost no difference in the films that were made using the XeCl laser compared with the, more standard, KrF laser. Three films were made from each target at each of three different oxygen pressures in the PLD chamber. The three conditions were base pressure, 2×10^{-5} Torr and oxygen pressures of 2×10^{-4} Torr and 2×10^{-3} Torr. This was to allow us to compare the effects of adding oxygen to the target from the precursor with that of adding oxygen to the PLD growth chamber.

3. Results

3.1. Structural Characterisation of the Films

The films' structural and chemical characteristics were obtained using X-ray diffraction XRD, (Rigaku Corporation, Tokyo, Japan and Bruker D2 Phaser, Coventry, UK), XANES and EXAFS techniques. These techniques gave us information on the lattice constant and grain size of the In_2O_3 matrix and the presence of any nanoparticles that existed in the films.

The XRD data, shown in Figure 1, were measured using Cu Kα radiation (λ = 1.5406 Å) using a θ–2θ scan. For the samples grown at base pressure, the data showed that the samples had diffraction peaks corresponding to (222) and (400) of the pure cubic bixbyite In_2O_3; the (006) peak is from the sapphire substrate. A small peak at ~36° (shown in red) indicated the presence of the secondary phase of FeO; however, no peaks from metallic iron were detected. The insets in Figure 1 show an enlarged plot of the (222) reflection; at base pressure, all three lattice constants were 10.18 ± 0.02 Å but there are real differences for the films grown at 2×10^{-3} Torr where the lattice constants for FeO, Fe_3O_4 and Fe_2O_3 were 10.17 ± 0.02 Å, 10.14 ± 0.02 Å and 10.12 ± 0.02 Å, respectively. This is in agreement with earlier results where it was found that the lattice constant increased slowly with increasing oxygen due to the elimination of isolated oxygen vacancies, but that at higher oxygen pressure it decreased rapidly due to the removal of the oxygen vacancy being accompanied by the oxidation of the large Fe^{2+} ion to the much smaller Fe^{3+} ion [15,26]. The size of the observed lattice contraction increased with the amount of oxygen in the target. Hence, the data indicated that the total amount of oxygen in the films depends on both the oxygen in the precursor as well as the oxygen in the PLD chamber.

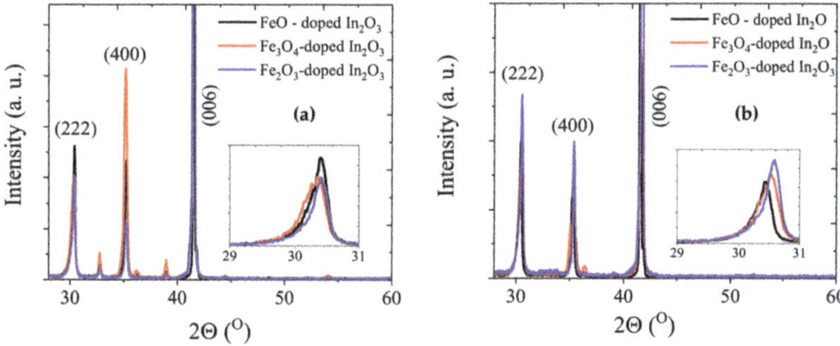

Figure 1. XRD data of the Fe-doped In_2O_3 thin films grown from different oxide precursors at (**a**) base pressure of 2×10^{-5} Torr and (**b**) 2×10^{-3} Torr. The insets demonstrate the effect of the precursor on the position of the (222) peak.

To have a more accurate estimate of the presence of defect phases and the ionisation state and environment of the Fe ion, K-edge XANES and EXAFS spectra have been measured. The films grown at base pressure and the powder oxide standards were measured on beamline 20-BM, and the films grown at higher oxygen pressure were measured on beamline 20-ID at the Advanced Photon Source. The setups on both beamlines were similar with Si (111) monochromators providing 1 eV energy resolution at the Fe K edge. The measurements were made at room temperature at a glancing angle of ~5° with the X-ray polarization normal to the surface of the films. Multielement solid-state detectors (4 element silicon drift detector on 20-ID and 13 element Ge detector on 20-BM) were used for fluorescence detection, and the samples were spun at a few Hz to avoid Bragg reflection interference from the single-crystal substrates. Typically, 4–8 scans were averaged for improved signal to noise. Data were analysed using the Demeter analysis package [28].

In the XANES spectra, as shown in Figure 2, the signals from wüstite (FeO), magnetite (Fe_3O_4) and hematite (Fe_2O_3) have been plotted alongside Fe-doped In_2O_3 films to be used as references. Figure 2a displays the data from the samples grown at base pressure and indicates an absorption at ~7117 eV (marked with an arrow) that is at a lower energy than for wüstite. This means that they all contain a small percentage of metallic iron. Such results are caused by the increased number of oxygen vacancies that are generated by PLD at base pressure [29] and all evidence of metallic iron has vanished from the films made at 2×10^{-4} Torr and 2×10^{-3} Torr. The XANES data for the films deposited at higher oxygen pressure, 2×10^{-3} Torr, are close to that of hematite, Fe_2O_3, indicating that most of the Fe ions are in the state Fe^{3+} although a small fraction of the Fe ions may be present as Fe^{2+}.

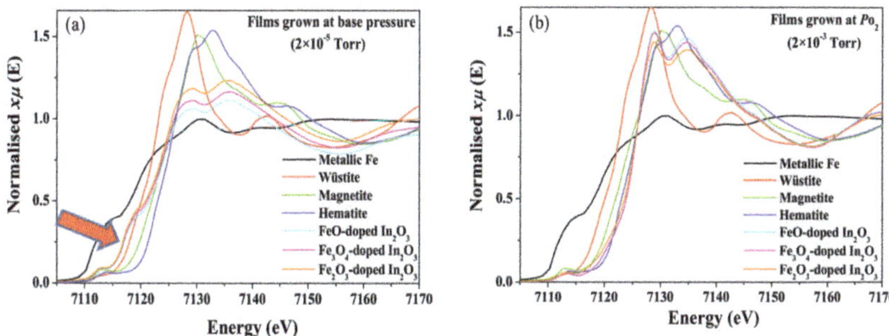

Figure 2. K-edge X-ray absorption near edge structure (XANES) spectra of reference compounds of metallic Fe, FeO, Fe_3O_4 and Fe_2O_3 and the Fe-doped In_2O_3 films grown from different precursors at (**a**) at base pressure of 2×10^{-5} Torr; where an arrow at 7117 eV indicates an additional absorption and (**b**) higher partial oxygen pressure of 2×10^{-3} Torr.

The environment of the Fe ions is obtained from an analysis of the Fourier transform of the EXAFS and the results are shown in Figure 3. The results for the films grown at base pressure are shown in Figure 3a and are compared with the EXAFS spectrum from a sample of Fe_2O_3-doped In_2O_3 sample that is believed to be pure substitutional [5]. All of the data shows a strong peak near $R = 1.6$ Å from near neighbour oxygens. Note that the transforms are not phase corrected, so the peaks are shifted ~0.3 Å lower from their actual distances. The spectra from the films that were grown at base pressure have similar peaks in the region $R = 2$–4 Å as the fully substitutional spectrum (shown in red) except near R~2.1 Å where they show small additional peaks that are likely characteristic of Fe metal. If the metallic fraction is very small, it is difficult to detect it in diffraction. Attempts at fitting with a metal site combined with a substitutional site were only moderately successful. A good fit was achieved with about 12% of the doped Fe in the metallic environment, but the substitutional site parameters had to be modified more than seems reasonable to accommodate the Fe. This indicates the possibility of a third type of Fe oxide site, in agreement with the XRD where a small signal from FeO was also detected. Unfortunately, the data range was such that it is difficult to reliably fit a three-site model. The reduction in the intensity for $2 \leq R \leq 3$ Å, also suggests the presence of some Fe oxide secondary phase in addition to the metallic Fe clusters.

Detection of an impurity phase by XRD depends on the ability to measure the square of the concentration. In this case, this is given by the percentage of Fe in the whole sample, 5%, and the percentage of these atoms that are in a metallic environment, about 12%. These combine to give a percentage of metallic Fe in the sample that is about 0.6% which was not measurable.

The data shown in Figure 3b are from the samples grown at 2×10^{-3} Torr. A plot of a reference sample of hematite has been included because the XANES data from the films had indicated that all the iron was in the Fe^{3+} state which is also characteristic of hematite. These data show that the atomic arrangement in the films is very close to that expected for Fe ions substituted on the In sites and distinct from that of hematite. At higher oxygen pressures, the Fe ions appear to be substitutional for all the targets and show no evidence of any metallic iron or any iron oxide. There are some differences in the edge data between our sample and the reference sample that may be due to the result of better structural order in our samples. While the basic structure of the peaks looks similar, there is an increasing difference in the height of the peaks at larger distances that could be a characteristic of disorder. The In_2O_3 has two different In sites and the disorder arises if the Fe substitutes randomly on both sites [15,27].

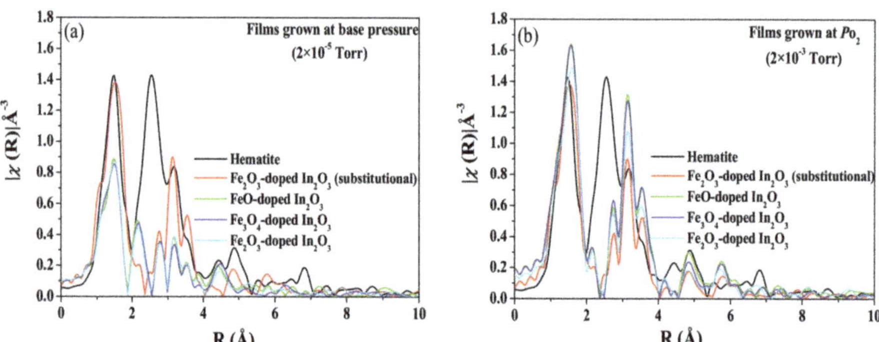

Figure 3. X-ray absorption fine structure (EXAFS) Fourier transform of Fe_2O_3 and substitutional Fe_2O_3-doped In_2O_3 deposited at base pressure as reference compounds and the Fe-doped In_2O_3 films grown from different precursors at (**a**) base pressure of 2×10^{-5} Torr and (**b**) O_2 pressure of 2×10^{-3} Torr.

In summary, we find evidence of defect phases, metallic iron from the XANES and FeO from the XRD and the EXAFS in films grown at base pressure. The results of the XANES and the EXAFS measurements of films grown with oxygen in the chamber indicated that substantially all the Fe^{3+} ions were situated on In^{3+} sites.

3.2. Magnetic Properties of the Films

Magnetic hysteresis loops were taken at 5 and 300 K for the substrates and all the Fe-doped In_2O_3 films using a Quantum Design SQUID magnetometer (San Diego, CA, USA). It was found that all the films displayed room temperature ferromagnetism; examples of the loops obtained are shown in Figure 4 and the values of M_s and H_c measured at room temperature for the different precursors and oxygen pressure in the PLD chamber are summarised in Table 2. A magnetic field of 10,000 Oe was applied parallel to the plane of the film during the magnetisation measurements. The diamagnetic contribution from the sapphire substrate was subtracted, as was the paramagnetic contribution from the film from the data shown in Figure 4.

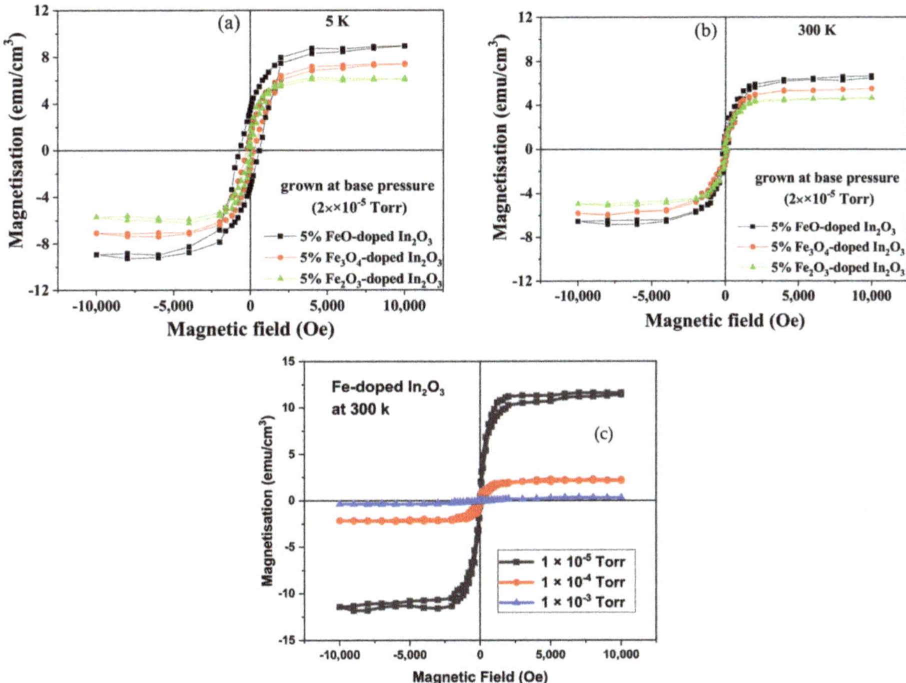

Figure 4. (**a**) and (**b**) are respectively the magnetic hysteresis loops measurements at 5 and 300 K for the Fe-doped In_2O_3 films from different oxide precursors at a base pressure of 2×10^{-5} Torr, (**c**) shows the room temperature data for films that were made with Fe metal in the precursor. The diamagnetic and paramagnetic terms have been subtracted from all the data shown here.

The paramagnetic contribution from the film, seen at 5 K, was consistent with it being due to free spins of all the Fe^{3+} ions because the value of p_{eff} was found to be 4.5 ± 0.3 compared with the expected value for Fe^{3+} of 4.9 [30]. The observed strong temperature dependence of the saturation magnetisation was observed previously in films of Fe-doped In_2O_3 that had semiconducting behaviour [3,15,31]. Part of the magnetisation observed from the films deposited at base pressure could be due to the 12% of the Fe ions that were in a metallic environment as observed by EXAFS leading to a metallic concentration of about 0.6%. If each of these contributed 2.2 μ_B to the bulk magnetisation, this would contribute about 1.3 emu/cm^3 to the magnetisation. This is comparable to the difference between the magnetisation observed for films grown at base pressure and those grown at 2×10^{-4} Torr for the films grown with the oxide precursors.

Previous results had found that oxygen vacancies and Fe^{2+} ions were necessary for ferromagnetism to be observed in $Fe\text{-}In_2O_3$ and these results are consistent with this earlier work [3,31,32]. This pattern was seen for all films and all oxygen content because it was found that both the saturation magnetisation and coercive field decreased with oxygen content whether the oxygen was in the target or in the PLD chamber. The decrease in magnetisation for added oxygen was larger for the films deposited at base pressure. The large value of M_s, seen only at base pressure, may be due to the existence of metallic iron as seen by XANES as well as a larger number of oxygen vacancies [3,31,32]. The coercive field of the films deposited at base pressure from the FeO target was increased significantly at 5 K compared with that measured at 300 K, leading to the deduction that blocked magnetic nanoparticles, of probably Fe metal, existed at low temperatures in that film. This increase was not seen for the films deposited at higher pressure. The magnetisation of the films made from the target that contained metallic iron as the precursor showed a stronger dependence on oxygen content than the films made from the oxide precursors. Interestingly, the coercive field seen at 5 K for the film grown from the Fe metal target at base pressure was higher than with those grown from the oxide targets grown at base pressure. This implies that if, as expected, the film grown with a Fe-target does contain a larger percentage of metallic iron, the nanoparticles are so small that their blocking temperatures are close to 5 K, or below, whereas significant blocking of nanoparticles has occurred for the film ablated from the target that had been made with FeO.

Temperature-dependent plots of the magnetisation were measured under the conditions of zero field cooled (ZFC) and field cooled (FC) to investigate further the relative importance of nanoparticles. This magnetisation was obtained for all the samples grown from FeO, Fe_3O_4 and Fe_2O_3 precursors and deposited at base pressure; this was the condition where the XANES measurement had indicated the presence of about 12% of the iron atoms in a metallic environment. A magnetic field of 100 Oe was applied in parallel to the plane of the samples [33–35]. The diamagnetic contribution from the sapphire substrate was subtracted from all ZFC and FC curves shown in Figure 5. The separation of the ZFC and the FC curves is due to the increase of the anisotropy field to become comparable or larger than the measuring field, 100 Oe, as the temperature is reduced. If the magnetisation had been dominated by nanoparticles, the curves should vary as $1/T$ in the reversible regime, but this is not observed here.

Table 2. The values of M_s and H_c measured at 300 K and at 5 K for the different precursors and oxygen pressure in the pulsed laser deposition (PLD) chamber.

Precursor	Base Pressure				2×10^{-4} Torr				2×10^{-3} Torr			
	Fe	FeO	Fe_3O_4	Fe_2O_3	Fe	FeO	Fe_3O_4	Fe_2O_3	Fe	FeO	Fe_3O_4	Fe_2O_3
M_s 300 K (5 K)emu/cm^3	11.0 ± 0.4 (12.0 ± 0.4)	6.8 ± 0.3 (9.2 ± 0.4)	6.0 ± 0.2 (7.8 ± 0.3)	4.2 ± 0.3 (6.1 ± 0.3)	2.6 ± 0.1 (3 ± 0.2)	5.1 ± 0.2 (7.3 ± 0.3)	4.4 ± 0.3 (6.1 ± 0.2)	3.2 ± 0.2 (5.2 ± 0.2)	0.3 ± 0.2 (5 ± 0.1)	3.6 ± 0.3 (4.8 ± 0.2)	2.5 ± 0.2 (3.9 ± 0.2)	2.0 ± 0.2 (3.0 ± 0.1)
H_c 300 K (5 K)Oe	25 ± 3 (150 ± 17)	135 ± 15 (535 ± 32)	113 ± 1 (271 ± 23)	100 ± 12 (153 ± 18)	100 ± 12 (120 ± 16)	123 ± 14 (127 ± 17)	106 ± 12 (111 ± 14)	94 ± 12 (96 ± 14)	50 ± 7 (80 ± 9)	118 ± 14 (110 ± 15)	97 ± 10 (98 ± 14)	92 ± 10 (87 ± 12)

The FC/ZFC curves shown in Figure 5a,b for the films made from oxide precursors and grown at base pressure are consistent with the increase in the coercive field at the temperature measured at 5 K as given in Table 1. The large coercive field observed at 5 K for the film grown from the FeO could be due to shape anisotropy of the metallic inclusions but even in this case, the $1/T$ dependence in the reversible region was not observed, indicating that the magnetic contributions from the nanoparticles are not dominating the overall magnetisation.

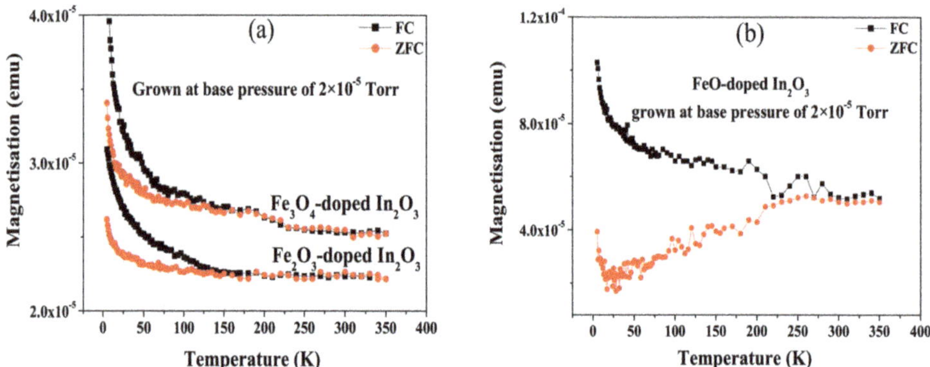

Figure 5. Field cooled (FC) and zero field cooled (ZFC) magnetisation curves of the Fe-doped In$_2$O$_3$ from Fe$_3$O$_4$ and Fe$_2$O$_3$ precursors in (**a**) and from FeO precursor in (**b**) where all samples were grown at a base pressure of 2×10^{-5} Torr.

The increase of the magnetisation below 50 K is a characteristic of all DMS materials. This behaviour arises from isolated paramagnetic ions which are not contributing to the long-range ferromagnetic order. Existence of these ions was already discussed because they give rise to a paramagnetic contribution to the hysteresis loops [30].

3.3. Optical Absorption

The optical properties of the Fe-doped In$_2$O$_3$ films were investigated by carrying out transmission and reflection measurements at room temperature. From these measurements, absorption data were obtained to gain an insight into the electronic structure and to estimate the density of gap states and the band gap of the films.

The optical properties of this material are sensitive to different targets and film preparation parameters, including the amount of oxygen, the following results will show the effect of changing the oxygen content. Figure 6 illustrates the absorption data at energies close to the band edge for all the Fe-doped In$_2$O$_3$ films grown at the base and different oxygen pressures. In doped In$_2$O$_3$ there are two dominant effects that can change the band gap. Isolated oxygen vacancies are donors and will be ionised to increase the band gap due to the Burstein-Moss effect, however, a lattice contraction will increase the band gap. Both effects are relevant here. The values of the band gap are summarised in Table 3. At low oxygen pressure, 2×10^{-5} Torr, the films have essentially the same lattice constant and the band gap is highest for the films with the lowest amount of oxygen in the target due to the Burstein-Moss effect. At the higher oxygen pressure, 2×10^{-3} Torr, the lattice contraction for the films containing the most oxygen is the dominant effect in determining the lattice constant.

All the spectra show a substantial amount of absorption below the energy gap due to energy states in the gap. We note that the highest density of gap states occurs for the three films grown at base pressure, which were known to contain about 12% of the iron atoms in a metallic environment.

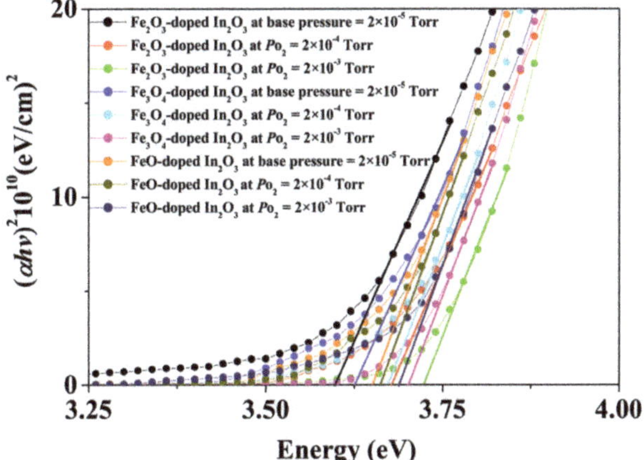

Figure 6. Absorption data of Fe-doped In_2O_3 samples grown from FeO, Fe_3O_4 and Fe_2O_3 precursors deposited at a base pressure of 2×10^{-5} Torr and the two higher oxygen pressures of 2×10^{-4} Torr and 2×10^{-3} Torr.

3.4. The Magneto-Optical Properties

The MCD spectra for all Fe-doped In_2O_3 samples were measured in the energy range between 1.7 and 4 eV at room temperature in Faraday geometry using an applied magnetic field of 18000 Oe, as displayed in Figure 7. The MCD is a very powerful technique because it indicates the amount of spin polarisation that is present in the quantum states that are involved in transitions at a particular energy [36]. The MCD signal from the sapphire substrate has been subtracted from the data shown in Figure 7.

Table 3. Summary of the band gap values of Fe-doped In_2O_3 thin films deposited at the base and higher oxygen pressures.

Sample	E_g (eV)		
	Base Pressure (2×10^{-5} Torr)	Oxygen Pressure (2×10^{-4} Torr)	Oxygen Pressure (2×10^{-3} Torr)
FeO-doped In_2O_3	3.65 ± 0.01	3.66 ± 0.01	3.69 ± 0.02
Fe_3O_4-doped In_2O_3	3.63 ± 0.02	3.67 ± 0.02	3.70 ± 0.01
Fe_2O_3-doped In_2O_3	3.60 ± 0.02	3.68 ± 0.02	3.72 ± 0.01

The results taken at base pressure are shown in Figure 7a; these are characteristic of films that contain metallic iron where the MCD may be calculated using the Maxwell–Garnett theory [5,15,37]. The signals indicate that the most metal is in the film made from the FeO target and the least in the one made from the Fe_2O_3 target, but in both the percentage of the volume occupied by metal is small, approximately 0.5% or less. In contrast, there is no sign of any metallic iron seen in the MCD spectra of the films made at the higher oxygen pressure shown in Figure 7b, as expected from the XANES results shown in Section 3.1. The MCD spectra varied between positive for the film with the lowest oxygen content, FeO-In_2O_3 grown 2×10^{-4} Torr, to negative for the film with the largest amount of oxygen, Fe_2O_3-In_2O_3 grown at 2×10^{-3} Torr, as the oxygen content was increased. It had been found earlier that the MCD was positive for films with a high density of carriers produced by oxygen vacancies and small or negative for those in the semiconducting regime [15]. These results are also consistent with the values of the saturation magnetisation obtained from the magnetic hysteresis loops measured by the SQUID shown in Table 1.

The dip observed in all the spectra shown in Figure 7b just above 3.4 eV and below the band gap energy stated in Table 2 is typical of magnetic oxide semiconductors. It arises from transitions from the valance band to empty donor states of oxygen vacancies situated below the conduction band edge and is a clear indication that these donor states are spin polarised [35]. The absorption from these states is seen in Figure 6 in the region of about 0.25 eV below the band gap.

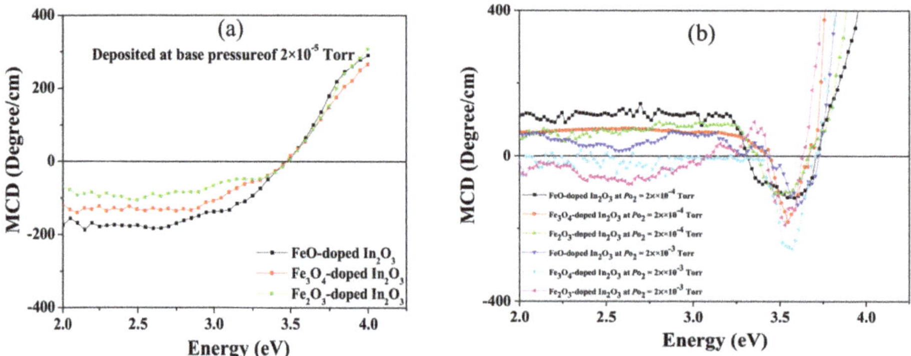

Figure 7. Magnetic circular dichroism (MCD) spectral shapes of the Fe-doped In_2O_3 samples deposited from different precursors at (**a**) base pressure and (**b**) at 2×10^{-4} Torr shown with black, red and green symbols for FeO, Fe_3O_4 and Fe_2O_3, respectively and at 2×10^{-3} Torr with blue, turquoise and pink symbols for FeO, Fe_3O_4 and Fe_2O_3, respectively.

4. Discussion

In this work, we studied a range of films that have differed in two aspects: The amount of oxygen in the PLD chamber and the precursor that was used to add Fe ions to the target material. We anticipated that films that were made with FeO in the target would contain less oxygen than those that were made with Fe_3O_4, which in turn would have less oxygen than those made with Fe_2O_3, however, we speculated that there may be subtle questions of chemistry of the PLD process that come into play, as we found in Co ions incorporated in ZnO [14].

The aim of this work was to see if the extra oxygen that is incorporated into the target enters the films in a similar way as the oxygen included in the PLD chamber. We investigated this by keeping all other variables constant. All the films had the same percentage of iron (5%), the same procedures for mixing and annealing the targets, the same substrate temperature and, as far as possible, the same thickness of the films. The processing of the targets was done in air, so it was interesting to observe that the amount of oxygen in the target still depended on the precursor and was not equalised during the process of fabrication.

The three precursors have different magnetic properties: FeO and Fe_2O_3 are antiferromagnetic, or very weakly ferromagnetic, whereas Fe_3O_4 is strongly ferromagnetic. Hence, it was important to check if any of the Fe_3O_4 precursor had survived in the films and the XAFS data showed clearly that it had not. Nor was there any suggestion that the magnetic properties of the film made with this precursor were significantly inconsistent with the other films.

We had evidence of defect phases of both metallic iron and FeO appearing in all our films that had been grown at base pressure. The appearance of FeO in these films occurred for all three oxide precursors as is clear from the XAFS shown in Figure 2a and was a consequence of the low oxygen pressure and not dependent on FeO being in one of the targets. Figure 8a shows the change of the magnetisation measured at room temperature as a function of the oxygen pressure for all of the films. In this case, the amount of oxygen in the target and the PLD chamber produced effects of similar magnitude. It is clear that the largest magnetisation in films without metallic iron was made from a target that had been produced using FeO. This is interesting because the precursor of choice had been assumed to be Fe_2O_3 which was found to be the worst performing precursor in this study.

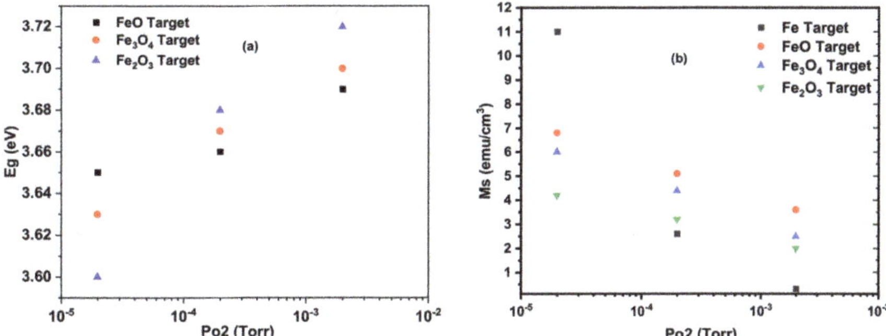

Figure 8. The dependence of (**a**) the band gap and (**b**) the saturation magnetisation on the oxygen pressure.

The results of this study are summarised in Figure 8 and Table 1. These show that as the level of oxygen in the PLD chamber is increased the energy of the band edge increases as shown in Figure 8a and the saturation magnetisation reduces as shown in Figure 8b. These results are in agreement with earlier work that deduced that the magnetisation in $(In_{1-x}Fe_x)_2O_3$ is due to oxygen vacancies and the compensating Fe^{2+} ions that are removed by the addition of oxygen [15].

The measurements described here give a coherent account of PLD films made with iron oxide precursors and ablated in different oxygen pressures. The measurements of the lattice constant, XANES and EXAFS spectra, band gap and magnetic measurements were combined to give a clear description of these films. The results fit the general pattern that the oxygen in the target, generated from the precursor, had a similar effect as adding oxygen to the PLD growth chamber. The films grown at base pressure contained some metallic iron as indicated by X-ray absorption and larger coercive fields at 5 K, however, the density of oxygen vacancies was smaller for the precursors, with more oxygen in the target as indicated by the band gap. At higher oxygen pressures, both the density of isolated oxygen vacancies and the density of Fe^{2+} ions were reduced with corresponding drops in the magnetisation, more details are in [38]. The magnetism of the films grown with metallic Fe decreased more rapidly as oxygen was added to the growth chamber. More work should be done on this interesting system.

It has been customary to fabricate targets using Fe_2O_3 mixed with In_2O_3 because it was assumed that this would naturally be a combination that would be best suited to incorporate the Fe into the In_2O_3 lattice. The work done here suggests that FeO would be a better choice because films made with this precursor have a higher magnetisation than those made with Fe_2O_3.

Author Contributions: Conceptualisation H.B.A. and G.A.G.; Methodology, H.B.A., A.M.A.S., S.M.H., A.M.F. and M.S.A.; Validation, H.B.A. and G.A.G.; Investigation, H.B.A., A.M.A.S., F.-X.J., S.M.H. and K.Y.M.; Resources, H.B.A., G.A.G., S.M.H. and M.S.A.; Writing—original draft preparation, H.B.A.; Writing—review and editing, G.A.G. and M.S.A.; Supervision, G.A.G., A.M.F.; Project administration, G.A.G. and M.S.A.

Funding: This research was funded by Saudi Cultural Attaché-London, UK Engineering and Physical Sciences Research EP/D070406/1, Advanced Photon Source, an Office of Science User Facility operated for the US Department of Energy (DOE) Office of Science by Argonne National Laboratory and was supported by the US DOE under Contract No. DE-AC02-06CH11357 and the Canadian Light Source and its funding partners and King Abdulaziz City for Science and Technology (KACST), Saudi Arabia.

Acknowledgments: H.B.A. acknowledges the receipt of a studentship from the Saudi Cultural Attaché-London. The SQUID and MCD measurements were taken on apparatus initially funded by the UK Engineering and Physical Sciences Research EP/D070406/1. This research used resources of the Advanced Photon Source, an Office of Science User Facility operated for the US Department of Energy (DOE) Office of Science by Argonne National Laboratory and was supported by the US DOE under Contract No. DE-AC02-06CH11357 and the Canadian Light

Source and its funding partners. In addition, resources for target preparation provided by King Abdulaziz City for Science and Technology (KACST), Saudi Arabia.

Conflicts of Interest: The authors have no conflicts of interest.

References

1. Pearton, S.J.; Norton, D.P.; Heo, Y.W.; Tien, L.-C.; Ivill, M.P.; Li, Y.; Kang, B.S.; Ren, F.; Kelly, J.; Hebard, A.F.; et al. ZnO spintronics and nanowire devices. *J. Electron. Mater.* **2006**, *35*, 862–868. [CrossRef]
2. Kim, H.; Gilmore, C.M.; Piqué, A.; Horwitz, J.S.; Mattoussi, H.; Murata, H.; Kafafi, Z.H.; Chrisey, D.B. Electrical, optical, and structural properties of indium–tin–oxide thin films for organic light-emitting devices. *J. Appl. Phys.* **1999**, *86*, 6451–6461. [CrossRef]
3. Mukherji, R.; Mathur, V.; Mukherji, M. A Perspective on Indium Oxide based diluted magnetic semiconductors. *Mater. Focus* **2018**, *7*, 842–850. [CrossRef]
4. Straumal, B.B.; Mazilkin, A.A.; Protasova, S.G.; Myatiev, A.A.; Straumal, P.B.; Schuetz, G.; Van, P.A.; Goering, E.; Baretzky, B. Magnetisation study of nanograined pure and Mn-doped ZnO films: Formation of a ferromagnetic grain-boundary foam. *Phys. Rev. B* **2009**, *79*, 205–206. [CrossRef]
5. Feng, Q.; Blythe, H.J.; Fox, A.M.; Gehring, G.A.; Jiang, F.-X.; Xu, X.-H.; Heald, S.M. Contrasting behavior of the structural and magnetic properties in Mn- and Fe-doped In_2O_3 films. *App. Phys. Lett. Mater* **2013**, *1*, 022107. [CrossRef]
6. Feng, Q.; Blythe, H.J.; Fox, A.M.; Qin, X.-F.; Xu, X.-H.; Heald, S.M.; Gehring, G.A. Grain boundary ferromagnetism in vanadium-doped In_2O_3 thin films. *EPL Europhys. Lett.* **2013**, *103*, 67007. [CrossRef]
7. An, Y.; Wang, S.; Feng, D.; Wu, Z.; Liu, J. Correlation between oxygen vacancies and magnetism in Fe-doped In_2O_3 films. *Appl. Surf. Sci.* **2013**, *276*, 535–538. [CrossRef]
8. Marotta, V.; Orlando, S.; Parisi, G.; Giardini, A. Indium and tin oxide polycrystalline thin films as NO gas sensors produced by reactive pulsed laser ablation and deposition. *Appl. Phys. A* **1999**, *69*, S675–S677. [CrossRef]
9. Zheng, B.; Lian, J.; Zhao, L.; Jiang, Q. Optical and electrical properties of In-doped CdO thin films fabricated by pulse laser deposition. *Appl. Surf. Sci.* **2010**, *256*, 2910–2914. [CrossRef]
10. Alshammari, M.S.; Alqahtani, M.S.; Albargi, H.B.; Alfihed, S.A.; Alshetwi, Y.A.; Alghihab, A.A.; Alsamrah, A.M.; Alshammari, N.M.; Aldosari, M.A.; Alyamani, A.; et al. Magnetic properties of In_2O_3 containing Fe_3O_4 nanoparticles. *Phys. Rev. B* **2014**, *90*, 144433. [CrossRef]
11. Zaki, A.M.; Blythe, H.J.; Heald, S.M.; Fox, A.M.; Gehring, G.A. Growth of high quality yttrium iron garnet films using standard pulsed laser deposition technique. *J. Magn. Magn. Mater.* **2018**, *453*, 254–257. [CrossRef]
12. Zaki, A.M. A study of thin magnetic films of Iron Oxides and Yttrium Iron Garnet. Ph.D. Thesis, University of Sheffield, Sheffield, UK, 2018.
13. Feng, Q.; Dizayee, W.; Li, X.; Score, D.S.; Neal, J.R.; Behan, A.J.; Mokhtari, A.; Alshammari, M.S.; Al-Qahtani, M.S.; Blythe, H.J.; et al. Enhanced magnetic properties in ZnCoAlO caused by exchange-coupling to Co nanoparticles. *New J. Phys.* **2016**, *18*, 113040. [CrossRef]
14. Ying, M.; Blythe, H.J.; Dizayee, W.; Heald, S.M.; Gerriu, F.M.; Fox, A.M.; Gehring, G.A. Advantageous use of metallic cobalt in the target for pulsed laser deposition of cobalt-doped ZnO films. *Appl. Phys. Lett.* **2016**, *109*, 072403. [CrossRef]
15. Jiang, F.-X.; Xu, X.-H.; Zhang, J.; Fan, X.-C.; Wu, H.-S.; Alshammari, M.; Feng, Q.; Blythe, H.J.; Score, D.S.; Addison, K.; et al. Room temperature ferromagnetism in metallic and insulating $(In_{1-x}Fe_x)_2O_3$ thin films. *J. Appl. Phys.* **2011**, *109*, 53907. [CrossRef]
16. Hagleitner, D.R.; Menhart, M.; Jacobson, P.; Blomberg, S.; Schulte, K.; Lundgren, E.; Kubicek, M.; Fleig, J.; Kubel, F.; Puls, C.; et al. Bulk and surface characterization of In_2O_3(001) single crystals. *Phys. Rev. B* **2012**, *85*, 115441. [CrossRef]
17. Kaleemulla, S.; Readdy, A.; Uthanna, S.; Reddy, P. Physical properties of pure In_2O_3 thin films. *Optoelectron. Adv. Mater. Rapid Commun.* **2008**, *2*, 782–787.
18. King, P.D.C.; Veal, T.D.; Payne, D.J.; Bourlange, A.; Egdell, R.G.; McConville, C.F.; King, P.; Veal, T.; Payne, D. Surface electron accumulation and the charge neutrality level in In_2O_3. *Phys. Rev. Lett.* **2008**, *101*, 116808. [CrossRef]

19. Fahed, C.; Qadri, S.B.; Kim, H.; Piqué, A.; Miller, M.; Mahadik, N.A.; Rao, M.V.; Osofsky, M. Transport, magnetic and structural properties of bulk $In_{2-x}Fe_xO_3$. *Phys. Status Solidi* **2010**, *7*, 2298–2301. [CrossRef]

20. Kohiki, S.; Sasaki, M.; Murakawa, Y.; Hori, K.; Okada, K.; Shimooka, H.; Tajiri, T.; Deguchi, H.; Matsushima, S.; Oku, M.; et al. Doping of Fe to In_2O_3. *Thin Solid Films* **2006**, *505*, 122–125. [CrossRef]

21. Fallah, H.R.; Ghasemi, M.; Hassanzadeh, A.; Steki, H. The effect of annealing on structural, electrical and optical properties of nanostructured ITO films prepared by e-beam evaporation. *Mater. Res. Bull.* **2007**, *42*, 487–496. [CrossRef]

22. Ho, H.W.; Zhao, B.C.; Xia, B.; Huang, S.L.; Wu, Y.; Tao, J.G.; Huan, A.C.H.; Wang, L. Magnetic and magnetotransport properties in Cu and Fe co-doped bulk In_2O_3 and ITO. *J. Physics Condens. Matter* **2008**, *20*, 475204. [CrossRef]

23. Jiang, F.-X.; Xu, X.-H.; Zhang, J.; Wu, H.-S.; Gehring, G. High temperature ferromagnetism of the vacuum-annealed $(In_{1-x}Fe_x)_2O_3$ powders. *Appl. Surf. Sci.* **2009**, *255*, 3655–3658. [CrossRef]

24. Kim, H.; Osofsky, M.; Miller, M.M.; Qadri, S.B.; Auyeung, R.C.Y.; Piqué, A. Room temperature ferromagnetism in transparent Fe-doped In_2O_3 films. *Appl. Phys. Lett.* **2012**, *100*, 32404. [CrossRef]

25. He, J.; Xu, S.; Yoo, Y.K.; Xue, Q.; Lee, H.-C.; Cheng, S.; Xiang, X.-D.; Dionne, G.F.; Takeuchi, I. Room temperature ferromagnetic n-type semiconductor in $(In_{1-x}Fe_x)_2O_{3-\sigma}$. *Appl. Phys. Lett.* **2005**, *86*, 052503. [CrossRef]

26. Xu, X.-H.; Jiang, F.-X.; Zhang, J.; Fan, X.-C.; Wu, H.-S.; Gehring, G.A. Magnetic and transport properties of n-type Fe-doped In_2O_3 ferromagnetic thin films. *Appl. Phys. Lett.* **2009**, *94*, 212510. [CrossRef]

27. An, Y.; Wang, S.; Feng, D.; Wu, Z.; Liu, J. Local structure of Fe-doped In_2O_3 films investigated by X-ray absorption fine structure spectroscopy. *Appl. Phys. A* **2014**, *115*, 823–828. [CrossRef]

28. Ravel, B.; Newville, M. ATHENA, ARTEMIS, HEPHAESTUS: Data analysis for X-ray absorption spectroscopy using IFEFFIT. *J. Synchrotron Radiat.* **2005**, *12*, 537–541. [CrossRef]

29. Chen, A.; Zhu, K.; Zhong, H.; Shao, Q.; Ge, G. A new investigation of oxygen flow influence on ITO thin films by magnetron sputtering. *Sol. Energy Mater. Sol. Cells* **2014**, *120*, 157–162. [CrossRef]

30. Hakimi, A.M.H.R.; Blamire, M.G.; Heald, S.M.; Alshammari, M.S.; Alqahtani, M.S.; Score, D.S.; Blythe, H.J.; Fox, A.M.; Gehring, G.A. Donor-band ferromagnetism in cobalt-doped indium oxide. *Phys. Rev. B* **2011**, *84*, 085201. [CrossRef]

31. Jiang, F.; Xu, X.; Zhang, J.; Fan, X.; Wu, H.; Gehring, G. Role of carrier and spin in tuning ferromagnetism in Mn and Cr-doped In_2O_3 thin films. *Appl. Phys. Lett.* **2010**, *96*, 052503. [CrossRef]

32. An, Y.; Wang, S.; Duan, L.; Liu, J.; Wu, Z. Local Mn structure and room temperature ferromagnetism in Mn-doped In_2O_3 films. *Appl. Phys. Lett.* **2013**, *102*, 212411. [CrossRef]

33. Jayakumar, O.; Gopalakrishnan, I.; Kulshreshtha, S.; Gupta, A.; Rao, K.; Louzguine-Luzgin, D.; Inoue, A.; Glans, P.; Guo, J.; Samanta, K.; et al. Structural and magnetic properties of $(In_{1-x}Fe_x)_2O_3$ ($0.0 \leq x \leq 0.25$) system: Prepared by gel combustion method. *Appl. Phys. Lett.* **2007**, *91*, 052504. [CrossRef]

34. Karmakar, D.; Mandal, S.K.; Kadam, R.M.; Paulose, P.L.; Rajarajan, A.K.; Nath, T.K.; Das, A.K.; Dasgupta, I.; Das, G.P. Ferromagnetism in Fe-doped ZnO nanocrystals: Experiment and theory. *Phys. Rev. B* **2007**, *75*, 144404. [CrossRef]

35. Singhal, A.; Achary, S.N.; Manjanna, J.; Jayakumar, O.D.; Kadam, R.M.; Tyagi, A.K. Colloidal Fe-Doped Indium oxide nanoparticles: Facile synthesis, structural, and magnetic properties. *J. Phys. Chem. C* **2009**, *113*, 3600–3606. [CrossRef]

36. Neal, J.R.; Behan, A.J.; Ibrahim, R.M.; Blythe, H.J.; Ziese, M.; Fox, A.M.; Gehring, G.A. Room-temperature magneto-optics of ferromagnetic transition-metal-doped ZnO Thin Films. *Phys. Rev. Lett.* **2006**, *96*, 197208. [CrossRef] [PubMed]

37. Score, D.S.; Alshammari, M.; Feng, Q.; Blythe, H.J.; Fox, A.M.; Gehring, G.A.; Quan, Z.-Y.; Li, X.-L.; Xu, X.-H. Magneto-optical properties of Co/ZnO multilayer films. *J. Phys. Conf. Ser.* **2010**, *200*, 062024. [CrossRef]

38. Al-bargi, H.B. Magnetic and magneto-optic investigations of thin films of oxides. Ph.D. Thesis, University of Sheffield, Sheffield, UK, 2018.

© 2019 by the authors. Licensee MDPI, Basel, Switzerland. This article is an open access article distributed under the terms and conditions of the Creative Commons Attribution (CC BY) license (http://creativecommons.org/licenses/by/4.0/).

Communication

Synthesis of Iron Oxide Nanostructures via Carbothermal Reaction of Fe Microspheres Generated by Infrared Pulsed Laser Ablation

Jeffrey C. De Vero [1,*,†], Alladin C. Jasmin [1,2], Lean L. Dasallas [1], Wilson O. Garcia [1] and Roland V. Sarmago [1]

[1] National Institute of Physics, University of the Philippines, Diliman, Quezon City 1101, Philippines; acjasmin@up.edu.ph (A.C.J.); ldasallas@nip.upd.edu.ph (L.L.D.); wgarcia@nip.upd.edu.ph (W.O.G.); roland.sarmago@up.edu.ph (R.V.S.)
[2] Department of Physical Science, University of the Philippines, Baguio 2600, Philippines
* Correspondence: jeffrey-devero@aist.go.jp; Tel.: +81-8044338426
† Current address: National Institute of Advanced Industrial Science and Technology, Tsukuba 305-8665, Japan.

Received: 4 February 2019; Accepted: 2 March 2019; Published: 7 March 2019

Abstract: Iron oxide nanostructures were synthesized using the carbothermal reaction of Fe microspheres generated by infrared pulsed laser ablation. The Fe microspheres were successfully deposited on Si(100) substrates by laser ablation of the Fe metal target using Nd:YAG pulsed laser operating at $\lambda = 1064$ nm. By varying the deposition time (number of pulses), Fe microspheres can be prepared with sizes ranging from 400 nm to 10 μm. Carbothermal reaction of these microspheres at high temperatures results in the self-assembly of iron oxide nanostructures, which grow radially outward from the Fe surface. Nanoflakes appear to grow on small Fe microspheres, whereas nanowires with lengths up to 4.0 μm formed on the large Fe microspheres. Composition analyses indicate that the Fe microspheres were covered with an Fe_3O_4 thin layer, which converted into Fe_2O_3 nanowires under carbothermal reactions. The apparent radial or outward growth of Fe_2O_3 nanowires was attributed to the compressive stresses generated across the $Fe/Fe_3O_4/Fe_2O_3$ interfaces during the carbothermal heat treatment, which provides the chemical driving force for Fe diffusion. Based on these results, plausible thermodynamic and kinetic considerations of the driving force for the growth of Fe_2O_3 nanostructures were discussed.

Keywords: nanostructure; iron oxide; pulsed laser deposition

---

1. Introduction

Nanomaterials offer a new way of designing structural, functional, and electronic devices on the nanometer scale. These nanostructures show exceptional and dimension-dependent properties that are not readily observed in their bulk form and which can be exploited for biological and molecular applications [1–5]. In particular, Fe_2O_3 (iron oxide) nanostructures have been intensively studied because of their unique electrical and magnetic properties while maintaining their chemical compatibility with biological tissues for possible biological applications [1–6]. Their use extend to recording, ultrahigh density memory storage, and targeted drug delivery, including water splitting for energy applications [4–6]. Several techniques have been reported for the growth and preparation of iron oxide nanostructures, which aim at finding simple ways of controlling the morphology, size, and growth direction of these nanostructures for enhanced functionality. Most studied preparation techniques include hydrothermal synthesis [3,4], sol-gel techniques [5], thermal oxidation [6,7], electric arc gas discharge [8], and the pulsed laser deposition (PLD) method [9–11]. Common to all these methods, metal catalysts, such as Au (template), are necessary to assist in the production of nanostructures, which partially limits the controllability of the features, size, and functionality of the nanostructures.

The PLD method offers flexibility in the growth of thin films, including the size of nanoparticles, composition, and phase, by controlling the laser excitation source (wavelength), substrate temperature, oxygen partial pressure, and target composition [12,13]. The PLD techniques have been successfully applied to prepare different oxide materials, such as superconductors [14–18], magnetic materials [19,20], perovskite cathodes, and barrier layers for solid oxide fuel cells [21–25], including perovskite thin-film solar cells [26,27]. In particular, the preparation of nanostructure thin films, such as Au-TiO_2 [28,29], and nanostructure multilayered perovskite cathodes [30] under different PLD growth conditions were explored. These materials in nanostructured thin film forms showed a significant enhancement of the catalytic activities as compared to their bulk counterparts. As in the case of the laser-assisted growth of one-dimensional (1D) nanostructures, such as nanowires [31,32], gas-based lasers operating in the ultraviolet wavelength (UV) were commonly utilized. However, the cost of the gas sources limits their practical application as well as the apparent deviation of the composition of the grown layer, which has a complex stoichiometry (more than three elements), thus necessitating the use of an alternative laser source for PLD assisted growth [33,34].

One possible candidate for the excitation source for PLD is the Nd:YAG laser. It is all solid-state, thus requiring no toxic and expensive gas, is easy to maintain, and is environmentally friendly. The Nd:YAG laser is also highly tunable, which can be operated from its fundamental harmonic of λ = 1064 nm (infrared) to its fourth harmonic at λ = 266 nm (UV). Hence, the Nd:YAG laser is a promising excitation source for a safe and low-cost PLD process. The use of the fourth harmonics (λ = 266 nm) of the Nd:YAG for coated conductor thin film was demonstrated and showed comparable electrical properties with UV gas-based lasers [35]. Previously, we reported the use of the fundamental harmonic (λ = 1064 nm) of the Nd:YAG laser for the preparation of high-temperature superconducting thin film materials. The initial morphology of the as-prepared samples is spheroidal, which crystallize and form a relatively smooth and flat film layer after high-temperature post-annealing (>850 °C) [36–40]. This spheroidal feature of the ablated species was observed for all types of oxide materials we investigated, suggesting that the ablated materials from the target are molten when they arrive on the substrate. Since the substrate is kept at room temperature, the ablated particles solidify on the substrate layer upon cooling. Chemical analysis of the ablated species together with time-resolved optical emission spectroscopy of the laser produced plasma during infrared pulsed laser ablation showed a strong tendency of the spheroidal particles to maintain the stoichiometry of the target [37,38]. We extended the idea of using the Nd:YAG laser operating at λ = 1064 nm as the excitation source for the formation of nanoparticles on a Si(100) substrate coupled with the carbothermal reaction process to synthesize iron oxide nanostructures. The objective of this work is two-fold: (1) Synthesis of iron oxide nanostructures by carbothermal oxidation of Fe microspheres generated by infrared pulsed laser ablation, demonstrating the formation of iron oxide nanostructures without the need of catalysts (or self-assembly); (2) discussion of the plausible mechanism for the self-assembly of iron oxide nanostructures in terms of thermodynamics and kinetic processes.

2. Materials and Methods

The experiment consists of two parts: (1) Preparation of Fe microspheres by infrared ablation of Fe metal, and (2) carbothermal oxidation of Fe microspheres to generate the iron oxide nanostructures as shown in Figure 1. For the Fe microsphere preparation, a high power tunable Q-switch Nd:YAG pulsed laser (Quanta-Ray Pro Spectra Physics, Santa Clara, CA, USA) operating at the fundamental wavelength, λ = 1064 nm, with an 8 ns pulse duration, was used to ablate a rotating Fe metal target (99.9%, Kurt J. Lesker Company, Jefferson Hills, PA, USA) with a Si(100) substrate placed 30 mm away from target as shown in Figure 1a. The ablation was performed at a laser fluence of 4.32 J/cm^2 at a 10 Hz repetition rate. The deposition chamber was continuously evacuated to reach a pressure of

10^{-2} mbar. During the ablation process, no substrate heating and no background gas was employed. Herein, the deposition time was varied from 35 min (21,000 pulses) to 75 min (45,000 pulses) to control the density and size of nanoparticles on the Si(100) substrate. Since the substrate is kept at room temperature, the condensation of iron nanoparticles and microclusters aggregate on the Si substrate as spheroids, similar to our previous reports [37,38]. To grow the iron oxide nanowires/nanoflakes, the Fe microspheres were heat treated in a carbon-rich environment, as schematically shown in Figure 1b. Herein, the Si wafer with Fe microspheres was placed in a ceramic crucible containing 5.0 g of 99.99% activated carbon powders. The heat treatment was carried out in a box combustion furnace at temperatures between 750 to 800 °C for 145 min. To systematically evaluate the microstructural features and the composition of the Fe microspheres and Fe_2O_3 nanostructures, scanning electron microscopy coupled with energy dispersive x-ray spectroscopy (SEM-EDS, Hitachi S-3400N, Tokyo, Japan) was performed. Fourier transform infrared transmission spectroscopy (FTIR, Bio-Rad FTS-40A spectrometer (Cambridge, MA, USA) with KBr as the reference) was used to further confirm the iron oxide phase after thermal oxidation in the carbon-rich environment.

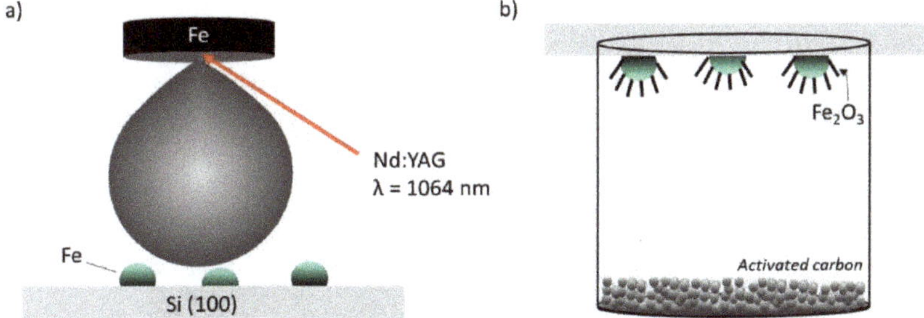

Figure 1. Schematic diagram of iron oxide nanostructure synthesis. The fabrication steps consist of two parts, namely (**a**) laser ablation of the Fe metal target using the Q-switch Nd:YAG pulsed laser operating at the fundamental harmonic, λ = 1064 nm. The Fe microspheres formed in the vicinity of unheated Si(100); (**b**) carbothermal oxidation of Fe microspheres at high temperatures (up to 800 °C) with activated carbon powder for 2 h. Iron oxide nanostructures grew radially outward from iron microspheres.

3. Results

Figure 2 shows the SEM micrographs of Fe microspheres generated by infrared laser ablation using the Nd:YAG laser. To control the size and density of the Fe microspheres, the deposition time was varied while keeping other deposition parameters fixed. Analysis of the SEM image reveals that the particle size density increases from 1.80×10^3 to 3.70×10^3 per μm^2 after fixing the deposition time to 75 min. The Fe particles are randomly distributed on the Si substrate with sizes ranging from 400 nm to 10 μm. Some Fe microspheres are smaller in size (~100 nm to 1.2 μm) at a shorter deposition time (35 min or 21,000 pulses, Figure 2a,b), whereas increasing the deposition time to 75 min (45,000 pulses) showed that large Fe microspheres up to 10 μm in diameter can be generated. At longer deposition times, the microspheres appear to coalesce together on the substrate (Figure 2c,d). The control of the microsphere density and size is possible by adjusting the laser pulses. However, the diameter of the particles has a tendency to saturate as evidenced by the distorted shape of the grains. It is possible that nanoparticles aggregated and further deposited on the microspheres as shown in Figure 2c,d. By changing the other PLD deposition conditions, such as the deposition pressure, substrate temperature, and target-to-substrate distance, it is possible to further tune the size, density, and morphological features of the grains [28,29,37]. It is interesting to note that the oxidation of the arriving particles can occur during the ablation process due to the minute amount of oxygen or during the cool down inside the growth chamber. This suggests that the Fe microspheres may have been covered with an oxide layer (oxidized) after the ablation process.

It can be seen that the characteristic feature of the surface particulates derived using infrared pulsed laser ablation is spherical or microspherical. This is because the nanoparticles, as well as the micron-sized droplets, are simultaneously generated by laser ablation. In general, to produce a thin film by PLD, the preparation condition is adjusted to decrease or suppress the droplets. More importantly, since no heat is induced on the site of arrival (usually by substrate heating), the droplets solidify, taking on a thermodynamically favored shape that in all the reported cases is spherical [36–38]. Previous studies also reveal that the ablation of a sintered target by an Nd:YAG laser operating at the infrared wavelength (λ = 1064 nm) produces ablation species consisting of molten-like spheroidal particles with stoichiometry resembling that of the target [36–38]. Longer wavelengths result in deeper target penetration, which causes subsurface evaporation and block-by-block ejection of larger clusters, which we observed as spheroids in the substrate [12,37,38,40]. Whereas, shorter wavelengths offer a higher photon energy, which is more suitable for efficient vaporization and ionization of the solid sample, resulting in atomized material delivery on a nearby substrate [12,33]. In particular, we reported UV pulsed laser ablation of iridate targets showing iridate nanoparticles (average size ~50 nm) on unheated MgO single crystal substrates [41]. The iridate nanoparticles are spheroidal with an apparent tendency to increase its particle density with increasing laser fluence [40]. The formation of large ZnO microspheres (sizes of more than 30 μm) was successfully synthesized using very high laser fluences up to 440 J/cm^2 (which is 10^3 higher than conventional laser density used in PLD) [41]. However, undesired severe target fragmentation was observed at this range of laser fluence. The systematic variation of the deposition conditions to create different features of iron microstructures is out of the scope of this work and is suggested for future studies.

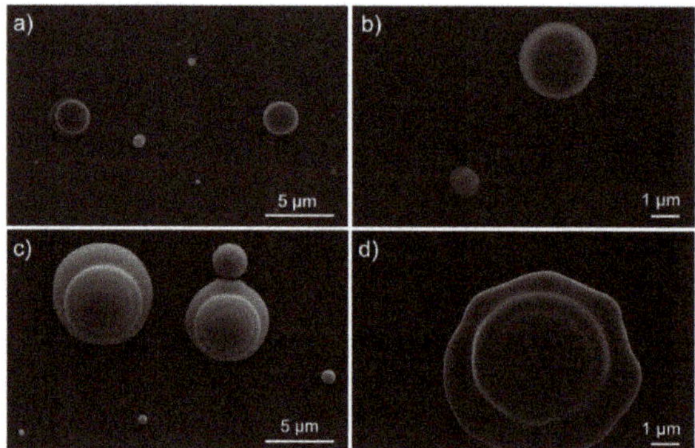

Figure 2. Fe microspheres on Si(100) generated by infrared pulsed laser ablation at deposition time of (**a,b**) 35 min (21,000 pulses); (**c,d**) 75 min (45,000 pulses). The surface features of Fe grains suggests that they are molten and cool when they arrive on the substrate.

Figure 3 shows the Fe microspheres (D = 10 μm) after carbothermal heat treatment at (a,b) 750 °C and (c,b) 800 °C for 145 min. It can be observed that iron oxide nanostructures grew radially outward from the surface of Fe microspheres. At 750 °C, iron oxide nanowires grew outward from the surface of Fe microspheres. The iron oxide nanostructures' diameters, D, vary from 500 nm (nanoflakes) to 2 μm (nanowires) as shown in Figure 3a. Nanosize pores on the surface of Fe microspheres are also evident at 750 °C (Figure 3a,b). On the other hand, carbothermal treatment at 800 °C decreases the size of the Fe microspheres, but results in longer nanowires extending up to L = 4.0 μm with a large aspect ratio (L/D = 44) as shown in Figure 3c. This suggests that increasing the carbothermal temperature assists in the growth of iron oxide nanowires. Pores are also present on the surface as well as some nanoflakes growing adjacent to the nanowires as shown in Figure 3d. A representative SEM micrograph of small microspheres (D = 1.8 μm) prepared at shorter laser pulses is also shown for comparison (Figure 3e). Nanoflakes appear to dominate the surface with no apparent growth of nanowires. The difference in the morphology of the iron oxide nanostructure indicates that the growth of nanowires sensitively changes with the size of the Fe microspheres. This suggests that large (>10.0 µm) microspheres are necessary to provide a sufficient supply of Fe during the carbothermal process for the growth of iron oxide nanowires. To illustrate the role of the carbon-rich environment in the self-assembly of iron oxide nanostructures, a reference Fe microsphere was oxidized in air (or without carbon) as shown in Figure 3f. It is clear from this result that the oxidation in air (or the trace amount of CO_2 in air) is insufficient to generate iron oxide nanowires, indicating that high vapor pressure carbon (or high density of carbon gaseous species) is critical for the synthesis of iron oxide nanostructures. The mechanistic growth of iron oxides under the carbothermal route will be discussed later. It is interesting to note that by changing the size/temperature, the resulting dimensions of the nanowires can be tuned, suggesting a flexibility of the infrared pulsed laser ablation assisted carbothermal growth of nanostructures. As the diameter of the wire decreases, the number of surfaces greatly increases. Hence, the large surface-to-volume ratio of the nanowire is expected to increase the number of active surfaces, which alters the physiochemical properties of the nanowires. A large aspect ratio is highly ideal for improving the sensitivity of the nanowires for real-time monitoring for sensor applications [1–7]. As a prospective future work, the micromanipulation and harvesting of iron oxide nanowires using the cantilevers of an atomic force microscope should be performed to isolate iron oxide nanowires for possible electronic applications (which our group has successfully demonstrated in the case of hydrothermally prepared ZnO nanowires exhibiting highly suitable scintillation properties, see [42,43]).

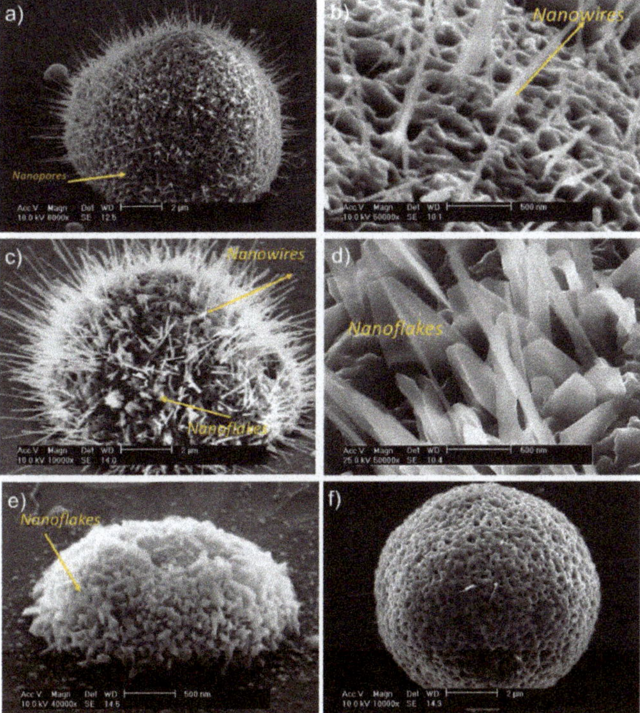

Figure 3. Fe microspheres (average diameter, D = 10 µm) after carbothermal heat treatment at (**a,b**) 750 °C and (**c,d**) 800 °C. The size and features of the iron oxide nanostructure sensitively change with temperature; (**e**) limited growth of nanoflakes on smaller Fe microspheres (D = 1.8 µm diameter); (**f**) reference Fe microsphere heat-treated in air.

Figure 4a shows the cross-sectional SEM image of iron oxide nanowires radially growing from the surface of Fe microspheres, which can extend up to 10 μm in length. The nanowires have a sharp tip with well-defined edges (290 nm in length) as shown in Figure 4b. The diameter of the iron oxide nanowires is typically around 440 nm. To elucidate the phase of the iron oxide nanowire generated by infrared pulsed laser ablation with the carbothermal technique, representative FTIR transmittance spectroscopy was performed as shown in Figure 4c. The FTIR spectra show absorbance peaks at 457.0 and 667.0 cm^{-1} as indicated in the figure. The characteristic absorption bands of the Fe_2O_3 phase are at 460 cm^{-1} (transverse-optical mode, TO) and 537 cm^{-1} (longitudinal-optical mode, LO). The FTIR spectra of the iron oxide nanostructure showed a small phase shift (3.0 cm^{-1}) in the TO mode and (5.0 cm^{-1}) LO mode as compared to the reported values of the TO and LO for Fe_2O_3 [44,45]. Representative EDS spectrum with a semi-quantitative analysis of Fe and O of the iron oxide nanostructure is shown in Figure 4d. The EDS confirmed that the nanowires contain both Fe and O in approximately a 2:3 ratio, indicating the formation of Fe_2O_3, which is consistent with the FTIR results. Hence, the iron oxide nanostructure formed can be assigned to the Fe_2O_3 phase. A trace amount of C was detected from the carbon used in the coating of the sample. Si was also detected since EDS may include some information from the substrate because of the penetration of the electron beam.

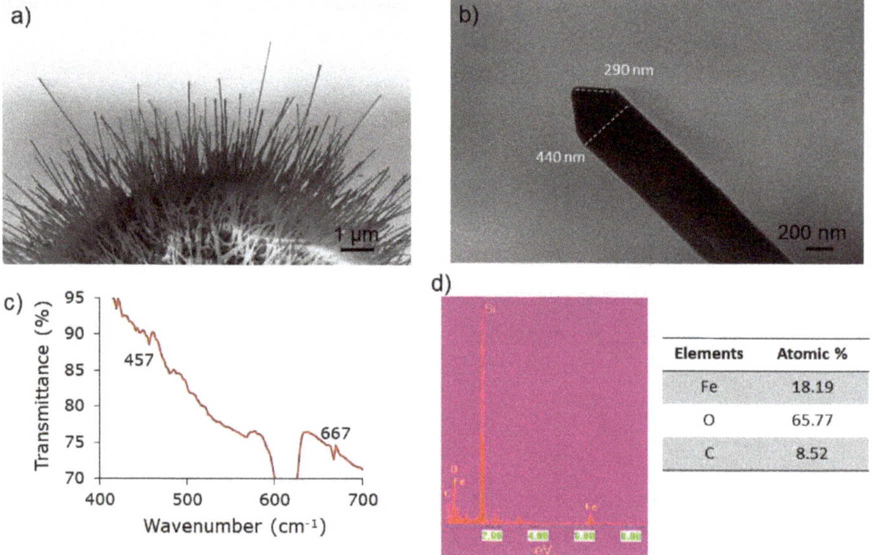

Figure 4. (**a**) SEM cross-sectional image of iron oxide nanostructures after carbothermal heat treatment showing the formation of iron oxide growing radially outward from the surface of Fe microspheres. (**b**) Representative high-resolution image of an iron oxide nanowire with well-defined edges. (**c**) FTIR spectra, (**d**) EDS spectra, and semi-quantitative analysis of the iron oxide nanostructures confirming the formation of the Fe_2O_3 phase.

4. Discussion

To describe the plausible growth mechanism of Fe_2O_3 nanostructures under the carbothermal reaction, the following chemical reactions are considered. The reaction of carbon and Fe with oxygen in air proceeds as follows:

$$C\ (s) + O_2\ (g) \rightarrow CO_2\ (g) \tag{1}$$

whereas:

$$3\ Fe\ (s) + 2\ O_2\ (g) \rightarrow Fe_3O_4 \tag{2}$$

Based on our results, the Fe_3O_4 appears to be in the intermediate phase of the nanoparticles or may be present in the walls of the microspheres and may react with carbon dioxide, forming Fe_2O_3 in the process:

$$2\ Fe_3O_4\ (s) + CO_2\ (g) = 3\ Fe_2O_3 + CO\ (g) \tag{3}$$

Alternatively, Fe_3O_4 may react with CO (g) following the reduction process:

$$Fe_3O_4\ (s) + CO\ (g) = 3\ FeO\ (s) + CO_2\ (g) \tag{4}$$

$$FeO\ (s) + CO\ (g) = Fe\ (s) + CO_2\ (g) \tag{5}$$

Note that gasification of solid carbon can also occur under large amounts of CO_2 gas via [46]:

$$C(s) + CO_2\ (g) = 2CO\ (g) \tag{6}$$

Hence, it possible that Equations (3) and (4) may occur simultaneously. The cycle proceeds until a sufficient supply of Fe is available to form Fe_2O_3 nanostructures, which in this case can be inferred from the change of the highly smooth surface of Fe microspheres to a rough porous like surface after the carbothermal reaction. In the case of the two stable oxide phases of Fe_3O_4 and Fe_2O_3, we can hypothesize that the surface contains two layers. The Fe_2O_3 may form at the outmost layer and subsurface of the less oxidized oxide phase containing Fe_3O_4. The ratio of these phases depends on the availability of Fe species present in bulk. It is highly improbable that Fe and C affect the Fe_3O_4-Fe_2O_3 layer and the reduction of Fe_3O_4 at the surface once both phases are established during the reaction. These reaction steps suggest that both supplies of Fe and carbon species are critically important to producing the Fe_2O_3 nanostructures. In other words, the availability of both Fe and C are the rate-limiting steps. In the absence of a sufficient carbon supply (low carbon vapor pressure), Fe_3O_4 may form around the walls of iron microspheres, which is consistent with our observation (Figure 3e) [47], and ceases to be converted to Fe_2O_3. Previously, we reported the spectroscopic analysis of the Fe layer on Si(100) carbothermally heat treated at 800 °C [47]. Raman spectroscopy reveals that the iron oxide nanostructures after carbothermal heat treatment were predominantly the Fe_2O_3 phase. A significantly broad Fe_3O_4 peak was also detected with respect to the literature value [48], suggesting that remnant Fe_3O_4 may be present in the sample, which did not fully convert to the Fe_2O_3 phase. The large shift in the observed Fe_3O_4 peak suggests that the Fe_3O_4 layer is less crystalline and more strained on the Fe microsphere than the highly faceted Fe_2O_3 nanowires. Based on the SEM analysis, the initially smooth surface of the Fe microspheres transformed into a rough and porous-like structure, which is a characteristic feature of the Fe_3O_4 film after thermal oxidation [46]. We take this to indicate that the walls of Fe microspheres contain a thin layer of Fe_3O_4. Kinetically, the solid–solid reaction is expected to be much slower compared to gas–solid. Hence, the driving force for Fe_2O_3 formation is the availability of carbon gas species surrounding the iron oxide nanoparticles [49].

To plausibly describe the apparent radial or outward growth of Fe_2O_3 nanowires on the surface of Fe microspheres generated by infrared pulsed laser ablation, we propose that compressive stresses are generated across the iron oxide interfaces during carbothermal processes, leading to the formation of iron nanostructures, as schematically shown in Figure 5. The atomic flux can be described using Fick's second law of diffusion [50] via:

$$J = cB\ \Delta\mu \tag{7}$$

where c is the concentration of the diffusion species (Fe), B is the mobility, and $\Delta\mu$ is the chemical potential gradient. The chemical potential gradient can be expressed in terms of the gradient of the hydrostatic pressure [51], σ, and can serve as the driving force, which can be written as:

$$J = \frac{cDV}{k_B T}\ \Delta\sigma \tag{8}$$

where the mobility, B, can be expressed as a factor of the diffusion coefficient, D, and the atomic (V) volume per unit of Boltzmann's constant, k_B, and the absolute temperature, T. The atomic flux is then directed from a more compressive/less tensile area to a less compressive/more tensile area. When Fe spheroids are heated in a carbon-rich environment, as described in the chemical reactions above, two thermodynamically stable oxide layers are formed, namely, Fe_3O_4 to Fe_2O_3 [49]. Since the molar volumes were in the order of Fe, or FeO < Fe_3O_4 < Fe_2O_3, hydrostatic stresses are generated across the interface. In this process of formation, Fe ions migrate from the Fe microsphere to the surface via bulk diffusion, leaving pores in the process. Fe ions can reach the Fe_3O_4/Fe_2O_3 via grain boundary diffusion. The Fe_3O_4 maybe in the transient state, transforming into Fe_2O_3 in the presence of a carbon-rich environment. The fast migration of Fe via surface diffusion can further promote the length of Fe_2O_3 nanowires. Hence, the migration of Fe ions via different diffusion pathways across the surface provides a continuous supply for the growth of the Fe_2O_3 nanowires. In this sense, larger sizes of Fe microspheres provide a larger volume for the mass transport of Fe ions. Large Fe microspheres (>10.0 µm) generate nanowires whereas small Fe microsphere result in the formation of nanoflakes. Our result is similar to the growth of CuO nanowires by hydrostatic stresses induced by the formation of different molar species of Cu across the Cu/Si interface under heat treatment [51]. The driving force for the growth of CuO nanowires was attributed to the stresses induced in the samples during annealing in air. Cu flux, which tends to migrate to some specific sites through surface and grain boundaries, provides for the growth of CuO nanowires [51]. The stress gradient determines the direction of Fe_2O_3.

The sufficient presence of CO_2 during heat treatment (or a carbon rich-environment) allows for the simultaneous reaction and reduction of Fe species, which is essential for the self-assembly of Fe_2O_3 nanostructures. Oxidation in air of the Fe microsphere did not produce an Fe_2O_3 nanostructure, indicating that the supply of both Fe and carbon are the rate-limiting steps for the growth of Fe_2O_3 nanostructures. Various stages of the growth formation of iron oxide nanostructures are schematically shown in Figure 5. Herein, Fe may diffuse from the bulk (microsphere) to the surface and may react with the oxygen and or carbon forming intermediate phase, Fe_3O_4 (stage 1 to stage 2). Fe ions may further diffuse at the Fe_3O_4/Fe_2O_3 interface via the grain boundary, whereas Fe may be transported through the nanowires via surface diffusion (stage 3). The diffusion of Fe ions through the nanowires was also reported to occur during the thermal oxidation of Fe metal at 600 °C, resulting in the growth of Fe_2O_3 nanowires [52], which is consistent with our results. In fact, the diffusivity of Fe ions is enhanced at high-temperatures (as in our case, which is 800 °C). Hence, the length of iron oxide nanowires strongly depends on the diffusion of Fe through the iron oxide interface. However, for growth to proceed, a carbon-rich environment is necessary as described in the chemical reactions presented in Equations (3)–(5). This is an important consideration in controlling the type of iron oxide nanostructure. Another possible growth route is via diffusion assisted seed crystal growth, however, the evidence suggests the contrary. As shown in Figure 3f, although Fe (seed) is present, the Fe/air interface did not produce iron oxide nanowires. However, under the carbothermal reaction route, the self-assembly of iron oxide nanowires occurred. The size and length of these nanowires are kinetically and thermodynamically controlled by the diffusion of Fe and the availability of carbon in air. It is important to note that in most surface-diffusion induced growth of nanowires, such as ZnO, the nanowires are prepared with small Au clusters on GaN substrates [53,54] and/or in Zn acetate seeds on amorphous substrates, where the diffusion of Zn ions controls the length of the nanowires [55]. A thorough discussion on the growth of ZnO nanowires from Zn-based seeds can be found in [55]. However, iron oxide nanowires were observed to grow on the surface of microspheres even without a catalyst as long as a sufficient amount of carbon was present, which was supplied by heating the activated carbon at high temperatures. The influence of the strain in the iron oxide nanostructure may also be explored in future works by changing the type of oxide substrate.

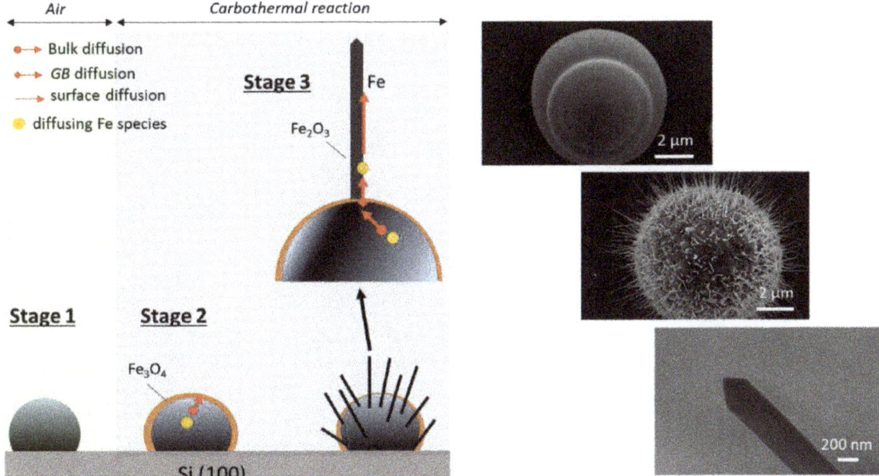

Figure 5. Plausible growth process of Fe_2O_3 nanowires under the carbothermal route. Key stages can be identified: Stage 1, deposition of the Fe microsphere; Stage 2, initial stage of the carbothermal annealing of Fe microspheres transform into Fe_3O_4 as the intermediate phase; Stage 3: final stage of carbothermal annealing, with the Fe_3O_4 transforming mostly into Fe_2O_3 nanowires. The stress gradient determines the direction of Fe_2O_3. Nanowires emerge from the Fe/Fe_3O_4 interface to outward stress-free Fe_2O_3 nanowire. Different types of pathways, such as bulk, grain boundary (GB), and surface diffusion, can assist in the transport of Fe across the interfaces. Representative SEM images of the Fe microsphere generated by infrared pulsed laser ablation and the carbothermally synthesized iron oxide nanostructures are also shown in the figure.

5. Summary

In summary, the formation of iron oxide nanostructures was successfully demonstrated using combined infrared pulsed laser ablation and carbothermal heat treatment of Fe microspheres. Results show that iron oxide nanowires and nanoflakes can be synthesized without the need of a catalyst. The iron oxide nanostructures grew radially outward from the Fe microspheres, suggesting a possible growth via self-assembly. Chemical and morphological analyses indicate that the synthesized iron oxide nanostructures can be attributed to the Fe_2O_3 phase. The plausible growth mechanism of iron oxide nanowires was described in terms of the chemical driving force for the diffusion of Fe species through the microsphere, as well as Fe_3O_4/Fe_2O_3 interfaces, which are both thermodynamically and kinetically controlled by the amounts of Fe and carbon gaseous species. The features of the nanoparticles can be easily tuned by simply adjusting the number of laser pulses or depositions during infrared pulsed laser ablation. The features of the iron oxide nanostructure (nanoflakes or nanowires) appeared to strongly depend on the size of the Fe microspheres. We also demonstrated that a sufficient amount of carbon during the heat treatment of Fe microspheres is necessary to generate iron oxide nanowires. An investigation of the electrical and magnetic properties of Fe_2O_3 is recommended for future work. It is interesting to note that the iron oxide nanostructures from infrared pulsed laser ablation result in the growth of Fe_2O_3 nanowires with a high aspect ratio. Iron oxide nanostructures with a high aspect ratio are highly suitable for bioengineering applications. Hence, this method is highly advantageous in fabricating iron oxide nanostructures with a high aspect ratio. It is important to point out that the use of an all solid-state laser, such as the Nd:YAG pulsed laser, allows for the controllable synthesis of nanostructure materials. It is envisioned that infrared laser ablation in combination with carbothermal heat-treatment could be a viable method to grow other nanostructured functional materials.

Author Contributions: Conceptualization, J.C.D.V. and A.C.J.; Methodology, J.C.D.V., A.C.J., L.L.D. and W.O.G.; Investigation, J.C.D.V. and A.C.J.; Data curation, J.C.D.V. and A.C.J.; Writing, J.C.D.V. and A.C.J.; Visualization, J.C.D.V. and A.C.J.; Supervision, R.V.S.; Funding acquisition, R.V.S.

Funding: This research received no external funding.

Acknowledgments: A. Jasmin acknowledges UP Baguio Local Faculty Fellowship Grant. University of the Philippines assisted in meeting the publication costs of this article.

Conflicts of Interest: The authors declare no conflict of interest.

References

1. Bery, C. Progress in functionalization of magnetic nanoparticles for applications in biomedicine. *J. Phys. D Appl. Phys.* **2009**, *42*, 224003. [CrossRef]
2. Xie, Y.; Ju, Y.; Toku, Y.; Morita, Y. Synthesis of single-crystal Fe_2O_3 nanowire array based on stress-induced atomic diffusion used for solar water splitting. *R. Soc. Open Sci.* **2018**, *5*, 171226. [CrossRef]
3. Takami, S.; Sato, T.; Mousava, T.; Ohara, S.; Umetsu, M.; Adschiri, T. Hydrothermal synthesis of surface-modified iron oxide nanoparticles. *Mater. Lett.* **2007**, *61*, 4769–4772. [CrossRef]
4. Li, J.; Shi, X.X.; Shen, M. Hydrothermal synthesis and functionalization of iron oxide nanoparticles for MR imaging applications. *Part. Part. Syst. Charact.* **2014**, *31*, 1223–1237. [CrossRef]
5. Walker, J.; Tannenbaum, R. Characterization of the sol–gel formation of iron (III) oxide/hydroxide nanonetworks from weak base molecule. *Chem. Mater.* **2006**, *18*, 4793–4801. [CrossRef]
6. Hiralal, P.; Unalan, H.; Wijayantha, K.; Kursumovic, A.; Jefferson, D.; MacManus-Driscoll, J.; Amaratunga, G. Growth and process conditions of aligned and patternable films of iron(III) oxide nanowires by thermal oxidation of iron. *Nanotechnology* **2008**, *19*, 455608. [CrossRef] [PubMed]
7. Yu, W.; Falker, J.; Yavuz, C.; Colvin, V. Synthesis of monodisperse iron oxide nanocrystals by thermal decomposition of iron carboxylate salts. *Chem. Commun.* **2004**, 2306–2307. [CrossRef] [PubMed]
8. Xue, D.; Gao, C.; Liu, Q.; Zhang, L. Preparation and characterization of hematite nanowire arrays. *J. Phys. Condens. Matter* **2003**, *15*, 1455. [CrossRef]
9. Shi, W.; Zheng, Y.; Peng, H.; Wang, N.; Lee, C.S.; Lee, S.T. Laser ablation synthesis and optical characterization of silicon carbide nanowires. *J. Am. Ceram. Soc.* **2000**, *83*, 3228–3330. [CrossRef]
10. Yu, D.; Sun, X.; Lee, C.; Bello, I.; Lee, S.; Gu, H.; Leung, K.; Zhou, G.; Dong, Z.; Zhang, Z. Synthesis of boron nitride nanotubes by means of excimer laser ablation at high temperature. *Appl. Phys. Lett.* **1998**, *72*, 1966–1967. [CrossRef]
11. Wang, N.; Zhang, Y.F.; Tang, Y.H.; Lee, C.S.; Lee, S.T. SiO_2-enhanced synthesis of Si nanowires by laser ablation. *Appl. Phys. Lett.* **1998**, *73*, 3902–3904. [CrossRef]
12. Chrisey, D.B. *Pulsed Laser Deposition of Thin Films*; Wiley-Interscience: New York, NY, USA, 1994; ISBN 978-0471592181.
13. Eason, R. *Pulsed Laser Deposition of Thin Films Applications-Led Growth of Functional Materials*; Wiley-Interscience: New York, NY, USA, 2006; ISBN 978-0471447092.
14. Dijkkamp, D.; Venkatesan, T.; Wu, X.D.; Shaheen, S.A.; Jiswari, N.; Min-lee, Y.H.; McLean, W.L.; Croft, M. Preparation of Y-Ba-Cu oxide superconductor thin films using pulsed laser evaporation from high T_c bulk material. *Appl. Phys. Lett.* **1987**, *51*, 619–621. [CrossRef]
15. de Vero, J.; Lee, D.; Shin, H.; Namuco, S.; Hwang, I.; Sarmago, R.; Song, J.H. Influence of deposition conditions on the growth of micron-thick highly *c*-axis textured superconducting $GdBa_2Cu_3O_{7-\delta}$ films on $SrTiO_3$ (100). *J. Vac. Sci. Technol.* **2018**, *36*, 031506. [CrossRef]
16. de Vero, J.; Hwang, I.; Santiago, A.; Chang, J.; Kim, J.; Sarmago, R.; Song, J. Growth of $Bi_2Sr_2CaCu_2O_{8+\delta}$ thin films with enhanced superconducting properties by incorporating $CaIrO_3$ nanoparticles. *Appl. Phys. Lett.* **2014**, *104*, 172603. [CrossRef]
17. Wu, X.D.; Dye, R.C.; Muenchausen, R.E.; Foltyn, S.R.; Maley, M.; Rollett, A.D.; Garcia, A.R.; Nogar, N.S. Epitaxial CeO_2 films as buffer layers for high-temperature superconducting thin films. *Appl. Phys. Lett.* **1991**, *58*, 2165. [CrossRef]
18. Singh, R.; Kumar, D. Pulsed laser deposition and characterization of high-T_c $YBa_2Cu_3O_{7-x}$ superconducting thin films. *Mater. Sci. Eng. Rep.* **1998**, *22*, 113–185. [CrossRef]

19. Shen, J.; Gai, Z.; Kirschner, J. Growth and magnetism of metallic films and multilayers by pulsed laser deposition. *Surf. Sci. Rep.* **2004**, *52*, 163–218. [CrossRef]

20. Chen, X.; Chien, C. Magnetic properties of epitaxial Mn-doped ZnO thin films. *J. Appl. Phys.* **2003**, *93*, 7876–7878. [CrossRef]

21. Plonczak, P.; Søgaard, A.B.M.; Ryll, T.; Martynczuk, J.; Hendriksen, P.; Gauckler, L. Tailoring of $La_xSr_xCo_yFe_{1-y}O_{3-\delta}$ nanostructure by Pulsed Laser Deposition. *Adv. Funct. Mater.* **2011**, *21*, 2764–2775. [CrossRef]

22. de Vero, J.C.; Develos-Bagarinao, K.; Kishimoto, H.; Ishiyama, T.; Yamaji, K.; Horita, T.; Yokokawa, H. Enhanced stability of solid oxide fuel cells by employing a modified cathode- interlayer interface with a dense $La0_.6Sr_{0.4}Co_{0.2}Fe_{0.8}O_{3-\delta}$ thin film. *J. Power Sources* **2018**, *377*, 128–135. [CrossRef]

23. de Vero, J.C.; Develos-Bagarinao, K.; Kishimoto, H.; Ishiyama, T.; Yamaji, K.; Horita, T.; Yokokawa, H. Optimization of GDC interlayer against $SrZrO_3$ formation in LSCF/GDC/YSZ triplets. In Proceedings of the 12th European SOFC and SOEC Forum, Lucerne, Switzerland, 5–8 July 2016; Volume 41, p. B1513.

24. Morales, M.; Pesce, A.; Slodczyk, A.; Torrell, M.; Piccardo, P., II; Montinaro, D.; Tarancón, A.; Morata, A. Enhanced performance of gadolinia-doped ceria diffusion barrier layers fabricated by pulsed laser deposition for large-area solid oxide fuel cells. *ACS Appl. Energy Mater.* **2018**, *1*, 1955–1964. [CrossRef]

25. Yan, J.; Matsumoto, H.; Akbay, T.; Yamada, T.; Ishihara, T. Preparation of $LaGaO_3$-based perovskite oxide film by pulsed-laser ablation method and application as a solid oxide fuel cell electrolyte. *J. Power Sources* **2006**, *157*, 714–719. [CrossRef]

26. Liang, Y.; Yao, Y.; Zhang, X.; Hsu, W.; Gong, Y.; Shin, J.; Wachsman, E.; Dagenais, M.; Takeuchi, I. Fabrication of organic-inorganic perovskite thin films for planar solar cells via pulsed laser deposition. *AIP Adv.* **2016**, *6*, 05001. [CrossRef]

27. Park, J.; Seo, J.; Park, S.; Shin, S.; Kim, Y.; Jeon, N.; Shin, H.; Noh, T.A.J.; Yoon, S.; Hwang, C.; et al. Efficient $CH_3NH_3PbI_3$ perovskite solar cells employing nanostructured p-Type NiO electrode formed by a pulsed laser deposition. *Adv. Mater.* **2015**, *27*, 4013–4019. [CrossRef] [PubMed]

28. Ghidelli, M.; Mascaretti, L.; Bricchi, B.; Zapelli, A.; Russo, V.; Casari, C.; Bassi, A.L. Engineering plasmonic nanostructured surfaces by pulsed laser deposition. *Appl. Surf. Sci.* **2018**, *433*, 1064–1073. [CrossRef]

29. Bricchi, B.; Ghidelli, M.; Mascaretti, L.; Zapelli, A.; Russo, V.; Casari, C.; Terraneo, G.; Alessandri, I.; Ducati, C.; Bassi, A.L. Intergration of plasmonic Au nanoparticles in TiO_2 hierarichal structures in a single-step pulsed laser co-deposition. *Mater. Des.* **2018**, *156*, 311–319. [CrossRef]

30. Develos-Bagarinao, K.; de Vero, J.; Kishimoto, H.; Yamaji, K.; Horita, T.; Yokokawa, H. Multilayered LSC and GDC: An approach for designing cathode materials with superior oxygen exchange properties for solid oxide fuel cells. *Nano Energy* **2018**, *52*, 369–380. [CrossRef]

31. Yang, R. One-Dimensional Nanostructures by Pulsed Laser Ablation. *Sci. Adv. Mater.* **2012**, *4*, 401–406. [CrossRef]

32. Morales, A.; Lieber, C. A laser ablation method for synthesis of crystalline semiconductor nanowires. *Science* **1998**, *279*, 208–211. [CrossRef] [PubMed]

33. Schou, J. Physical aspects of the pulsed laser deposition technique: The stoichiometric transfer of material from target to film. *Appl. Surf. Sci.* **2009**, *10*, 5191–5198. [CrossRef]

34. Arnold, C.B.; Aziz, M.J. Stoichiometry issues in pulsed laser deposition of alloys grown from multicomponent targets. *Appl. Phys. A* **1999**, *69*, S23–S27. [CrossRef]

35. Ichino, Y.; Yoshida, Y.; Yoshimura, T.; Takai, Y.; Yoshizumi, M.; Izumi, T.; Shiohara, Y. Potential of Nd:YAG pulsed laser deposition method for coated conductor production. *Phys. C* **2010**, *470*, 1234–1237. [CrossRef]

36. de Vero, J.; Blanca, G.R.S.; Vitug, J.; Garcia, W.; Sarmago, R. Stoichiometric transfer of material in the infrared pulsed laser deposition of yttrium doped Bi-2212 films. *Phys. C* **2011**, *471*, 378–383. [CrossRef]

37. de Vero, J.; Gabayno, J.F.; Garcia, W.O.; Sarmago, R.V. Growth of $Bi_2Sr_2CaCu_2O_{8+\delta}$ thin films deposited by infrared (1064 nm) pulsed laser deposition. *Phys. C* **2010**, *470*, 149–154. [CrossRef]

38. Vitug, J.; de Vero, J.; Blanca, G.R.S.; Sarmago, R.; Garcia, W. Stoichiometric transfer by infrared pulsed laser deposition of y-doped Bi–Sr–Ca–Cu–O investigated using time-resolved optical emission spectroscopy. *J. Appl. Spectrosc.* **2012**, *78*, 855–860. [CrossRef]

39. de Vero, J.; Lopez, R.A.; Garcia, W.O.; Sarmago, R.V. Post deposition heat treatment effects of ceramic superconducting films produced by infrared Nd:YAG pulsed laser deposition. In *Heat Treatment*; Czerwinski, F., Ed.; InTechOpen: Winchester, UK, 2012; pp. 197–205. ISBN 978-953-51-0768-2.

40. de Vero, J.C.; Hwang, I.; Shin, H.; Santiago, A.; Lee, D.; Chang, J.; Kim, J.; Sarmago, R.; Song, J.H. Growth and superconducting properties of $Bi_2Sr_2CaCu_2O_{8-\delta}$ thin films incorporated with iridate nanoparticles. *Phys. Status Solidi A* **2014**, *211*, 1787–1793. [CrossRef]

41. Nakamura, D.; Tanaka, T.; Ikebuchi, T.; Ueyama, T.; Higashihata, M.; Okada, T. Synthesis of Spherical ZnO Microcrystals by Laser Ablation in Air. *Electro Commun. Jpn.* **2016**, *99*, 58–63. [CrossRef]

42. Santos-Putungan, A.; Singidas, B.; Sarmago, R. Manipulation of low temperature grown ZnO rigid structures via Atomic Force Microscope. *HCTL Open Int. J. Technol. Innov. Res.* **2014**, *11*, 1–8.

43. Santos-Putungan, A.B.; Empizo, M.J.F.; Yamanoi, K.; Vargas, R.M.; Arita, R.; Minami, Y.; Shimizu, T.; Salvador, A.A.; Sarmago, R.V.; Sarukura, N. Intense and fast UV emitting ZnO microrods fabricated by low temperature aqueous chemoical growth method. *J. Lum.* **2016**, *169*, 216–219.

44. Chen, Z.; Cvelbar, U.; Mozetic, M.; He, J.; Sunkara, M. Long-range ordering of oxygen-vacancy planes in Fe_2O_3 nanowires and nanobelts. *Chem. Mater.* **2008**, *20*, 3224–3228. [CrossRef]

45. Jaeger, R.C. Thermal Oxidation of Silicon. In *Introduction to Microelectronic Fabrication*; Prentice Hall Inc.: New York, NY, USA, 2001; p. 30.

46. Arthur, J.R. Reaction between C and O_2. *Trans. Faraday Soc.* **1951**, *47*, 164–177. [CrossRef]

47. Jasmin, A.; Rillera, H.; Semblante, O.; Sarmago, R. Surface morphology, microstructure, raman characterization and magnetic ordering of oxidized Fe-sputtered films on silicon substrate. *AIP Conf. Proc.* **2012**, *1482*, 572–577.

48. Maslar, J.; Hurst, W.; Bowers, W.; Hendricks, J.; Aquino, M. In situ raman spectroscopic investigation of aqueous iron corrosion at elevated temperatures and pressures. *J. Electrochem. Soc.* **2000**, *147*, 2532–2542. [CrossRef]

49. Moon, J.; Sahajwalla, V. Kinetic model for the uniform conversion of self-reducing iron oxide carbon briquettes. *ISIJ Int.* **2003**, *43*, 1136–1142. [CrossRef]

50. Crank, J. *Mathematics of Diffusion*, 2nd ed.; Oxford Science Publications: Oxford, UK, 1975; p. 36.

51. Chen, M.; Yue, Y.; Jun, Y. Growth of metal and metal oxide nanowires driven by the stress-induced migration. *J. Appl. Phys.* **2012**, *11*, 104305. [CrossRef]

52. Yuan, L.; Wang, Y.; Cai, R.; Jiang, Q.; Wang, J.; Li, B.; Sharma, A.; Zhou, G. The origin of hematite nanowire growth during thermal oxidation of iron. *Mater. Eng. B* **2012**, *177*, 327–336. [CrossRef]

53. Kim, D.; Gosele, U.; Zacharias, M. Surface-diffusion induced growth of ZnO nanowires. *J. Cryst. Growth* **2009**, *311*, 3216–3219. [CrossRef]

54. Shih, P.-H.; Wu, S. Growth mechanism studies of ZnO nanowires: Experimental observations and short-circuit diffusion analysis. *Nanomaterials* **2017**, *7*, 188. [CrossRef]

55. Cutinho, J.; Chang, B.S.; Oyola-Reynoso, S.; Chen, J.; Akhter, S.S.; Tevis, I.; Bello, N.; Martin, A.; Foster, M.; Thuo, M. Automous thermal-oxidative composition inversion and texture tuning of liquid metal surfaces. *ACS Nano* **2018**, *12*, 4744–4753. [CrossRef] [PubMed]

© 2019 by the authors. Licensee MDPI, Basel, Switzerland. This article is an open access article distributed under the terms and conditions of the Creative Commons Attribution (CC BY) license (http://creativecommons.org/licenses/by/4.0/).

Article

Pulsed Laser Deposition of Aluminum Nitride Films: Correlation between Mechanical, Optical, and Structural Properties

Lilyana Kolaklieva [1], Vasiliy Chitanov [1], Anna Szekeres [2], Krassimira Antonova [2], Penka Terziyska [2], Zsolt Fogarassy [3], Peter Petrik [3], Ion N. Mihailescu [4] and Liviu Duta [4,*

[1]   Central Laboratory of Applied Physics, Bulgarian Academy of Sciences, 61 St. Petersburg Blvd., 4000 Plovdiv, Bulgaria; ohmic@mbox.digsys.bg (L.K.); vchitanov@gmail.com (V.C.)
[2]   Institute of Solid State Physics, Bulgarian Academy of Sciences, Tzarigradsko Chaussee 72, 1784 Sofia, Bulgaria; szekeres@issp.bas.bg (A.S.); krasa@issp.bas.bg (K.A.); penka@issp.bas.bg (P.T.)
[3]   Centre for Energy Research, Hungarian Academy of Sciences, Konkoly-Thege út 29-33, H-1121 Budapest, Hungary; fogarassy.zsolt@energia.mta.hu (Z.F.); petrik.peter@energia.mta.hu (P.P.)
[4]   National Institute for Lasers, Plasma, and Radiation Physics, 409 Atomistilor Street, 077125 Magurele, Romania; ion.mihailescu@inflpr.ro
*   Correspondence: liviu.duta@inflpr.ro

Received: 9 February 2019; Accepted: 13 March 2019; Published: 17 March 2019

Abstract: Aluminum nitride (AlN) films were synthesized onto Si(100) substrates by pulsed laser deposition (PLD) in vacuum or nitrogen, at 0.1, 1, 5, or 10 Pa, and substrate temperatures ranging from RT to 800 °C. The laser parameters were set at: incident laser fluence of 3–10 J/cm^2 and laser pulse repetition frequency of 3, 10, or 40 Hz, respectively. The films' hardness was investigated by depth-sensing nanoindentation. The optical properties were studied by FTIR spectroscopy and UV-near IR ellipsometry. Hardness values within the range of 22–30 GPa and Young's modulus values of 230–280 GPa have been inferred. These values were determined by the AlN film structure that consisted of nanocrystallite grains, strongly dependent on the deposition parameters. The values of optical constants, superior to amorphous AlN, support the presence of crystallites in the amorphous film matrix. They were visualized by TEM and evidenced by FTIR spectroscopy. The characteristic Reststrahlen band of the *h*-AlN lattice with component lines arising from IR active phonon vibrational modes in AlN nanocrystallites was well detectable within the spectral range of 950–500 cm^{-1}. Control X-ray diffraction and atomic force microscopy data were introduced and discussed. All measurements delivered congruent results and have clearly shown a correlation between the films' structure and the mechanical and optical properties dependent on the experimental conditions.

Keywords: aluminum nitride; pulsed laser deposition; nanoindentation testing; TEM imaging; FTIR spectroscopy; ellipsometry; complex refractive index

---

1. Introduction

Pulsed laser-assisted coatings represent a clean and fast route applied for surface modification and controlled micro-structuring of a wide range of materials. When compared to other physical vapor deposition methods, i.e., thermal evaporation or sputtering, pulsed laser deposition (PLD) stands out as a simple, versatile, rapid, and cost-effective method, which can enable precise control of thickness and morphology for the fabrication of high-quality thin films [1,2]. Amorphous or crystalline, extremely adherent, stoichiometric, dense, or porous structures from various complex materials can be synthesized, even at relatively low deposition temperatures, by simply varying the experimental parameters, mainly related to the (*i*) laser (fluence, wavelength, pulse-duration, and repetition rate) and (*ii*) deposition conditions (target-to-substrate distance, substrate temperature, nature, and pressure of the environment) [2–4].

Thin though hard coatings have proven invaluable for the production of mechanical parts or tools due to their hardness and wear-resistance characteristics [5,6]. In this respect, for the last couple of years, a great interest in using nitride-based films as protective coatings, due to their physical, chemical, electronic, thermal, or mechanical properties, has been reported [7–10]. In particular, aluminum nitride (AlN) coatings possess such characteristics, which make them suitable candidates for a wide range of applications, including insulating and buffer layers, photodetectors, light-emitting diodes, laser diodes, acoustic devices, or designs of self-sustainable opto- and micro-electronical devices [11–16]. Hard protective AlN coatings in multi-layered systems as AlN/TiN and CrN/AlN were intensively studied for tribological applications [8,9,17,18]. AlN is also commonly used in piezoelectric thin films [19,20], for the fabrication of micro-electro-mechanical system (MEMS) devices [21].

Depending on the deposition techniques and technological protocols, the AlN film structure can vary from fully-amorphous to nanocrystalline, with a tendency to decrease the volume fraction of grain boundaries [22–25]. This may significantly modify the physical, chemical, and mechanical properties of films with nano-sized crystalline structure in comparison to polycrystalline materials, which have grain sizes usually in the range of 100–300 μm [26]. Highly *c*-axis-oriented AlN films exhibit a large piezoelectric coefficient and are attractive for electroacoustic devices via surface acoustic waves [12,13]. Therefore, the fabrication of hard coatings based on properly-oriented nanocrystalline AlN layers requires a good understanding of their microstructure as a function of deposition conditions. However, obtaining AlN films with a definite structure and crystalline quality still remains a challenge for most deposition techniques. The PLD method has the main advantage of ensuring the growth of thin AlN films with good crystallinity and stoichiometry at relatively low temperatures [27]. Furthermore, PLD for AlN film synthesis proved to be one of the methods resulting in superior mechanical properties of the material [28]. There is still no straightforward theoretical or experimental model of the processes during deposition and the resulting film properties. Hence, the characterization of film growth and the mechanisms governing the film synthesis are important tasks in all application areas of AlN films.

Thin AlN films synthesis by the PLD technique is also the subject of our research. We focused during the years on the influence of the technological parameters, such as the assisting nitrogen gas pressure, incident laser fluence, repetition rate of laser pulses, substrate temperature, and the presence of an additional matching sub-layer, on the physical properties of AlN films synthesized by PLD [23,29–36] onto Si(100) substrates. Physical properties, such as surface roughness, microstructure, composition, amorphous-to-polycrystalline phase ratio, and optical constants appropriate for various applications, have been systematically studied. A systematization of the experimental results and finding the correlation between the structure and properties of the PLD AlN films and their preparation conditions would allow for the optimization of the deposition process in order to fabricate AlN films with the desired quality.

We resume with this paper the investigations with special attention to new, previously-unstudied phenomena, in the trial to better understand the quite complicated physical and chemical PLD process. Thus, by depth-sensing nanoindentation, the mechanical properties of the PLD AlN films, fabricated at substrate temperatures ranging from room temperature (RT) up to 800 °C and, varying other deposition parameters such as ambient environment, gas pressure, laser incident fluence, and laser pulse frequency (LPF), were studied. Complementary results obtained by transmission electron microscopy (TEM), Fourier transform infrared (FTIR) spectroscopy, and UV-near IR ellipsometry are also reported, with the aim of finding the relationship between the structural properties of films and their mechanical properties.

2. Experimental Details

2.1. AlN Film Preparation

AlN films were synthesized onto Si(100) substrates by laser ablation of a polycrystalline stoichiometric AlN target using a pulsed KrF* excimer laser source COMPex Pro205 (Coherent, Göttingen, Germany, λ = 248 nm, $\nu_{FWHM} \leq 25$ ns). The laser beam was oriented at 45° with respect to the target surface. The laser pulse energy was ~360 mJ, and the incident laser fluence was set at ~3, 4, 4.8, or 10 J/cm^2, respectively. The separation distance between the target and Si substrate was 5 cm. The PLD process was performed in vacuum (~10^{-4} Pa) or at different N$_2$ gas pressures of 0.1, 1, 5, or 10 Pa, respectively. Before each experiment, the irradiation chamber was evacuated down to a residual pressure of ~10^{-5} Pa.

Prior to deposition, the Si substrates were cleaned in diluted (5%) hydrogen fluoride solution in order to eliminate the native oxide layer. The target was cleaned by baking at 800 °C for 1 h in a vacuum followed by a short multipulse laser ablation with 1000 pulses. A shutter was interposed in this case between the target and substrate to collect the expelled impurities.

During deposition, the target was continuously rotated with 0.4 Hz and translated along two orthogonal axes to avoid piercing and allow for the growth of uniform thin films. The substrate was heated either at 800, 450, 400, and 350 °C or was maintained at RT. The chosen temperature was kept constant with the help of a PID-EXCEL temperature controller (Excel Instruments, Gujarat, India).

For the deposition of one thin film, 15,000, 20,000, or 25,000 consecutive laser pulses were applied, with a corresponding LPF of 40, 10, or 3 Hz, respectively.

2.2. Nanoindentation Testing

The mechanical properties of the synthesized AlN films were investigated by a depth-sensing indentation method using Compact Platform CPX-MHT/NHT equipment (CSM Instruments/Anton-Paar, Peseux, Switzerland). Nanoindentation was performed with a triangular diamond Berkovich pyramid having a facet angle of 65.3° $\pm$ 0.3° (CSM-Instruments SA certificate B-N 41), in the loading interval starting from 5–100 mN. The nanohardness and elastic modulus were determined from the load/displacement curves applying the Oliver and Phar method [37].

2.3. Transmission Electron Microscopic Measurements

The structure of the PLD AlN films was investigated by transmission electron microscopy (TEM) and high-resolution transmission electron microscopy (HR-TEM) with a Philips CM-20 (Amsterdam, The Netherlands) operated at a 200-kV accelerating voltage and a JEOL 3010 (Tokyo, Japan) operated at a 300-kV accelerating voltage. The cross-sectional TEM samples were prepared by ion beam milling.

2.4. Optical Measurements

The influence of the deposition conditions on the films' complex refractive index ($\tilde{n} = n - jk$, where n is the refractive index and k is the extinction coefficient) was studied by spectroscopic ellipsometry (SE) measurements on an M1000D ellipsometer from J.A. Woollam Co., Inc. (Lincoln, NE, USA) working in the spectral range of 193–1000 nm. In the SE data analysis, the Complete EASE J.A. Woollam Co., Inc. software (version 5.08) was used [38]. The experimental SE spectra were taken at RT and different angles of light incidence of 60°, 65°, and 70°. In data simulation, a two-layer optical model (substrate–1st layer (film bulk)–2nd layer (surface roughness)) was applied. In the spectral range of 400–1000 nm, the data were fitted by the Cauchy model to obtain the films' thickness values. The ellipsometric data were fitted by a Tauc–Lorentz general oscillator model. The surface roughness layer was modeled as a mixture of 50% material (film) and 50% voids (air) and was calculated by applying Bruggeman's effective medium approximation theory.

FTIR reflectance spectra were obtained in a linearly-polarized incidence beam by using a Bruker Vertex 70 instrument (Billerica, MA, USA) equipped with a reflectance accessory A513/Q. Both *s* and *p* irradiation polarizations were exploited at an incident angle of 70°. In this geometry, it is more correct to consider the orientation of E with respect to the normal to the film surface z instead of the optical nanocrystalline axis c, which could be oriented in a certain direction with a probability depending on the deposition conditions. Furthermore, it should be underlined that during every measurement, the components of the electric vector E oriented along the x, y, and z directions were presented with different weights at different temperatures. Thus, all electric field components contributed to the phonon-polariton modes in randomly-oriented AlN nanocrystallites. The spectral resolution was 2 cm^{-1}, and the total number of scans per each measurement was 64.

3. Results and Discussion

3.1. Nanoindentation Testing

For all AlN films, the measured load-penetration depth curves with maximum indentation loads were smooth, with no discontinuities. The smooth loading nature testifies to the good film uniformity and adherence to the Si substrate. Even for the highest displacement load of 100 mN, when the indentation depth was close to the film thickness, there were no signs of cracking or peeling, which demonstrates the good interface quality. In Figure 1, a typical load versus indenter displacement curve is presented, corresponding to a test performed on an AlN film deposited at 800 °C, in 0.1 Pa N$_2$ pressure and at a LPF of 40 Hz. The main parameters used for the analysis are marked on the graph. F_m is the peak load corresponding to a maximum nominal penetration depth h_{max}, which depends on the hardness and, consequently, on the film structure. The stiffness S results from the slope of the tangent to the unloading curve. The measured depth h verifies the relation $h = h_s + h_c/\varepsilon$, where h_s is the displacement of the surface at the perimeter of the contact, h_c is the vertical distance along which the contact is made, and ε is an indenter constant.

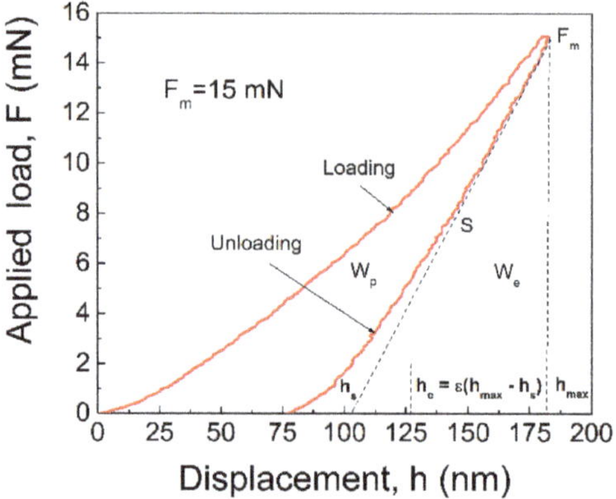

Figure 1. Typical load-displacement curve at the maximum load of 15 mN in the case of an AlN film deposited at 800 °C, in 0.1 Pa N$_2$ pressure and with a laser pulse frequency of 40 Hz.

The area between the loading and unloading curves defines the plastic deformation work W_p, while that between the unloading curve and perpendicular to the maximum penetration depth, h_{max}, is a measure of the elastic deformation work, W_e. The ratio $W_e/(W_e + W_p)$ defines the elastic recovery of the coating after indentation and is associated with the coating ability to go back after deformation. For the studied films, this ratio varied from 49–67%, depending on the deposition conditions. This implies a very good coating recovery after mechanical deformation.

The measured load and displacement curves were analyzed, and the nanohardness H and elastic modulus E were evaluated [37]. The hardness was estimated from the relation $H = F_{max}/A$, where A is the projected contact area of the indentation. By fitting the unloading curve, i.e., the stiffness

$S = \mathrm{d}F/\mathrm{d}h$, the projected area A can be determined. The Young's modulus E is determined from the relation $(1 - v^2)/E = (2A^{1/2}/S\pi^{1/2}) - (1-v_i^2)/E_i$ [37], where E_i and v_i are the elastic modulus and Poisson's ratio of the indenter (v was assumed equal to 0.22).

The dependence of the nanohardness and elastic modulus on the indentation depth corresponding to the applied load in the interval from 5–100 mN is presented in Figure 2. Below the loading value of 5 mN, the hardness determination with sufficient accuracy is limited by the surface roughness [39]. Our PLD AlN films exhibited a considerably smooth and uniform surface morphology, with a root mean squared roughness in the range of 0.24–2.5 nm, depending on the deposition conditions [35,36,40].

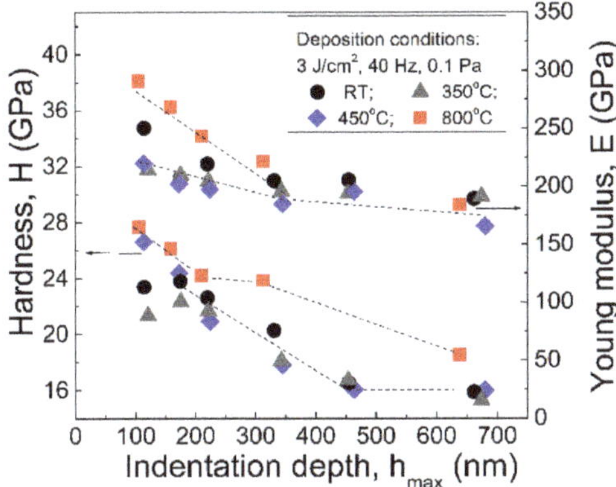

Figure 2. Nanohardness H and elastic modulus E as a function of the maximum nominal penetration depth, h_{max}, of the PLD AlN films obtained using the deposition conditions given in the inset.

The variation of hardness with the indenter penetration depth points to a region of $\Delta h_{\mathrm{max}} \approx (150\text{–}200)$ nm, corresponding to 10–15 mN loading, where the H values of the films could be recorded with the weak influence of the Si substrate on the test measurements. With the further increase of the applied load, i.e., the increase of the maximum penetration depth, the nanohardness value dropped rapidly below 20 GPa, followed by a smooth decrease to values that approached the Si substrate hardness of $\approx$15 GPa. The latter implies an increasing influence of the substrate [41]. Taking the observed dependence into account, the further presented results correspond to the load of 15 mN, for which the influence of the Si substrate on the H values was similar. The observed variation of H values with substrate temperature can be assigned to a change in the microstructure of films. Elevated temperature facilitates the crystallization process, and thus, a less defective structure with a larger amount and size of h-AlN crystallite grain boundaries was growing, characterized by higher nanohardness values.

Our recent investigations on PLD AlN films have established that the variation of the nitrogen pressure, on one hand [23,30,33], and LPF, on the other [24,31,36], had the strongest influence on the formation of the AlN microstructure. The effect of laser incident fluence can be compensated by the variation of those two parameters. Correspondingly, in Figure 3, the H values are represented as a function of LPF (Figure 3a) and N_2 pressure (Figure 3b) at other PLD parameters given in the insets. The AlN films were deposited under different conditions as either the N_2 pressure was kept constant at 0.1 Pa while varying the laser fluence, LPF, and substrate temperature (Figure 3a) or the substrate temperature was kept at 800 °C (Figure 3b) while varying the nitrogen pressure, laser fluence, and LPF. The observed hardness behavior is closely related to the processes of film growth and the resulting film microstructure, which yielded variation in the film hardness values. Nevertheless, all H values were within 22–30 GPa range, superior to the ones registered in the case of films obtained by other deposition techniques [22,42–47].

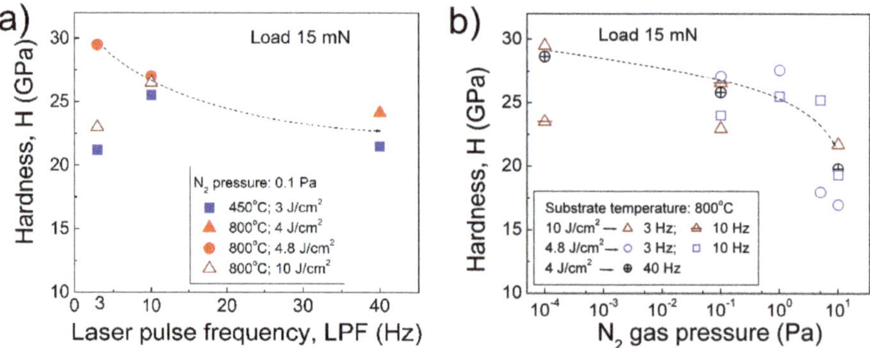

Figure 3. Nanohardness as a function of laser pulse repetition frequency (**a**) and N_2 gas pressure (**b**) for PLD AlN films deposited with variation of other PLD parameters (as given in the insets).

In general, a higher deposition temperature enhances the reaction at the surface of the substrate and promotes the formation of crystallites in the growing film [48]. As a result, AlN films deposited at 800 °C possessed higher hardness values (Figure 3a). When the deposition was performed at low N_2 pressure, a high laser fluence of 10 J/cm^2, and a low LPF of 3 Hz, the species evaporated from the polycrystalline AlN target acquired a much higher kinetic energy. This excess energy was transferred to adatoms when reaching the surface of the growing film, obstructing the ordering in a crystalline network. AlN films formed in these conditions were amorphous, as previously revealed by our TEM and XRD studies [23].

With increasing the LPF from 3 to 10 and 40 Hz, the multiple, consecutive vaporization "cleaned up" the space between the target and substrate. Consequently, the atoms ejected from the target had much more energy when reaching the substrate, contributing to the boost of the mobility of adatoms and surface diffusion. As one can observe in Figure 3a, the forming microstructure could be however more defective with lowered microhardness. When increasing the nitrogen pressure (Figure 3b), the particles ejected from the target in the plasma plume lost their energy in collisions with nitrogen particles. Accordingly, they could not significantly contribute to the thermally-induced mobility promoted by heating the substrate, but bound to their impinging sites without further surface migration. As a result, the formed film structure was less crystalline and more defective, which was reflected in the lower hardness values (Figure 3b).

As known [49,50], the hardness and elastic modulus are important material parameters that indicate the resistance to elastic/plastic deformation and could be used for the estimation of the coating wear behavior. The H/E ratio characterizes the elastic strain to failure resistance, while the H^3/E^2 ratio evaluates the coating resistance to plastic deformation at sliding contact load. Both ratios are associated with the coating toughness, a key parameter for the evaluation of the tribological properties of materials [50]. Hence, the improvement of the tribological behavior can be achieved by increasing the coating hardness and decreasing the elastic modulus. In Figure 4, the resistance for elastic strain to failure (H/E) and to plastic deformation (H^3/E^2) of AlN films versus deposition temperature is presented.

From the dependence of these ranking parameters, one can state that the studied PLD AlN films had a very high H^3/E^2 ratio compared to other AlN coatings [43,44].

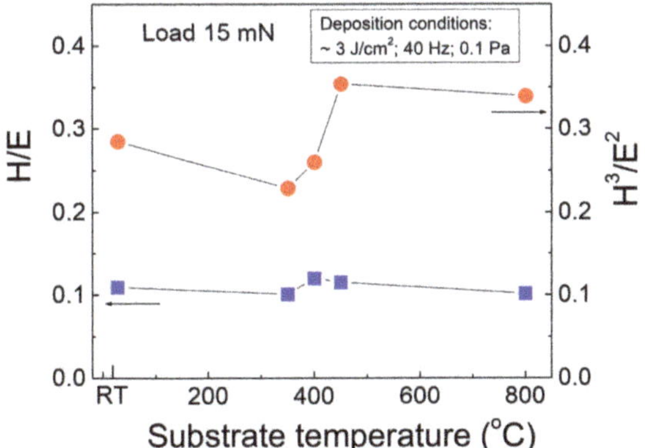

Figure 4. Resistance for elastic strain to failure (H/E) and plastic deformation (H^3/E^2) of the PLD AlN films as a function of substrate temperature during deposition.

3.2. TEM Observations

Four types of significantly different AlN structures were revealed in previous TEM studies of PLD films [23,30,36]. Amorphous AlN layers are mostly forming at RT or in a growth environment where the mobility of the atoms after reaching the substrate surface is limited. When increasing the temperature, nano-sized crystalline grains in an amorphous matrix emerged. This case is well visible in Figure 5a, where the HR-TEM image of the AlN film, deposited at 450 $^\circ$C, 0.1 Pa N$_2$ pressure, LPF of 40 Hz, and incident laser fluence of 3 J/cm^2, revealed hexagonal nanocrystallites surrounding with amorphous AlN. The reduced crystallinity was due to the relatively low substrate temperature of 450 $^\circ$C. Here, AlN crystallites were hexagonal (h-AlN), but the metastable cubic (c-AlN) phase can also grow in the amorphous matrix [23,33,36]. The hardness of such AlN films may vary significantly due to the variation in the thickness of the amorphous matrix between the crystalline particles [51] and/or voids possibly incorporated into the layer, which may significantly reduce the film's hardness.

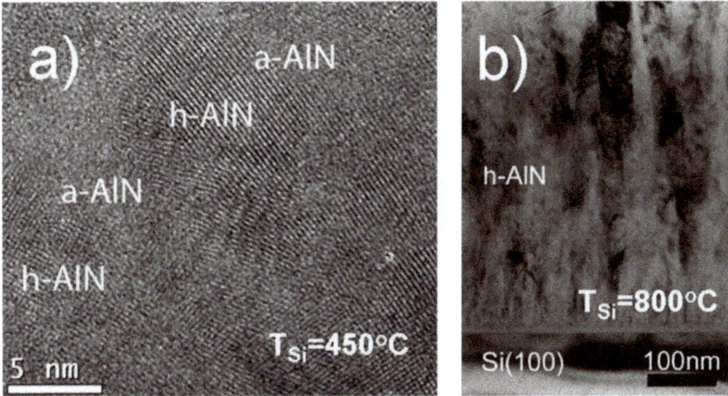

Figure 5. HR-TEM image of nano-sized crystalline grains in amorphous matrix (**a**) and bright field cross-sectional TEM image (**b**) of the PLD AlN films deposited at 450 and 800 $^\circ$C, respectively. The other PLD parameters were identical: N$_2$ pressure of 0.1 Pa, LPF of 40 Hz, and laser fluence of $\approx$3 J/cm^2.

The third type of AlN layer consists of columnar crystals with a highly crystalline h-AlN structure, mostly with the (001) texture [48]. A similar crystalline structure was observed for the AlN films grown at 800 $^\circ$C. This is illustrated in Figure 5b, where the bright-field (BF) cross-sectional TEM image of AlN film deposited at 800 °C, 0.1 Pa N_2 pressure, LPF of 40 Hz, and incident laser fluence of 3 J/cm^2 is shown. The columnar grains with a crystalline *h*-AlN structure are well seen. In the case of AlN films deposited at a higher temperature (800 °C), but in vacuum [23], a highly-ordered crystalline film structure was observed, where the *h*-AlN crystallites had grown epitaxially in a columnar orientation perpendicular to the Si substrate (Figure 6a,b). Although an epitaxial growth is achieved (as shown in Figure 6c), the layer is not a single crystal because *h*-AlN crystals grow with two preferred orientations, rotated from each other by 30° due to the growth of the *h*-AlN (001) plane onto the cubic Si lattice. The dark-field cross-sectional TEM image in Figure 6b was prepared from two dark-field images (separated from each other by color), which were recorded from spots with two possible epitaxial orientations. The selected area electron diffraction patterns in Figure 6c were taken from the cross-sectional TEM image in Figure 6a. In the first pattern (Figure 6c$_1$), the Si(100) substrate is shown, while the other two patterns (Figure 6c$_2$,c$_3$) show two possible epitaxially-oriented areas in the AlN film.

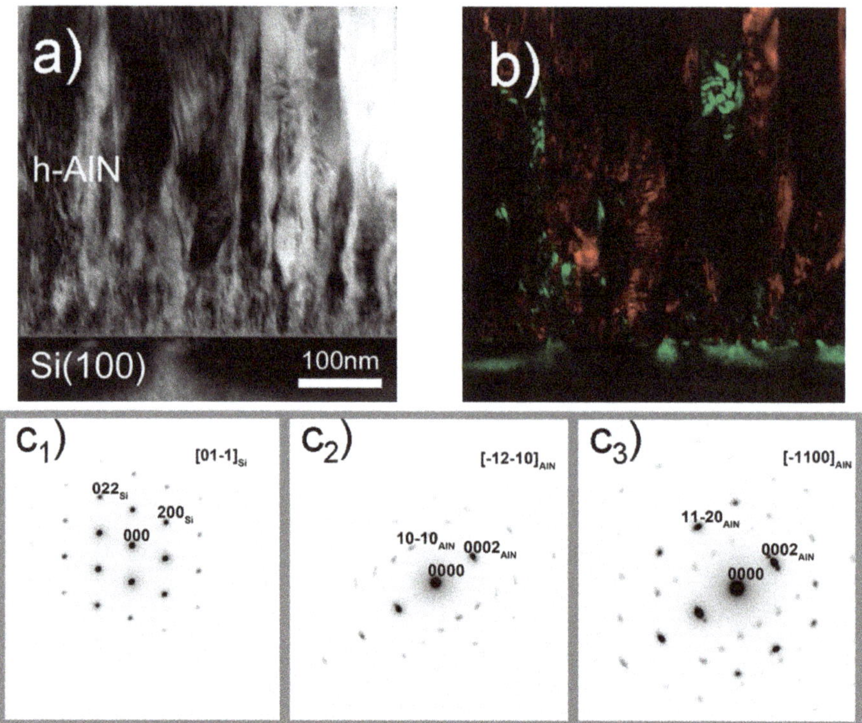

Figure 6. Bright-field (**a**) and dark-field (**b**) cross-sectional TEM images of the PLD AlN film deposited in vacuum (10^{-4} Pa) at a temperature of 800 °C, laser fluence of 10 J/cm^2, and LPF of 3 Hz. In (**c**), the corresponding selected area electron diffraction (SAED) patterns from (**a**) are shown: SAED pattern of the Si(100) substrate (**c$_1$**) and SAED patterns of AlN films (**c$_2$**,**c$_3$**) taken from two possible epitaxially-oriented areas.

TEM observations correlated well with the results of our earlier studies of PLD AlN films by X-ray diffraction (XRD, Bruker Corporation, Billerica, MA, USA) [23,24,30,35,52]. Our analysis revealed that a stable *h*-AlN phase was forming with predominant (002) *c*-axis orientation, for films deposited at 450 and 800 °C, low laser fluence (<10 J/cm^2), small nitrogen pressure (vacuum or 0.1 Pa), and high LPF (10 or 40 Hz). For a higher laser fluence of 10 J/cm^2, nitrogen pressure of 0.1 Pa, and LPF of 3 Hz, films were amorphous. At intermediate values of PLD parameters, the coexistence of hexagonal and cubic AlN crystallites occurred [52]. The average grains size was 10–60 nm, as determined with the Scherrer equation. We mention that high-quality AlN (002) films were synthesized by PLD on (La,Sr)(Al,Ta)O$_3$ substrates [53]. According to [54], higher laser fluence and substrate temperature and lower ambient pressure are beneficial for PLD synthesis of AlN thin films with the (002) orientation.

The structural changes ensuing from the variation of the PLD conditions were reflected in the alteration of the surface morphology of the AlN films. The latter has been studied by atomic force microscopy (AFM) and discussed in detail elsewhere [35,36,40]. The obtained results can be briefly summarized as follows. The smoothest surface (RMS roughness of ~0.46 nm) was found in the case of AlN films deposited in nitrogen at low pressure (0.1 Pa), 450 °C, and a LPF of 3 Hz, for which the TEM imaging detected the amorphous AlN phase only. On the other hand, the highest surface roughness (RMS roughness of ~2.5 nm) was obtained in the case of films deposited at 800 °C, for which better crystallinity and larger-sized crystallites coming up to the surface were detected [36]. The influence of nitrogen pressure on the surface roughness of the PLD AlN films has been reported in [40]. It was shown that deposition at a substrate temperature of 800 °C in vacuum (~10^{-4} Pa) resulted in considerably high surface roughness (RMS roughness of ~1.8 nm), while increasing the nitrogen pressure up to 10 Pa yielded minimal roughness values (RMS roughness of ~0.24 nm).

The hardness values of the AlN films as a function of the film structure are shown in Figure 7. The data demonstrate well the sensitivity of the AlN film structure to the PLD conditions. As can be seen, the PLD AlN films with the amorphous structure possessed the lowest hardness values. The reason is that the amorphous material is characterized by a short-range order with a distribution in bond lengths that generally results in lower stiffness, as compared to the corresponding crystalline phase [55]. The higher the stiffness of the atomic bondings, the higher the material's hardness is. This explains the observed increased hardness of the PLD AlN films when the degree of crystallinity increased for example by enhancing the substrate temperature from 350 to 800 °C or increasing the LPF from 3 to 40 Hz, respectively.

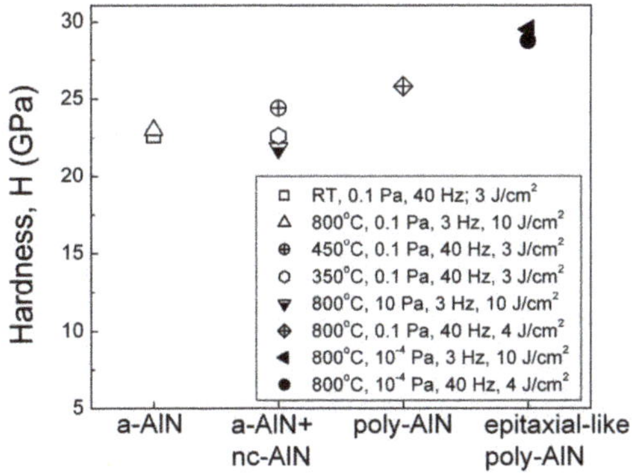

Figure 7. Variation of the hardness values with the film structure obtained at different PLD conditions, given in the inset.

One can observe in Figure 7 that the appearance of nanocrystallites in the amorphous matrix increased the AlN film's hardness. The size and amount of crystallite grains are determinative in the hardness level of coatings [56–58]. However, when two phases coexist in films, the hardness values can be greatly influenced by the thickness of the amorphous matrix separating the nanocrystals. Moreover, when the crystalline particles are forming in the gas space, it is easier to involve cavities (voids) from their environment, which can greatly reduce the hardness of the layer.

The highest hardness values were registered for the PLD films with epitaxial-like growth of AlN on the Si(100) substrates, i.e., when the PLD process proceeded in vacuum at the highest temperature (800 °C) (see Figure 7). In this case, the largest size of nanocrystallites (10–20 nm), growing in a columnar grain structure with preferred grain orientations and in a negligible amount of amorphous matter, was observed (Figures 5b and 6). Such an ordered structure is characterized by a strongly-reduced amount of defects in grain boundaries and, consequently, a higher H value, as was observed.

3.3. FTIR Reflectance Spectra Analysis

FTIR reflectance spectra are given in Figure 8 for the case of p-polarized (Figure 8a) and s-polarized (Figure 8b) incident beams recorded at a radiation angle of 70°. The results in Figure 8 correspond to AlN films deposited with a laser fluence of 3 J/cm^2 and different temperatures and LPF of 40 Hz. For higher temperatures, the spectra exhibited a complex and broad band within the 950–500 cm^{-1} region. The complexity of the spectral envelope can be assigned to the nanocrystallites' disorientation. The Berreman effect was registered in p-polarization, which allows for identification of the longitudinal (LO) phonon vibrational modes [59]. This gives the possibility to characterize thin films' microstructure directly from IR spectroscopy. A comparison of the spectra taken in both s- and p-polarization points to a clear difference in the high frequency end of the band (Figure 8).

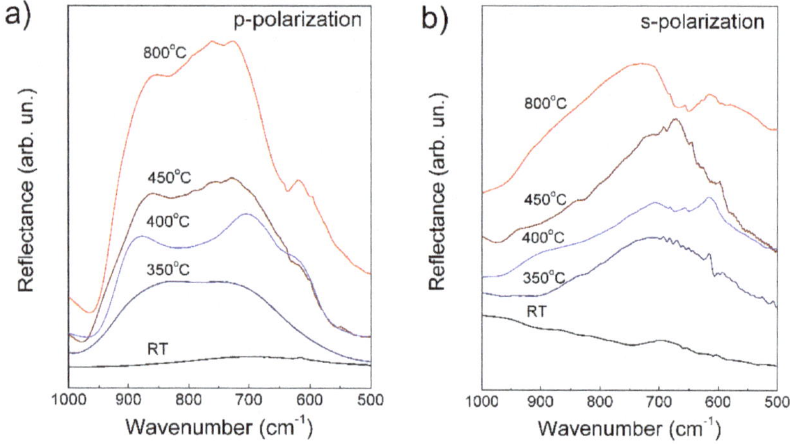

Figure 8. FTIR reflectance spectra of PLD AlN films, investigated in a linearly p-polarized (**a**) and s-polarized (**b**) incident beam.

The 950–500 cm^{-1} region is characteristic for the Reststrahlen band of the h-AlN crystal with component lines peaking around 611, 670, 890, and 912 cm^{-1}, arising from A$_1$(TO), E$_1$(TO), A$_1$(LO), and E$_1$(LO) IR active phonon vibrational modes, respectively [60–63]. For samples prepared at low substrate temperatures (RT and 350 °C), the deconvolution of the measured Reststrahlen band in p-polarized radiation was not possible. For higher substrate temperatures, the position of peaks was determined by the Levenberg–Marquardt deconvolution method with a fitting mean square error of 10^{-3}. The peaks and their assignments are collected in Table 1. When decreasing the substrate temperature, a major decrease of frequencies was observed for the E$_1$(TO) and A$_1$(LO) phonon-polariton modes. At a large angle of p-polarized incidence radiation such as 70°, the A$_1$(LO) mode, which is polarized parallel to the nanocrystallite c-axis, will be the most sensitive to the orientation of the crystal phase (Figure 8a). Any deviation of c-axis from the surface normal leads to a structure disorientation that is equivalent to a dumping of the phonon-polariton resonance vibration [64]. The enhanced structure disordering at lower temperatures also influences the E$_1$(TO) mode, which is polarized parallel to the a-axis, i.e., is parallel to the substrate surface in a good c-axis-oriented layer. Consequently, the resonance frequency decrease was more evident in the spectra measured in s-polarization (see Figure 8b). Besides, this mode is two-fold degenerated, i.e., it cumulates vibrations of two sets of atoms with the same frequency [65]. Thus, an increasing disorder with the temperature decrease will cause the peak's widening (as observed for all components), which leads to an increase of the entire Reststrahlen band half width. This is illustrated in Figure 9 for the AlN films deposited at 0.1 Pa N_2 pressure and LPF of 40 Hz. The incident laser fluence was kept within the range of 3–4 J/cm^2.

It should be mentioned that the features around 620–610 cm^{-1} in both sets of spectra in Figure 8 could hardly be assigned to the phonon mode A_1(TO) of *h*-AlN only. Indeed, the vibrational modes of Si substrate [66] and those of possible AlO$_x$ phases [67] were also present in the above-mentioned region. Possible AlO$_x$ bonds could be formed either during film preparation or storage of the samples under atmospheric conditions. In our opinion, the latter assumption is more likely to occur.

Table 1. Peak position of the phonon-polariton modes in the Reststrahlen band, registered with *p*-polarized radiation in the AlN films (Figure 8a). TO, transverse.

Substrate Temperature (°C)	A_1(TO) Mode (cm^{-1})	E_1(TO) Mode (cm^{-1})	A_1(LO) Mode (cm^{-1})	E_1(LO) Mode (cm^{-1})
800	618	740	890	920
450	615	730	885	925
400	616	704	870	922

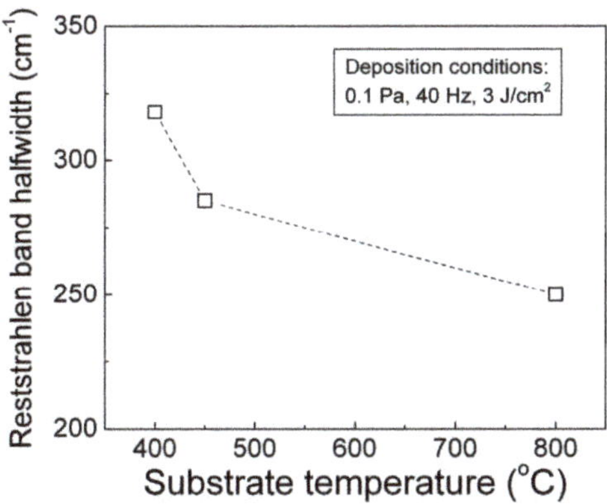

Figure 9. Reststrahlen band half width in *p*-polarized radiation (Figure 6a) as a function of substrate temperature during AlN film deposition at N_2 pressure of 0.1 Pa, LPF of 40 Hz, and laser fluence of 3–4 J/cm^2.

From the presented results, one can conclude that despite the poor crystalline phase, revealed by TEM, the FTIR spectra of AlN thin films deposited at temperatures higher than 350 °C clearly exhibited the characteristic Reststrahlen band of the AlN crystal with a hexagonal lattice. This band was originating from the *h*-AlN nanocrystallites, the size and ordering of which were increasing with the substrate temperature. For the AlN films synthesized at a substrate temperature of 350 °C, the spectra did not preserve the shape of a Reststrahlen band, and therefore, if nanocrystals were formed, their contribution could be negligible. At RT, a completely amorphous layer was grown. According to the SE results, the optical thickness of this layer was relatively small with respect to the wavelengths of the measured spectral region ($\sim\lambda/20$), and the recorded FTIR spectrum was flat. In such a thin amorphous film, neither a Reststrahlen band, nor the multiple interference effect could be observed in the FTIR spectra [68].

3.4. Spectroscopic Ellipsometry

The ellipsometric results revealed a clear dependence on technological conditions, in good agreement with TEM and FTIR investigations. We note that each AlN film yielded a certain thickness, which was within the 400–1000-nm range (corresponding to an estimated deposition rate of ~2.8 × 10^{-2}–7 × 10^{-2} nm/pulse), depending on the PLD technological protocol. For illustration purposes, in Figure 10, the optical constants n and k are shown for AlN films deposited in ambient nitrogen at a pressure of 0.1 Pa and laser fluence of 3–4 J/cm^2, by varying the substrate temperature and LPF. These values are characteristic for the corresponding AlN structures and correlated well with TEM observations. The refractive index values either coincided or were superior to those of amorphous AlN and remained inferior to those of high-quality polycrystalline *h*-AlN films. This suggests the coexistence of crystalline and amorphous AlN phases. Independently of the substrate's deposition temperature, films deposited at LPF of 3 Hz (Figure 10) possessed n values characteristic to an amorphous AlN structure. In accordance with the TEM results, larger LPF yielded nanostructured films with better ordering at LPF of 10 Hz, which reflects slightly higher index values. The exception is the AlN film deposited at RT (data represented by black dots in Figure 10), which was completely amorphous, as revealed by TEM, but its n values were close to those of *nc*-AlN. Additional compositional study of this sample by energy dispersive spectroscopy (EDS), performed in a scanning electron microscopy (SEM) system, has disclosed an over-stoichiometric AlN with an average Al/N ratio of 1.14. One can notice from the SEM-EDS results in Table 2 that at elevated temperatures, the films' composition was close to the stoichiometric AlN. When deposited at RT, AlN films contained an excess amount of Al atoms, which could contribute to the observed higher index values.

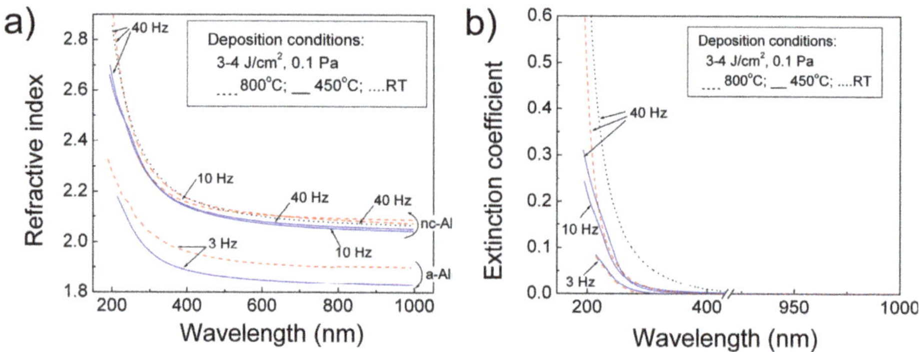

Figure 10. Dispersion curves of the refractive index (**a**) and extinction coefficient (**b**) of the studied AlN films deposited in the conditions presented in the insets.

Table 2. SEM-EDS data for AlN films deposited at different substrate temperatures in ambient N$_2$ at a pressure of 0.1 Pa, incident laser fluence of 3 J/cm^2, and LPF of 40 Hz.

Substrate Temperature (°C)	Al/N Atomic Ratio
800	0.98
450	0.98
350	0.97
RT	1.14

PLD AlN films were transparent in the 400–1000-nm spectral region, as the k values, dependent on substrate temperature and LPF (Figure 10b), approached zero. Below 400 nm, because of reaching the absorption edge, the extinction coefficient increased, and its value varied with the deposition conditions. A large shift of the absorption edge to higher wavelengths was observed for the RT deposited AlN film, suggesting a strong reduction of the optical bandgap in comparison to those deposited at elevated temperatures.

4. Conclusions

Aluminum nitride (AlN) films with different structural features were synthesized onto Si(100) substrates by pulsed laser deposition in vacuum and ambient nitrogen, at various pressures, substrate temperatures, laser incident fluences, and laser pulse frequencies. From the results of nanoindentation tests, transmission electron microscopy, X-ray diffraction, atomic force microscopy, Fourier transform infrared spectroscopy, and spectroscopic ellipsometry, the correlation between the mechanical properties, film structure, and optical parameters, dependent on deposition conditions, was studied.

The growth process and resulting film microstructures yielded variation in the film hardness within 22–30 GPa. Elevated substrate temperatures facilitated the crystallization process and, thus, a less defective structure for which increased nanohardness values were reached. Enhanced hardness values, in the range of 22–27 GPa, were observed for AlN films with a structure that consisted of nanocrystallite grains of 5–50 nm embedded in an amorphous matrix, strongly dependent on the deposition conditions. These values were superior to those obtained by other deposition techniques or reported for crystalline AlN. The refractive index value, superior to that of amorphous AlN, supported the existence of crystallites inside the film volume. In the case of PLD AlN films deposited at temperatures higher than 350 °C, the FTIR results evidenced vibrational bands within the characteristics Reststrahlen band of 950–500 cm^{-1}, which were assigned to hexagonal AlN crystallites. For lower temperatures, the Reststrahlen band gradually vanished, and the PLD film at room temperature exhibited an FTIR spectrum characteristic of a completely-amorphous AlN material.

The mechanical and optical properties of the synthesized AlN films conformed to the applied PLD technological parameters.

Author Contributions: Conceptualization, A.S. and L.D.; methodology, P.T., Z.F., and L.D.; validation, L.K., A.S., K.A., P.P., and L.D.; formal analysis, K.A., and P.T.; investigation, L.K., V.C., K.A., Z.F., and L.D.; resources, I.N.M.; writing, original draft preparation, L.K., A.S., K.A., and L.D.; writing, review and editing, A.S., P.P., I.N.M., and L.D.; visualization, L.D.; supervision, A.S.; project administration, A.S. and L.D.

Funding: The Bulgarian co-authors thank the European Regional Development Fund, the Ministry of Economy of Bulgaria, Operational Programme "Development of the Competitiveness of the Bulgarian economy" 2007–2013, Contract No. BG161PO003-1.2.04-0027-C0001. The Romanian co-authors acknowledge the support of the Core Programme, Contract 16N/2019. Liviu Duta thanks the support from the grant of the Ministry of Research and Innovation, CNCS-UEFISCDI, Project Number PN-III-P1-1.1-PD-2016-1568 (PD 6/2018), within PNCDI III. Peter Petrik is grateful for the support from OTKA Grant No. K115852.

Acknowledgments: All authors acknowledge with thanks the support of this work by the Bulgarian, Hungarian, and Romanian Academies of Sciences under the 2014–2017 Collaboration Agreements.

Conflicts of Interest: The authors declare no conflict of interest.

References

1. Chrisey, D.B.; Hubler, G.K. *Pulsed Laser Deposition of Thin Films*; John Wiley & Sons: Hoboken, NJ, USA, 1994.
2. Eason, R. *Pulsed Laser Deposition of Thin Films—Applications-Led Growth of Functional Materials*; Wiley-Interscience: Hoboken, NJ, USA, 2007.
3. Koh, A.T.T.; Foong, Y.M.; Chua, D.H.C. Cooling rate and energy dependence of pulsed laser fabricated graphene on nickel at reduced temperature. *Appl. Phys. Lett.* **2010**, *97*, 114102. [CrossRef]
4. Yang, Z.; Hao, J. Progress in pulsed laser deposited two-dimensional layered materials for device applications. *J. Mater. Chem. C* **2016**, *4*, 8859–8878. [CrossRef]
5. Rodriguez, R.J.; Garcia, J.A.; Medrano, A.; Rico, M.; Sanchez, R.; Martinez, R.; Labrugère, C.; Lahaye, M.; Guette, A. Tribological behaviour of hard coatings deposited by arc-evaporation PVD. *Vacuum* **2002**, *67*, 559–566. [CrossRef]
6. Huang, Z.P.; Sun, Y.; Bell, T. Friction behaviour of TiN, CrN, (TiAl)N coatings. *Wear* **1994**, *173*, 13–20. [CrossRef]
7. Jianxin, D.; Aihua, L. Dry sliding wear behavior of PVD TiN, Ti$_{55}$Al$_{45}$N, and Ti$_{35}$Al$_{65}$N coatings at temperatures up to 600 °C. *Int. J. Refract. Met. Hard Mater.* **2013**, *41*, 241–249. [CrossRef]

8. Jianxin, D.; Fengfang, W.; Yunsong, L.; Youqiang, X.; Shipeng, L. Erosion wear of CrN, TiN, CrAlN, and TiAlN PVD nitride coatings. *Int. J. Refract. Met. Hard Mater.* **2012**, *35*, 10–16. [CrossRef]

9. Liang, C.L.; Cheng, G.A.; Zheng, R.T.; Liu, H.P. Fabrication and performance of TiN/TiAlN nanometer modulated coatings. *Thin Solid Films* **2011**, *520*, 813–817. [CrossRef]

10. Cecchini, R.; Fabrizi, A.; Cabibbo, M.; Paternoster, C.; Mavrin, B.N.; Denisov, V.N.; Novikova, N.N.; Haïdopoulo, M. Mechanical, microstructural and oxidation properties of reactively sputtered thin CrN coatings on steel. *Thin Solid Films* **2011**, *519*, 6515–6521. [CrossRef]

11. Chen, Y.; Zhang, Z.; Jiang, H.; Li, Z.; Miao, G.; Song, H. The optimized growth of AlN templates for back-illuminated AlGaN-based solar-blind ultraviolet photodetectors by MOCVD. *J. Mater. Chem. C* **2018**, *6*, 4936–4942. [CrossRef]

12. Gao, J.; Hao, Z.; Luo, Y.; Li, G. Frequency response improvement of a two-port surface acoustic wave device based on epitaxial AlN thin film. *IOP Conf. Ser. Mater. Sci. Eng.* **2018**, *284*, 012028. [CrossRef]

13. Maouhoub, S.; Aoura, Y.; Mir, A. FEM simulation of AlN thin layers on diamond substrates for high frequency SAW devices. *Diam. Relat. Mater.* **2016**, *62*, 7–13. [CrossRef]

14. Schubert, E.F. *Light-Emitting Diodes*, 3rd ed.; Cambridge University Press: Cambridge, UK, 2018.

15. Galca, A.C.; Stan, G.E.; Trinca, L.M.; Negrila, C.C.; Nistor, L.C. Structural and optical properties of c-axis oriented aluminum nitride thin films prepared at low temperature by reactive radio-frequency magnetron sputtering. *Thin Solid Films* **2012**, *524*, 328–333. [CrossRef]

16. Dumitru, V.; Morosanu, C.; Sandu, V.; Stoica, A. Optical and structural differences between RF and DC Al_xN_y magnetron sputtered films. *Thin Solid Films* **2000**, *359*, 17–20. [CrossRef]

17. Pankov, V.; Evstigneev, M.; Prince, R.H. Room-temperature fabrication of hard AlN/TiN superlattice coatings by pulsed laser deposition. *J. Vac. Sci. Technol. A* **2002**, *20*, 430–436. [CrossRef]

18. Kim, G.S.; Lee, S.Y.; Hahn, J.H.; Lee, S.Y. Synthesis of CrN/AlN superlattice coatings using closed-field unbalanced magnetron sputtering process. *Surf. Coat. Technol.* **2003**, *171*, 91–95. [CrossRef]

19. Abid, I.; Faisal, M.-Y. Reactive sputtering of aluminum nitride (002) thin films for piezoelectric applications: A review. *Sensors* **2018**, *18*, 1797. [CrossRef]

20. Nan, L.; Satyesh, K.Y.; Jian, W.; Xiang-Yang, L.; Amit, M. Growth and stress-induced transformation of Zinc blende AlN layers in Al-AlN-TiN multilayers. *Sci. Rep.* **2015**, *5*, 18554. [CrossRef]

21. Abels, C.; Mastronardi, V.M.; Guido, F.; Dattoma, T.; Qualtieri, A.; Megill, W.M.; De Vittorio, M.; Rizzi, F. Nitride-based materials for flexible MEMS tactile and flow sensors in robotics. *Sensors* **2017**, *17*, 1080. [CrossRef]

22. Oliveira, I.C.; Grigorov, K.G.; Maciel, H.S.; Massi, M.; Otani, C. High textured AlN thin films grown by RF magnetron sputtering; composition, structure, morphology and hardness. *Vacuum* **2004**, *75*, 331–338. [CrossRef]

23. Szekeres, A.; Fogarassy, Z.; Petrik, P.; Vlaikova, E.; Cziraki, A.; Socol, G.; Ristoscu, C.; Mihailescu, I.N. Structural characterization of AlN films synthesized by pulsed laser deposition. *Appl. Surf. Sci.* **2011**, *257*, 5370–5374. [CrossRef]

24. Antonova, K.; Duta, L.; Szekeres, A.; Stan, G.E.; Mihailescu, I.N.; Anastasescu, M.; Stroescu, H.; Gartner, M. Influence of laser pulse frequency on the microstructure of aluminum nitride thin films synthesized by pulsed laser deposition. *Appl. Surf. Sci.* **2017**, *394*, 197–204. [CrossRef]

25. Giba, A.E.; Pigeat, P.; Bruyère, S.; Easwarakhanthan, T.; Mücklich, F.; Horwat, D. Controlling refractive index in AlN films by texture and crystallinity manipulation. *Thin Solid Films* **2017**, *636*, 537–545. [CrossRef]

26. Meyers, M.A.; Mishra, A.; Benson, D.J. Mechanical properties of nanocrystalline materials. *Prog. Mater. Sci.* **2006**, *51*, 427–556. [CrossRef]

27. Ristoscu, C.; Mihailescu, I.N. Thin Films and Nanoparticles by Pulsed Laser Deposition: Wetting, Adherence, and Nanostructuring. In *Pulsed Laser Ablation: Advances and Applications in Nanoparticles and Nanostructuring Thin Films*, 1st ed.; Mihailescu, I.N., Caricato, A.P., Eds.; Taylor&Francis Group: Singapore, 2018; pp. 245–276. [CrossRef]

28. Kumar, A.; Chan, H.L.; Weimer, J.J.; Sanderson, L. Structural characterization of pulsed laser-deposited AlN thin films on semiconductor substrates. *Thin Solid Films* **1997**, *308–309*, 406–409. [CrossRef]

29. Bakalova, S.; Szekeres, A.; Anastasescu, M.; Gartner, M.; Duta, L.; Socol, G.; Ristoscu, C.; Mihailescu, I.N. VIS/IR spectroscopy of thin AlN films grown by pulsed laser deposition at 400 °C and 800 °C and various N_2 pressures. *J. Phys. Conf. Ser.* **2016**, *514*, 012001. [CrossRef]

30. Bakalova, S.; Szekeres, A.; Fogarassy, Z.; Georgiev, S.; Ivanov, T.; Socol, G.; Ristoscu, C.; Mihailescu, I.N. Synthesis of nanostructured PLD AlN films: XRD and Surface-enhanced Raman scattering studies. *Micro Nanosyst.* **2014**, *6*, 9–13. [CrossRef]

31. Antonova, K.; Szekeres, A.; Duta, L.; Stan, G.E.; Mihailescu, N.; Mihailescu, I.N. Orientation of the nanocrystallites in AlN thin film determined by FTIR spectroscopy. *J. Phys. Conf. Ser.* **2016**, *682*, 012024. [CrossRef]

32. Simeonov, S.; Bakalova, S.; Kafedjiiska, E.; Szekeres, A.; Socol, G.; Grigorescu, S.; Mihailescu, I.N. Admittance study of MIS structures with pulsed plasma deposited AlN films. *J. Optoelectron. Adv. Mater.* **2007**, *9*, 323–325.

33. Bakalova, S.; Szekeres, A.; Cziraki, A.; Lungu, C.P.; Grigorescu, S.; Socol, G.; Axente, E.; Mihailescu, I.N. Influence of in-situ nitrogen pressure on crystallization of pulsed laser deposited AlN Films. *Appl. Surf. Sci.* **2007**, *253*, 8215–8219. [CrossRef]

34. Simeonov, S.; Bakalova, S.; Szekeres, A.; Kafedjiiska, E.; Grigorescu, S.; Socol, G.; Mihailescu, I.N. Extended analysis of the admittance frequency dependence of MIS structures with pulsed laser deposited AlN films as gate dielectric. *J. Phys. Conf. Ser.* **2008**, *113*, 012050. [CrossRef]

35. Duta, L.; Stan, G.E.; Stroescu, H.; Gartner, M.; Anastasescu, M.; Fogarassy, Z.; Mihailescu, N.; Szekeres, A.; Bakalova, S.; Mihailescu, I.N. Multi-stage pulsed laser deposition of aluminum nitride at different temperatures. *Appl. Surf. Sci.* **2016**, *374*, 143–150. [CrossRef]

36. Fogarassy, Z.; Petrik, P.; Duta, L.; Mihailescu, N.; Anastasescu, M.; Gartner, M.; Antonova, K.; Szekeres, A. TEM and AFM studies of aluminum nitride films synthesized by pulsed laser deposition. *Appl. Phys. A* **2017**, *123*, 756. [CrossRef]

37. Oliver, W.C.; Pharr, G.M. An improved technique for determining hardness and elastic modulus using load and displacement sensing indentation experiments. *J. Mater. Res.* **1992**, *7*, 1564–1583. [CrossRef]

38. Software CompleteEASE® 5.08 supplied by J. A. Woollam Co., Inc. Available online: https://www.jawoollam.com/ellipsometry-software/completeease (accessed on 15 March 2019).

39. Hay, J.L.; Pharr, G.M. Instrumented Indentation Testing. In *ASM Handbook: Mechanical Testing and Evaluation*; Khun, H., Medlin, D., Eds.; ASM International: Novelty, OH, USA, 2000; Volume 8, pp. 232–243. [CrossRef]

40. Bakalova, S.; Szekeres, A.; Huhn, G.; Havancsak, K.; Grigorescu, S.; Socol, G.; Ristoscu, C.; Mihailescu, I.N. Surface morphology studies of AlN films synthesized by pulsed laser deposition. *Vacuum* **2009**, *84*, 155–157. [CrossRef]

41. Buckle, H.; Westbrook, J.H.; Conrad, H. *The Science of Hardness Testing and Its Research Applications*; ASTM: Philadelphia, PA, USA, 1971.

42. Panda, P.; Ramaseshan, R.; Ravi, N.; Mangamma, G.; Jose, F.; Dash, S.; Suzuki, K.; Suematsu, H. Reduction of residual stress in AlN thin films synthesized by magnetron sputtering technique. *Mater. Chem. Phys.* **2017**, *200*, 78–84. [CrossRef]

43. Wei, Q.P.; Zhang, X.W.; Liu, D.Y.; Jie, L.I.; Zhou, K.C.; Zhang, D.; Yu, Z.M. Effects of sputtering pressure on nanostructure and nanomechanical properties of AlN films prepared by RF reactive sputtering. *Trans. Nonferrous Met. Soc. China* **2014**, *24*, 2845–2855. [CrossRef]

44. Sippola, P.; Perros, A.P.; Ylivaara, O.M.E.; Ronkainen, H.; Julin, J.; Liu, X.; Sajavaara, T.; Etula, J.; Lipsanen, H.; Puurunen, R.L. Comparison of mechanical properties and composition of magnetron sputter and plasma enhanced atomic layer deposition aluminum nitride films. *J. Vac. Sci. Technol. A* **2018**, *36*, 051508. [CrossRef]

45. Guillaumot, A.; Lapostolle, F.; Dublanche-Tixier, C.; Oliveira, J.C.; Billard, A.; Langlade, C. Reactive deposition of Al-N coatings in Ar/N_2 atmospheres using pulsed-DC or high power impulse magnetron sputtering discharges. *Vacuum* **2010**, *85*, 120–125. [CrossRef]

46. Kohout, J.; Qian, J.; Schmitt, T.; Vernhes, R.; Zabeida, O.; Klemberg-Sapieha, J.; Martinu, L. Hard AlN films prepared by low duty cycle magnetron sputtering and by other deposition techniques. *J. Vac. Sci. Technol. A* **2017**, *35*, 061505. [CrossRef]

47. Besleaga, C.; Dumitru, V.; Trinca, L.M.; Popa, A.-C.; Negrila, C.-C.; Kołodziejczyk, Ł.; Luculescu, C.R.; Ionescu, G.C.; Ripeanu, R.G.; Vladescu, A.; et al. Mechanical, corrosion and biological properties of room-temperature sputtered aluminum nitride films with dissimilar nanostructure. *Nanomaterials* **2017**, *7*, 394. [CrossRef]

48. Barna, P.B.; Adamik, M. Fundamental structure forming phenomena of polycrystalline films and the structure zone models. *Thin Solid Films* **1998**, *317*, 27–33. [CrossRef]

49. Oberle, T.L. Wear of metals. *J. Met.* **1951**, *3*, 438–439. [CrossRef]

50. Leyland, A.; Matthews, A. On the significance of the H/E ratio in wear control: A nanocomposite coating approach to optimised tribological behavior. *Wear* **2000**, *246*, 1–11. [CrossRef]

51. Musil, J. Hard and superhard nanocomposite coatings. *Surf. Coat. Technol.* **2000**, *125*, 322–330. [CrossRef]

52. Szekeres, A.; Cziraki, A.; Huhn, G.; Havancsak, K.; Vlaikova, E.; Socol, G.; Ristoscu, C.; Mihailescu, I.N. Laser technology for synthesis of AlN films: Influence of the incident laser fluence on the films microstructure. *J. Phys. Conf. Ser.* **2012**, *356*, 012003. [CrossRef]

53. Wang, W.; Yang, W.; Liu, Z.; Lin, Y.; Zhou, S.; Lin, Z.; Wang, H.; Qian, H.; Li, G. Synthesis of high-quality AlN films on (La,Sr)(Al,Ta)O$_3$ substrates by pulsed laser deposition. *Mater. Lett.* **2015**, *39*, 483–486. [CrossRef]

54. Hua, T.S.; Zhu, B.; Song, R.G. Characterisation of AlN nano thin films prepared by PLD. *Surf. Eng.* **2019**, 1–8. [CrossRef]

55. Music, D.; Hensling, F.; Pazur, T.; Bednarcik, J.; Hans, M.; Schnabel, V.; Hostert, C.; Schneider, J.M. Bonding and elastic properties of amorphous AlYB$_{14}$. *Solid State Commun.* **2013**, *169*, 6–9. [CrossRef]

56. Kobayashi, S.; Tsurekawa, S.; Watanabe, T. A new approach to grain boundary engineering for nanocrystalline materials. *Beilstein J. Nanotechnol.* **2016**, *7*, 1829–1849. [CrossRef]

57. Uberuaga, B.P.; Vernon, L.J.; Martinez, E.; Voter, A.F. The relationship between grain boundary structure, defect mobility, and grain boundary sink efficiency. *Sci. Rep.* **2015**, *5*, 9095. [CrossRef]

58. Suzuki, A.; Mishin, Y. Atomic mechanisms of grain boundary diffusion: Low versus high temperatures. *J. Mater. Sci.* **2005**, *40*, 3155–3161. [CrossRef]

59. Berreman, D.W. Infrared absorption at longitudinal optic frequency in cubic crystal films. *Phys. Rev.* **1963**, *130*, 2193–2198. [CrossRef]

60. Davydov, V.Y.; Kitaev, Y.E.; Goncharuk, I.N.; Smirnov, A.N.; Graul, J.; Semchinova, O.; Uffmann, D.; Smirnov, M.B.; Mirgorodsky, A.P.; Evarestov, R.A. Phonon dispersion and Raman scattering in hexagonal GaN and AlN. *Phys. Rev. B* **1998**, *58*, 12899–12907. [CrossRef]

61. Prokofyeva, T.; Seon, M.; Vanbuskirk, J.; Holtz, M.; Nikishin, S.A.; Faleev, N.N.; Temkin, H.; Zollner, S. Vibrational properties of AlN grown on (111)-oriented silicon. *Phys. Rev. B* **2001**, *63*, 125313. [CrossRef]

62. Lu, Y.F.; Ren, Z.M.; Chong, T.C.; Cheong, B.A.; Chow, S.K.; Wang, J.P. Ion-assisted pulsed laser deposition of aluminum nitride thin films. *J. Appl. Phys.* **2000**, *87*, 1540–1542. [CrossRef]

63. Taborda, J.A.P.; Caicedo, J.C.; Grisales, M.; Saldarriaga, W.; Riascos, H. Deposition pressure effect on chemical, morphological and optical properties of binary Al-nitrides. *Opt. Laser Technol.* **2015**, *69*, 92–103. [CrossRef]

64. Landau, L.D.; Lifshitz, E.M. Mechanics. In *Course of Theoretical Physics*, 2nd ed.; Landau, L.D., Lifshitz, E.M., Eds.; Pergamon Press: Oxford, UK, 1969; Volume 1.

65. Born, M.; Kun, H. *Dynamical Theory of Crystal Lattices*; Series Oxford Classic Texts in the Physical Sciences; Clarendon Press: Oxford, UK, 1998.

66. Kitamura, R.; Pilon, L.; Jonasz, M. Optical constants of silica glass from extreme ultraviolet to far infrared at near room temperature. *Appl. Opt.* **2007**, *46*, 8118–8133. [CrossRef]

67. Kazan, M.; Rufflé, B.; Zgheib, C.; Masri, P. Phonon dynamics in AlN lattice contaminated by oxygen. *Diam. Relat. Mater.* **2006**, *15*, 1525–1534. [CrossRef]

68. Klingshirn, C.F. *Semiconductor Optics*, 4th ed.; Springer: Berlin/Heidelberg, Germany, 2012; ISBN 978-3-642-28362-8.

 © 2019 by the authors. Licensee MDPI, Basel, Switzerland. This article is an open access article distributed under the terms and conditions of the Creative Commons Attribution (CC BY) license (http://creativecommons.org/licenses/by/4.0/).

 *coatings*

Review

Pulsed Laser Deposited Films for Microbatteries

Christian M. Julien * and Alain Mauger

Institut de Minéralogie, de Physique des Matériaux et de Cosmochimie (IMPMC), Campus Pierre et Marie Curie, Sorbonne Université, CNRS UMR 7590, 4 Place Jussieu, 75005 Paris, France; alain.mauger@upmc.fr
* Correspondence: christian.julien@upmc.fr; Tel.: +33-673-404-684

Received: 10 May 2019; Accepted: 10 June 2019; Published: 14 June 2019

Abstract: This review article presents a survey of the literature on pulsed laser deposited thin film materials used in devices for energy storage and conversion, i.e., lithium microbatteries, supercapacitors, and electrochromic displays. Three classes of materials are considered: Positive electrode materials (cathodes), solid electrolytes, and negative electrode materials (anodes). The growth conditions and electrochemical properties are presented for each material and state-of-the-art of lithium microbatteries are also reported.

Keywords: PLD films; energy storage; thin-film electrodes; thin-film solid electrolyte; lithium microbatteries

1. Introduction

It has been widely demonstrated that pulsed-laser deposition (PLD) based on the process of the transportation of a material (laser ablation) is a successful technique for the growth of stoichiometric multicomponent oxide films [1,2]. Indeed, PLD has shown unique advantages for the formation of dense films for energy storage and conversion, namely a high reproducibility, easy control of the growth rate, and a high film purity with a variety of substrates, such as amorphous glass, oriented silicon [3], stainless steel [4], $(001)Al_2O_3$ [5], indium tin oxide (ITO)- and ZnO-coated glass, and ITO-coated Upilex polymer [6]. Generally, the stoichiometry of the target phase is preserved in PLD films of oxides but a deviation is observed for lithiated material that implies an Li-enriched target. Consequently, the loss of volatile Li during deposition is compensated for by using about a 15 wt.% excess of Li_2O [7,8]. Accurate stoichiometry can be obtained by controlling several parameters of the process. The typical set-up for the fabrication of PLD films consists of a stainless-steel vacuum chamber evacuated down to a residual pressure less than 1×10^{-4} Pa before material deposition. Energy (laser fluence) in the range of 1.0 to 3.0 J·cm^{-2} is generated by a pulsed-laser beam, which falls onto the target surface with an incidence angle of approximately 45° (see Table 1 for laser characteristics). Indeed, four PLD parameters are of prime importance for the growth of films: The laser fluence; type of substrate; orientation and lattice parameters, which must match with those of the film for an efficient epitaxy process; substrate temperature (T_s); and the oxygen partial pressure (P_{O_2}). In addition, as the capacity of the microbattery depends on the electrode thickness, the duration of the deposition (t_p) must also be considered. The activity of a thin-film electrode, i.e., specific discharge energy, is proportional to the thickness, thus an increase of the film thickness leads to a power limitation because of the slow transport kinetic of Li$^+$ ions. Consequently, PLD is a popular technique due to the growth of a compact and dense film, which is replaced by a thick and porous film. Another advantage of the physical vacuum-like deposition techniques is the possibility of depositing a thin layer on top of the microbattery, which protects the device against a reactivity toward moisture. Moreover, due to the well-defined surface area of PLD films, a direct comparison of the electrochemical activity of materials can be done for different morphologies, from amorphous to single-crystalline materials [9].

Table 1. Typical laser beams for PLD films of transition-metal (TM) oxides for energy storage.

Laser	Wavelength (nm)	Pulse Width (ns)	Frequency (Hz)	Laser Fluence (J·cm^{-2})	Ref.
Excimer KrF	248	20	1	12	[10]
Excimer ArF	193	10	5	2	[11]
Excimer XeCl	308	–	–	–	[12]
Nd:YAG	532 [a]	8	10	6	[13]
Nd:YAG	266 [b]	–	–	1.6	[14]

[a] Frequency doubled; [b] Fourth harmonic.

Due to their high energy and power densities, lithium-ion batteries (LIBs) are initial power sources that are widely used in portable devices (laptops, mobile phones, cameras, etc.) and are now employed for sustainable transportation, such as full electric vehicles (EVs) and hybrid electric vehicles (HEVs). The fabrication of electrochemical cells with a thin-film architecture allows the development of microbatteries for powering micro-scaling devices, such as stand-alone sensor systems, medical implants and devices, labs-on-chip, credit cards, etc. In addition to technological applications, positive (cathode) and negative (anode) electrodes in the thin-film form are useful for studying the intrinsic properties of the material without the use of a polymeric binder and carbonaceous additive [15]. The use of thin-film technology may offer various advantages, such as: (i) Thin films are well suited for the design of devices; (ii) thinning of the layers provides a lower resistance in the transverse direction for weakly semiconducting materials; (iii) a reduction of the thickness of the solid electrolyte film allows the use of glassy materials with a low ionic conductivity; (iv) a reduction of the charge-transfer resistance of the electrolyte–electrode interface; (v) easy manufacture of microbatteries using the same technique that is currently used in the microelectronics industry; and (vi) the construction of microbatteries is realized in almost any two-dimensional shape. However, the fabrication of microbatteries also contains many difficulties, which are comprehensively discussed below [16].

In the present review paper, we present the properties of pulsed-laser deposited films used as components of energy storage devices (i.e., batteries, supercapacitors, electrochromics, etc.). The remainder of the article is organized as follows. First, the state-of-the-art of lithium microbatteries using PLD films are summarized in Section 2, providing the characteristics of the best lithium microbatteries fabricated so far. In Sections 3–5, the three classes of active materials constituting electrochemical microdevices, realized via the PLD technique, are considered: (i) Positive electrode materials (cathodes), (ii) electrolytes, and (iii) negative electrode materials (anodes). For each material, the growth conditions and electrochemical properties are presented. Finally, in Section 6, we compare and discuss the growth conditions that allow the best electrochemical performance of each electrochemically active component of microbatteries.

2. Lithium Microbatteries

The concept of a thin-film solid-state battery is quite old [17]. The subject of thin-film microbatteries has been discussed in the scientific literature for many years. The review by Kennedy is a good source for work prior to 1977 [18]. The concept and design of all-solid-state planar thin-film microbatteries have been patented by Bates et al. [19–24], who reported on micropower sources using lithium phosphate, lithium phosphorus oxynitride, and lithium phosphorus lithium oxide as a solid thin-film electrolyte. Julien investigated the electrochemical performance of individual layers in a microbattery in relation to the growth mechanism and thin-film structure [25]. Dudney addressed how to build a battery layer-by-layer by vapor deposition [26]. More recently, Oudenhoven et al. reviewed the concepts of three-dimensional (3D) microbatteries [27]. In 2015, Wang et al. discussed the choice of materials for lithium and lithium-ion microbatteries and reviewed the chemistry and electrochemistry for applications in microelectronic devices [28]. Ferrari et al. highlighted the importance of 3D microarchitecture electrodes to fabricate microgenerators for micro-electromechanical systems (MEMSs) [29]. In the presentation of in situ analytical microprobes, Meng et al. described PLD-produced thin-film lithium microbatteries using the PLD technique and showing the production of a multilayer structure with dense and smooth films [30].

There are many variations on the general scheme of microbatteries outlined in the literature. Two principal options are shown in Figure 1 [25]. Figure 1a shows a four-layer design on a conducting substrate (i.e., oriented silicon wafer) that can act as a current collector. Figure 1b shows a six-layer stack incorporating two metallic current collectors. There are two fast-ion conductor (FIC) layers in this design: A thick layer, which is the solid electrolyte itself, and a thin buffer film that acts as an electrolyte layer between the FIC and the Li metal film to prevent interface passivation. In 1992, a thin-film solid-state microbattery with an overall thickness of approximately 10 μm, including the TiS_2 cathode, oxide-sulfide solid electrolyte, LiI buffer, and Li metal anode, was developed at the Technology Laboratory of Eveready Battery Company (EBC) [31]. Laïk et al. evaluated the performance of three 4-V commercial all-solid-state lithium microbatteries (200-μm thick) with a nominal capacity of 700 μAh based on a $LiCoO_2$ cathode material [32]. Note that a typical 1-mWh battery weighs 2.5 mg and has a volume less than 1 μL, providing a specific energy and power of 400 Wh·kg^{-1} and 1 kWh·L^{-1}, respectively [26].

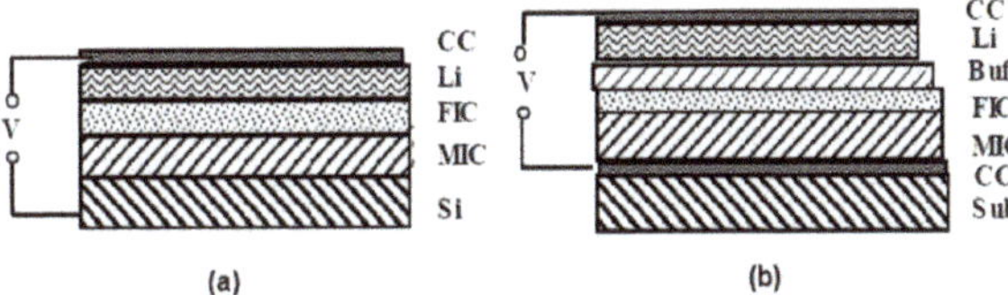

Figure 1. Design principles for lithium microbatteries composed of a lithium film (Li), fast-ion conductor (FIC), mixed ionic-electronic conductor (MIC), current collector(s) (CC), silicon substrate (Si), glass substrate (Sub), and buffer layer (Buf). (**a**) four-layer design on a conducting substrate and (**b**) six-layer stack incorporating two metallic current collectors (Reproduced with permission from [25]. Copyright 2000 Springer).

Regarding the manufacture of thin-film batteries, several start-up companies have marketed micropower sources. Enfucell developed a thin, printable, and flexible SoftBattery® used in various wearable electronics products [33]. Cymbet Corporation fabricates the EnerChip™ battery, which is a battery bare die and can be embedded with other integrated circuits [34]. Excellatron announced a pilot production line (10,000 cells/month) of thin-film solid-state batteries (approximately 0.3 μm thick) made of cathode films of $LiCoO_2$ or $LiMn_2O_4$, LiPON as the electrolyte, and Li metal or Sn_3N_4 as the anode based on the technology developed at Oak Ridge National Labs. Using a 2-μm thick positive electrode, these microbatteries have been cycled in excess of 2000 cycles [35]. Some industrial developments of thin film microbatteries are listed in Table 2.

Table 2. Industrial developments of thin film microbatteries.

Manufacturer	Electrochemical Chain	Specifications	Ref.
Cymbet Co.	EnerChip™ $LiCoO_2$/Li	60 μAh·cm^{-2}·μm^{-1}/5000 cycles	[34]
Infinite Power Solutions	$LiCoO_2$ or V_2O5/LiPON/Li	"Thinergy" 40 μAh·cm^{-2}·μm^{-1}	[36]
Front Edge Technology	$LiCoO_2$/LiPON/Li	"Nanoenergy" 0.9 mAh cm^{-2}	[37]
Ulvac Inc.	$LiCoO_2$/Li_3PO_4/Li	50 μAh·cm^{-2}·μm^{-1}	[38]
STMicroelectronics	$LiCoO_2$/LiPON/Li	700 μAh/discharge at 5 mA	[39]
Excellatron	$LiCoO_2$-$LiMnO_2$-LiPON-Sn_3N_4	0.3 μm thick/0.1 mAh/2000 cycles	[35]
Enfucell	MnO_2-based cell	voltage rating >3 V	[33]
GS Caltex	n/a	300 μm thick/3.9 V/ 8000 cycles	[40]

A sequential PLD technique was applied for the fabrication of a rechargeable thin-film lithium battery (2-μm thick, area of 0.23 cm^2) with partially crystallized LCO as the cathode, an Li$_{6.1}$V$_{0.61}$Si$_{0.39}$O$_{5.36}$ (LVSO) glassy electrolyte, and SnO film anode [41]. The ablation beam produced by a Q-switched Nd:YAG laser (λ = 266 nm, repetition rate of 10 Hz) was used at the fluence of 3.5 mJ·cm^{-2}. A single phase LCO film was obtained by post annealing at 600 °C for 1 h in air and the amorphous LVSO film exhibited an ionic conductivity of ca. 10^{-7} S·cm^{-1} at room temperature. Such a Li microbattery cycled at 44 μA·cm^{-2} in the voltage range of 0.7 to 3.0 V delivered a capacity of 9.5 Ah·cm^{-2}. After 100 cycles, the capacity retention was 45% of that of the first cycle. Characteristics of solid-state lithium microbatteries fabricated by the PLD technique reported in the literature are listed in Table 3. Most of the microcells use LiCoO$_2$ as the positive electrode, providing a nominal voltage of ~3.9 V vs. Li$^+$/Li. Thus, among the fabricated all-solid-state thin-film lithium batteries, the electrochemical chain of Li-In/80Li$_2$S–20P$_2$S$_5$/LiCoO$_2$ with an average potential of 3.5 V exhibits the best performance in terms of energy density.

Sakuda et al. reported the construction of an SSLMB based on an LCO positive electrode with an SE coating, a highly conductive 80Li$_2$S–20P$_2$S$_5$ solid electrolyte, and an Li-In alloy as the anode [42]. Such a microbattery delivered a specific capacity of 95 mAh·g^{-1} at a current density of 0.13 mA·cm^{-2}. By using LCO thin films prepared from a Li$_y$CoO$_\delta$ target containing 15% Li$_2$O, Xia et al. fabricated thin-film microbatteries by the successive deposition of an LCO cathode on a Pt/Ti/SiO$_2$ (amorphous)/Si composite substrate and amorphous Si anode [12]. Recently, the analysis by impedance spectroscopy of the microcell Li/LiPON/Li$_4$Ti$_5$O$_{12}$ (nominal voltage of 1.5 V), in which LiPON is an amorphous lithium phosphorus oxynitride (i.e., nitrogen-modified Li$_3$PO$_4$), has clarified the debate on the interface stability with lithium; it was clearly shown that LiPON forms a well-conducting solid electrolyte interface (SEI) layer [43]. Despite the low ionic conductivity (1 μS·cm^{-1}) and the rather large contribution to the internal cell resistance, LiPON can be used as a solid electrolyte film with a thickness of ~1 μm or less. Therefore, use of the LiPON-LiCoO$_2$ system is very popular in the construction of thin-film lithium microbatteries [26].

Table 3. Solid-state lithium microbatteries fabricated by the PLD technique.

Electrochemical Chain	Characteristics	Ref.
Li/Li$_3$PO$_4$/LiCoO$_2$	9.5 μAh·cm^{-2}	[14]
Li/Li$_4$SiO$_4$/LiCoO$_2$	10 μAh·cm^{-2}	[44]
SnO/Li$_{6.1}$V$_{0.61}$Si$_{0.39}$O$_{5.36}$/LiCoO$_2$	9.5 Ah·cm^{-2} at 44 μA·cm^{-2}	[41]
In/80Li$_2$S–20P$_2$S$_5$/LiCoO$_2$	95 mAh·g^{-1} at 0.13 mA·cm^{-2}	[42]
Li/Li$_3$PO$_4$/LiMnPO$_4$	12 mAh·g^{-1}	[45]
Li/LiPON/Li$_4$Ti$_5$O$_{12}$	32 μAh·cm^{-2} at 3.5 μA·cm^{-2}	[43]

3. Positive Electrode PLD Films

The main parameter for a microbattery is the delivered specific capacity. Rather than being expressed as the conventional unit of mAh·g^{-1}, due to the uncertainty in the film density, technologists prefer the stored charge, Q (expressed in μAh or in coulomb), per film surface area and the film thickness, i.e., μAh·cm^{-2}·μm^{-1} of mC·cm^{-2}·μm^{-1}. The relation between the gravimetric capacity, Q_m, of the material and the volumetric capacity of a film, Q_f, is given by:

$$Q_f = 0.36\, d\, Q_m \tag{1}$$

where Q_f is expressed in mC·cm^{-2}·μm^{-1}, Q_m in mAh·g^{-1}, and d is the density of the material in g·cm^{-3}. Table 4 summarizes the energetic quantities for the studied cathode compounds.

Table 4. Characteristics of the oxide materials used as a positive electrode in Li batteries. Δx_m is the quantity of electrons transferred (or Li uptake).

TMO Material	Density (g/cm^3)	C_m (mAh g^{-1}) for Δx_m	C_f (µA·cm^{-2}·µm^{-1})	C_f (mC·cm^{-2}·µm^{-1})
V$_2$O$_5$	3.35	294 ($\Delta x_m = 2$)	98.5	354
MoO$_3$	4.69	279 ($\Delta x_m = 2$)	130.8	471
LiCoO$_2$	5.03 [a]	137 ($\Delta x_m = 0.5$)	68.9	248
LiNi$_{0.5}$Co$_{0.5}$O$_2$	4.90 [a]	274 ($\Delta x_m = 1$)	134.2	483
LiMn$_2$O$_4$	4.32 [a]	148 ($\Delta x_m = 1$)	64.0	274

[a] after Ozuku, T.; Ueda, A. *J. Electrochem. Soc.* **1994**, *141*, 2972.

3.1. LiCoO$_2$ (LCO)

Having a lamellar structure, LiCoO$_2$ (LCO) is the prototypal positive electrode material commonly used in Li-ion batteries that yields a practical specific capacity of 135 mAh·g^{-1} in the voltage range from ~3.8 V (fully lithiated state) to ~4.2 V (charge state at Li$_{0.5}$CoO$_2$) [46]. Since the early work in 1996 by Berkeley's group [47], numerous studies have been devoted to the growth of LiCoO$_2$ thin films prepared by the PLD technique due in large to their high electrochemical performances. Further investigations of dense and well-defined PLD films described the phase evolution during Li extraction and the kinetics of Li$^+$ ions in the host lattice, which eventually found applications in the fabrication of the cathode element in microbattery stacks [48]. The two crystal forms, HT- and LT-LiCoO$_2$ phases, with the rock-salt (rhombohedral, *R-3m* space group) and spinel (cubic, *Fd3m* space group) structure, respectively, have been synthesized by pulsed-laser deposition. It was pointed out that the crystallographic texture for LCO films differs from one deposit technique to another, i.e., PLD versus sputtering, which influences the electrochemical properties due to the diffusion plane orientation [49]. Julien et al. stated that well-crystallized PLD-grown LCO thin films with a single layered structure can be obtained at substrate temperatures (T_s) as low as 300 °C [3].

The first growth of single phase LCO films by the PLD method was realized by Antaya et al. [4]. Films deposited on unheated stainless-steel substrates were amorphous but crystallized readily with heat treatment in air above 500 °C. Later, Striebel et al. [47] demonstrated the promise of PLD-grown films as cathodes for rechargeable lithium cells. Crystalline (003)-textured LCO films with thicknesses ranging from 0.2 to 1.5 µm were prepared without postdeposition treatment, which displayed a specific capacity of films of 62 µAh·cm^2·µm^{-1} and an Li diffusion coefficient of 1×10^{-10} cm^2·s^{-1}. Highly dense LCO films were first elaborated by the PLD process using a KrF laser under oxygen flow rates of 30 sccm and the pressure was maintained at 260 Pa on (200)-textured F-doped SnO$_2$ on fa used silica substrate maintained at $T_s = 700$ °C [49]. As-prepared LCO thin films were (00l) textured and had a density of 85% of the single crystal. The charge–discharge profile of the films was typical of the LCO bulk and presented an ~18% capacity loss for a single cycle to 4.15 V. In the potential range of 4.14 to 4.19 V, the measured chemical diffusion coefficients ranged from 1.7×10^{-12} to 2.6×10^{-9} cm^2·s^{-1} for as-deposited films and films annealed at 700 °C, respectively. Structural analysis of nanostructured LCO films prepared with PLD has been conducted by several research groups. Julien et al. [3,8,50] analyzed changes of the stoichiometry (i.e., the absence of the Co$_3$O$_4$ amorphous phase) as a function of the growth conditions using Raman spectroscopy. The inclusion of Co$_3$O$_4$ impurity is detected by analysis of the Raman intensity of the A_{1g} modes. Impurity-free films exhibit a specific capacity as high as 195 mC·cm^{-2}·µm^{-1} for polycrystalline films grown from an Li-rich target (i.e., excess of 15% Li$_2$O). The work by Okada et al. revealed that a decrease of the amount of inclusions can be obtained by a lower laser fluence and lower T_s [51]. Figure 2 presents the relationship between the impurity inclusions and growth conditions of PLD-grown LCO films established from spectroscopic Raman data. In this figure, the Co$_3$O$_4$ phase is grown under the conditions of high P_{O_2}, i.e., above the gray dashed line.

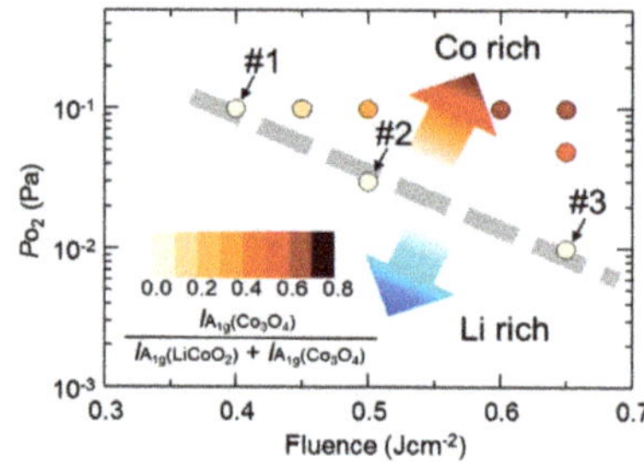

Figure 2. The relationship between impurity inclusions and growth conditions of PLD-grown LCO films established from spectroscopic Raman data (Reproduced with permission from [51]. Copyright 2017 AIP Publishing).

Ohnishi et al. [52,53] stated that suppression of the Co_3O_4 spinel phase can be ensured by the growth under a relatively low oxygen partial pressure. Zhang et al. [54] discussed the effect of the deposition conditions on the structure and morphology. The advantages of the preferential orientation of LCO films prepared by PLD has been discussed by numerous groups with the conclusion that dense uniaxial textured (003)-oriented films are obtained by a well-chosen substrate [6,53,55–59]. However, Xia et al. [59] stated that the fast transport of Li^+ ions is obtained for LCO films with a random orientation, in contrast with the results obtained with films having (003)-preferred orientation. Contrastingly, Nishio et al. claimed an excellent electrochemical performance for epitaxially grown LCO (77-nm thick) with a (104)-orientation on a (100) $Nb:SrTiO_3$ substrate that exhibited a discharge capacity of 26 mAh·g^{-1} even at high rates up to 100C [60]. Huo et al. stated that the film orientation is strongly dependent on the thickness and size of grains and demonstrated that films structured with parallel (003) planes are grown for thicknesses up to 300 nm [11]. Liu et al. showed that under certain PLD conditions, such as a high repetition rate of 35 Hz and low oxygen partial pressure of $P_{O_2} = 1$ Pa, LCO films tend to grow LCO films with a random orientation [61].

Epitaxial LCO thin films deposited on (001)-Al_2O_3 substrates remained in a single phase in a narrow range, $250 \leq T_s \leq 300\,°C$, whereas secondary phases appeared at $T_s > 300\,°C$, i.e., Co_2O_3, Co_3O_4, and $LiCo_2O_4$ [5]. Xia et al. established that thin LCO films can be easily grown with a (003) orientation because of the lowest surface energy for the (003) plane, while the minimized strain energy in thick LCO films allows preferential (101) and (104) textures. It seems that this last type of orientation favors the electrochemical performance of the LCO cathode [12]. The reduction of the laser fluence results in a decrease of the surface roughness of LCO films. With post annealing at 400 °C and optimized deposition conditions, LCO films exhibit an initial discharge capacity of 36 µAh·cm^{-2}·µm^{-1} and a cycleability of 94% [57]. Composition control was monitored to prepare stoichiometric LCO films using an Li-enriched target with a high-rate growth via an increase of the laser fluence to 0.29 J·cm^{-2} and an adjustment of the P_{O_2} to scatter the excess lithium. Ohnishi et al. showed that by using a $Li_{1.1}CoO_{2+\delta}$ target, the deposition of stoichiometric LCO with the highest crystallinity can be realized at the rate of 0.06 Å per pulse at the P_{O_2} of 0.1 Pa and $T_s = 800\,°C$ (Figure 3) [52,62].

Recently, Nishio et al. claimed that a high deposition rate of 1.2 Å·s^{-1} tends to form oxygen-deficient LCO films due to the destabilization of Co^{3+} cations and showed that post-annealing in air cancels the impurity phase [63]. Funayama et al. studied the effects of mechanical stress applied to LCO films (200 nm thick) deposited on Li-glass ceramic by the PLD method at 600 °C under a 20 Pa oxygen partial pressure for 1 h. Due to the lattice volume change, the generated electromotive force was 6.1×10^{-12} V·Pa^{-1} [64]. The electrode behavior shows an increase of the discharge capacity from 10 to 40 mAh·g^{-1} at a 2C rate with an increase of the T_s from 600 to 750 °C, whereas a $T_s = 800\,°C$ worsens the performance. Studies of the physico-chemistry of PLD-grown LCO thin films report the structural, surface morphology, optical, and electrical properties [65–67].

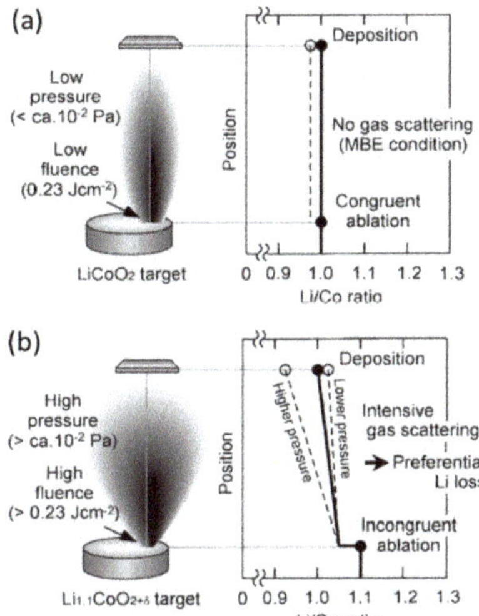

Figure 3. Schematic representation of the fast-laser ablation growth according to the Li/Co ratio variation of the plume for a stoichiometric (**a**) and Li-enriched target (**b**). (Reproduced with permission from [62]. Copyright 2012 IOP Publishing).

The electrochemical properties, i.e., thermodynamics and kinetics, of lithium intercalation in PLD LCO thin films have been widely investigated by the structure–electrochemistry relationship [59,68–70]; structural evolution upon charge-discharge cycles [71]; effect of doping by Ti, Al, and Mg [49,72–74]; characterization of the electrode/electrolyte interface [75]; and lithium-ion kinetics vs. basal plane orientation [76–80]. Highly (003)-oriented impurity-free LCO thin films grown by PLD on a stainless-steel substrate display an initial discharge capacity of 52.5 $\mu Ah \cdot cm^{-2} \cdot \mu m^{-1}$ and a capacity loss of 0.18% per cycle at a moderate current density of 12.7 $\mu A \cdot cm^{-2}$. These films show a very small lattice expansion upon charge, i.e., 0.09 Å for a charge of 4.2 V [81]. Figure 4 shows the typical electrochemical features of PLD LCO thin films grown on an Si wafer maintained at different temperatures. Both the specific discharge capacity and the mid-voltage increased with increasing T_s. For a film deposited at $T_s = 300\ °C$ under $P_{O_2} = 15$ Pa, the discharge capacity reached a value ~140 $mC \cdot cm^{-2} \cdot \mu m^{-1}$.

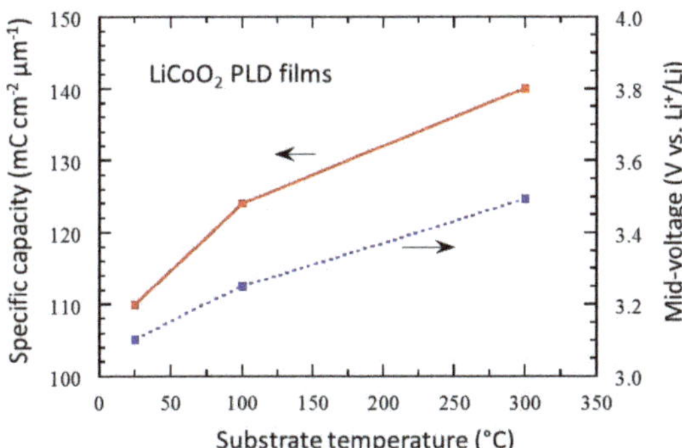

Figure 4. Electrochemical features of PLD-grown LCO thin films: specific discharge capacity and discharge mid-voltage vs. substrate temperature.

The electrochemical behavior of doped LCO thin films show that the voltage plateau at 3.65 V disappears in the charge curve of $LiTi_{0.05}Co_{0.95}O_2$ due to the doping effect, which cancels the semiconductor-metal-like transition of the LCO framework [81]. The Al-doped LCO film ($LiCo_{0.5}Al_{0.5}O_2$) exhibits a steady increase in the voltage vs. Li extraction with the absence of a voltage plateau as observed in stoichiometric LCO films; however, such films suffered from a limited capability and an upper bound of the diffusion coefficient of Li ($D^* = 9 \times 10^{-13}$ cm^2·s^{-1}) was observed [49]. Recent studies report on the improved electrochemical behavior of surface-modified LCO films using lithium tantalate (LTaO) and lithium niobite (LNbO). Coating with LNbO preserves the LCO surface and decreases the interfacial resistance, which indicates fast lithium transport [82]. LCO films modified by amorphous tungsten oxide (LWO) fabricated by PLD show a high capacity retention of $Q_r = 80\%$ at a high rate of 20C, against $Q_r = 0\%$ for bare LCO films cycled at the same C-rate. A slight increase of the superficial diffusion coefficient of Li$^+$ ions from 2.2×10^{-13} and 3.0×10^{-13} cm^2·s^{-1} was also observed, owing to the surface modification [83–86]. Note that LWO as well as LNBO are lithium ion conductors, which act as an efficient buffer between the electrolyte and LCO cathode. The structural degradation of cycled LCO films was investigated by Raman spectroscopy over 400 cycles, showing microstructural modification due to nanocrystallization and phase separation [87].

All-solid-state lithium microbatteries (SSLMB) using $LiCoO_2$ films have been developed using various inorganic solid electrolyte (SE) films, i.e., LiPON, Li_2S–P_2S_5, and amorphous Li_3PO_4. The thin-film battery with an electrochemical chain Li/amorphous Li_3PO_4/LCO/Pt shows a columnar-like LCO cathode (see the cross-sectional SEM image in Figure 5a) [88]. The excellent electrochemical performance is displayed in Figure 5b. This microcell exhibited an increase in capacity of up to 240 µAh·cm^{-2} when increasing the LCO thickness to 6.7 µm, which is 54% of the theoretical specific capacity of LCO (69 µAh·cm^{-2}·µm^{-1}). Shiraki et al. fabricated an SSLMB with an epitaxial LCO thin-film cathode (200 nm thick) by using PLD with a polycrystalline $Li_{1.1}CoO_2$ target ablated at a laser fluence of 1 J·cm^{-2}, LiPON solid electrolyte (2 µm thick), and Li film as the anode (0.5 µm thick) [89]. The authors reported cyclic voltammograms with six redox peaks, which drastically changed upon cycling but did not display the galvanostatic charge–discharge profile of the SSLMB.

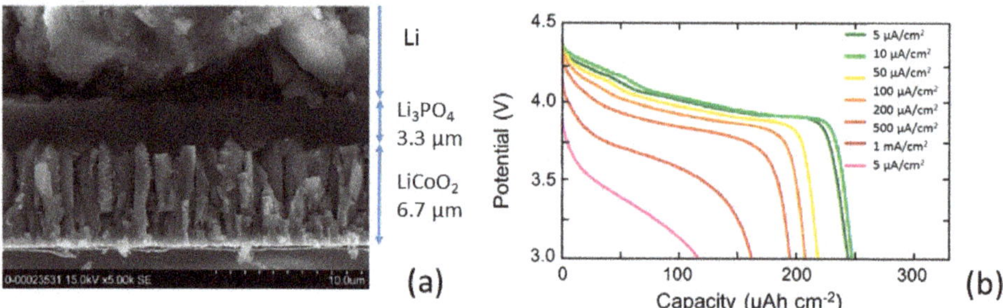

Figure 5. (a) SEM image cross-section of a Li/amorphous Li_3PO_4/LCO thin-film microbatteries. (b) Discharge curves for various current densities in the voltage range of 3.0 to 4.5 V vs. Li$^+$/Li. (Reproduced with permission from [88]. Copyright 2014 Elsevier).

3.2. LiNiO$_2$ (LNO)

PLD-grown thin films of lithium nickel oxide (LNO), i.e., $Li_xNi_{1-x}O$ and stoichiometric $LiNiO_2$, are applied as electrochromic and/or battery electrodes. In an early work, Rubin et al. established the complex relationship of the surface morphology and chemical composition of $Li_xNi_{1-x}O$ thin films vs. the deposition oxygen partial pressure, substrate temperature, and substrate–target distance as well [90]. LNO films produced at $T_s < 600$ °C immediately absorb CO_2 and H_2O when exposed to air, whereas they show long-term stability for $T_s = 600$ °C. LNO film with a composition of $Li_{0.5}Ni_{0.5}O$ (cubic rock-salt *Fd-3m* structure instead of the rhombohedral *R-3m* structure for $LiNiO_2$) was obtained under a deposition atmosphere of P_{O_2} = 60 mTorr. This film (150-nm thick) showed excellent electrochemical reversibility as an electrochromic item in the range of 1.0 to 3.4 V vs. Li$^+$/Li. An electrochromic device using WO$_3$ as the opposite electrode and PEO/LiTFSI as the solid polymer electrolyte (250-µm thick) showed an optical transmission range of ≈70% at 550 nm. Bouessay et al. optimized the PLD conditions to prepare NiO films, i.e., P_{O_2} = 0.1 mbar and T_s = 25 °C, and analyzed the electrochromic reversibility associated with the Ni^{3+}/Ni^{2+} redox couple [91]. Using a laser fluence of 1 to 2 J·cm^{-2}, which corresponded to an ablation rate of 0.9 Å·s^{-1}, NiO films with a cubic rock-salt structure (*Fm-3m* space group) were formed. Porous PLD NiO films were prepared using nickel foil as the target in a low oxygen atmosphere (P_{O_2} = 50 Pa) [92,93] and were applied as the electrode for supercapacitors, showing a high specific capacitance of 835 F·g^{-1} at a 1 A·g^{-1} current density.

Similarly to LCO, the PLD growth of stoichiometric LiNiO$_2$ with an α-NaFeO$_2$ layered structure requests an Li-enriched target, i.e., LiNiO$_2$ + 15%Li$_2$O. López-Iturbe and coworkers attempted to avoid Li loss by using an Ar atmosphere of P_{Ar} = 10 mTorr and laser fluence of 15 J·cm^{-2} [94], while Rao et al. introduced pure oxygen (P_{O_2} = 0.1 Torr) in the PLD chamber and ablated the target at a laser fluence of 10 J·cm^{-2} [95]. The LNO films prepared at T_s = 700 °C exhibited an initial discharge capacity of 175 mC·cm^{-2}·µm^{-1}. Yuki et al. used, as the oxygen evolution reaction (OER), electrocatalysts, which were prepared LNO films from a target composed of Li$_2$O and NiO$_2$ sintered at 1000 °C for 8.5 h in air ablated by a Nd:YAG laser operating at 532 nm [96]. Recently, PLD Li$_x$Ni$_{2-x}$O$_2$ thin films with $0.15 \leq x \leq 0.45$ deposited on a glass substrate under a pressure of 0.1 Pa and annealed at 350 °C were grown by PLD with an LiNiO$_2$ structure. The films appeared to be entirely made of particles even in the cross-section (grain size of 95 nm for x = 0.45). The average surface roughness estimated from the AFM measurements decreased with an increasing x, reaching a value of 0.615 nm for x = 0.45 [97].

3.3. LiNi$_{1-y}$Co$_y$O$_2$ (NCO)

The LiNi$_{1-y}$Co$_y$O$_2$ ($0 < y < 1$) system with a layered α-NaFeO$_2$ structure belongs to a LiCoO$_2$-LiNiO$_2$ solid solution with a higher reversible capacity than LCO and better cycleability than LNO. Among these substituted oxides, Ni-rich LiNi$_{0.8}$Co$_{0.2}$O$_2$ (NCO) has been identified as one the most attractive cathodes [98]. In this context, several works investigated the growth of NCO thin films using pulsed-laser deposition. Dense PLD NCO films grown at T_s > 400 °C exhibited a gravimetric density of 4.8 g·cm^{-3} [99]. Ramana et al. grew NCO films deposited on an Ni foil substrate at temperatures of $25 \leq T_s \leq 500$ °C under P_{O_2} = 6 to 18 Pa from Li-rich ceramic (15 mol% Li$_2$O excess to avoid NiO or Co$_3$O$_4$ impurity phases) [100]. At $T_s \leq 300$ °C, the PLD film showed the highest intensity of the (00*l*) reflection, which indicates that the *c*-axis was normal to the film surface. The XRD (003) diffraction peak at 2θ = 18.5° corresponds to an interplanar spacing of 0.145 nm. Phase diagram mapping (Figure 6) was proposed to highlight the effect of the growth temperature on the microstructure of PLD LiNi$_{0.8}$Co$_{0.2}$O$_2$ films. Galvanostatic titration carried out at a rate of C/30 in the potential range of 2.5 to 4.2 V showed a discharge capacity of 82 µAh·cm^{-2} µm^{-1}, which compares with the theoretical value of 136 µAh·cm^{-2}·µm^{-1} (490 mC·cm^{-2}·µm^{-1}) [101].

Hirayama et al. fabricated NCO films using the standard PLD conditions (Φ = 100–220 mJ, P_{O_2} = 3.3 Pa, target composition Li/Ni(Co) = 1.3, and T_s = 600–650 °C) at a deposition rate of 0.3 nm·min^{-1} on an oriented SrTiO$_3$ (STO) substrate. Microstructural analysis shows a misfit of ca. 5% and roughness of 1 to 3 nm for the film grown with an in-plane orientation at T_s = 600 °C. AFM imaging revealed the surface modification for films cycled in the voltage range of 2 to 5 V [102]. PLD NCO films (0.62 µm thick) were electrochemically characterized by galvanostatic titration (GITT), cyclic voltammetry (CV), and electrochemical impedance spectroscopy (EIS). Films grown at T_s = 600 °C under P_{O_2} = 13 Pa with a laser fluence of Φ = 450 mJ per pulse exhibited an average specific capacity of ~60 µAh·cm^{-2}·µm^{-1} and the Li$^+$ diffusion coefficient varied from 3×10^{-13} to 2×10^{-10} cm^2·s^{-1}. After 100 cycles, the electrode showed a capacity retention of 85% [103]. The kinetics of the Li-ion intercalation in PLD NCO films grown on an Nb-doped STO substrate at T_s = 600 °C under P_{O_2} = 3.3 Pa were investigated by EIS [104]. Nyquist plots showed changes of the electrode impedance as a function of the Li extraction/insertion with a larger value at a potential of 4.2 V. Baskaran et al. prepared NCO films on Pt and Si substrates heated at T_s = 500 °C under a low oxygen partial pressure of 0.21 Pa [105]. The 40-min deposited films (120-nm thick) displayed a specific discharge capacity of 69.6 μAh cm$^{-2}\cdot\mu$m^{-1} (145 mAh·g^{-1}) after 10 cycles. Based on these results, a Li-ion microbattery was fabricated with a LNCO/Li$_{3.4}$V$_{0.6}$Si$_{0.4}$O$_4$(LVSO)/SnO configuration with a thickness of ~1.5 μm. Such a microcell delivered a capacity of 16.1 μAh·cm$^{-2}\cdot\mu$m^{-1} after 20 cycles.

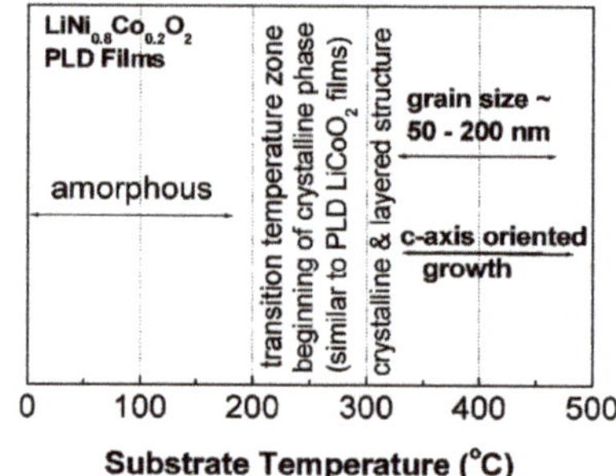

Figure 6. Phase diagram of microstructure development in PLD LiNi$_{0.8}$Co$_{0.2}$O$_2$ films as a function of the growth temperature. (Reproduced with permission from [100]. Copyright 2006 American Chemical Society).

3.4. LiNi$_{1-y}$Mn$_y$O$_2$ (NMO)

The substitution of Mn for Ni has been demonstrated to be beneficial for LiNiO$_2$ cathode materials. LiNi$_{0.5}$Mn$_{0.5}$O$_2$ (NMO) thin films were prepared on stainless steel and gold substrates from an Li-enriched target with an Li/(Ni + Mn) ratio of 1.5 in the NiO + MnO$_2$ mixture. Under standard conditions (Φ = 2 J·cm^{-2}, T_s = 550 °C, and P_{O_2} = 266 Pa), impurity-free and (001)-textured PLD films (300–500 nm thick) were obtained after an annealing process at 650 °C [106]. Galvanostatic charge–discharge tests showed that NMO films deposited on stainless steel displayed an electrochemical response, with a large voltage plateau between 2.5 and 3 V attributed to the presence of spinel phases (i.e., LiMn$_2$O$_4$ and LiNi$_{0.5}$Mn$_{1.5}$O$_4$), while NMO films prepared on Au substrate showed the typical fingerprint of the LiMO$_2$ layered compound with a single plateau at ca. 3.7 V. The analysis of the kinetics from CV measurements in the 2.5 to 4.5 voltage range led to a diffusion coefficient of the Li$^+$-ions of D^* = 3.13 × 10^{-13} cm^2·s^{-1} for the Li insertion and 7.44 × 10^{-14} cm^2·s^{-1} for the Li extraction. GITT results showed that D^* is highly dependent on the electrode potential in the range of 10^{-12} to 10^{-16} cm^2·s^{-1} [107]. Sakamoto et al. addressed the growth of epitaxial PLD NMO films with an orientation of the basal layered plane (BLP) that depends on the SrTiO$_3$ (STO) substrate plane: The BLP is parallel to the STO(110) substrate, while the BLP is perpendicular to the STO(111) substrate. The relationship between the lattice parameters and applied voltage highlights the charge–discharge processes for both orientations [108]. In a second article, Sakamoto et al. examined the structural properties of the surface and bulk of LiNi$_{0.5}$Mn$_{0.5}$O$_2$ epitaxial thin films during an electrochemical reaction using in situ X-ray scattering [109]. In normal conditions, the two-dimensional diffusion of Li during (de)intercalation proceeds for the (110) plane. However, 3D-diffusion activity can be observed for a high degree of cation mixing (Ni^{2+} in Li(3a) sites).

3.5. Li(Ni, Co, Al)O$_2$ (NCA)

The growth of LiNi$_{0.8}$Co$_{0.15}$Al$_{0.05}$O$_2$ (NCA) has been envisaged because Al-doping provides excellent structural and thermal stability for the electrode with the suppression of phase transitions. NCA thin films were prepared on Ni substrates at T_s = 500 °C by PLD with an energy and laser-pulse repetition rate of 300 mJ and 10 Hz, respectively, under P_{O_2} = 18 Pa. The PLD target (i.e., pellet pressed at 1.5 to 5.0 × 10^3 kg·cm^{-2}), optimized by using bulk NCA Li-enriched with 15 mol% Li$_2$O as the precursor, was sintered at 800 °C for 24 h [110]. The Li//NCA microcells delivered an initial specific capacity of 92 μAh·cm^{-2}·μm^{-1}. The kinetics of Li$^+$ ions in PLD films measured by the GITT method in the voltage range of 2.50 to 4.25 V vs. Li$^+$/Li revealed a diffusion coefficient of 4×10^{-11} cm^2·s^{-1} with a maximum of 1×10^{-10} cm^2·s^{-1} for the composition of Li$_{0.7}$Ni$_{0.8}$Co$_{0.15}$Al$_{0.05}$O$_2$. Figure 7 displays the specific capacity of Li//LiNi$_{0.8}$Co$_{0.15}$Al$_{0.05}$O$_2$ cells as a function of the substrate temperature. PLD films were grown onto various substrates at $25 \leq T_s \leq 500$ °C under a controlled O$_2$ atmosphere (P_{O_2} = 50 mTorr). NCA thin films prepared onto Ni foil at T_s = 500 °C exhibited the best specific discharge capacity of 100 μA·cm^{-2}·μm^{-1}.

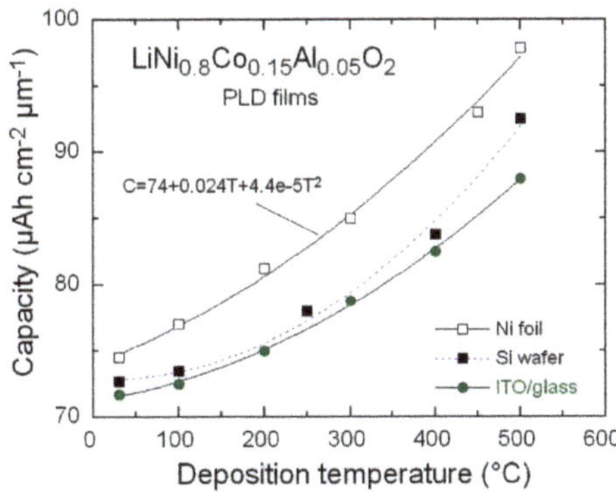

Figure 7. The specific capacity of LiNi$_{0.8}$Co$_{0.15}$Al$_{0.05}$O$_2$ thin films deposited onto Ni foil, Si wafer, and ITO-coated glass as a function of the substrate temperature.

3.6. Li(Ni, Mn, Co)O$_2$ (NMC)

Lithiated nickel-manganese-cobalt oxides, LiNi$_{1-y-z}$Mn$_y$Co$_z$O$_2$ (NMC), is a complex LiNiO$_2$-LiMnO$_2$-LiCoO$_2$ solid solution, which displays the same structure as rock salt, α-NaFeO$_2$, with a valence state of cations as Ni^{2+}, Mn^{4+}, and Co^{3+}. LiNi$_{1/3}$Co$_{1/3}$Mn$_{1/3}$O$_2$ (NMC333) thin film electrodes were prepared by pulsed laser deposition on a Pt/Ti/SiO$_2$/Si substrate (where the 150-nm thick Pt and the 100-nm thick Ti films act as the current collector and buffer layer, respectively) at room temperature under a P_{O_2} of 6.6 Pa. The deposition rate was about 4.4 nm·min^{-1}. Impurity-free NMC333 films were obtained using an Li-enriched target (15% excess Li$_2$O). The charge–discharge profiles strongly depended on the film morphology that was tuned by increasing the annealing temperature from 400 to 500 °C. The best electrochemical features were obtained for annealing at 450 °C, showing a discharge plateau of about 3.7 and 3.6 V [111]. Epitaxial and highly textured LiNi$_{0.5}$Mn$_{0.3}$Co$_{0.2}$O$_2$ (NMC532) thin film cathodes were deposited by a one-step PLD process [112]. Using a laser fluence of Φ = 6 J·cm^{-2}, T_s = 750 °C, and an Li-enriched target (20% Li$_2$CO$_3$ excess), the films were deposited on silicon (111), stainless steel (SS), and c-cut sapphire (0001) substrates. It was stated that highly dense and epitaxial films with a strong (003) reflection peak intensity are achieved at high T_s. Films grown on an Au-SS substrate delivered 125 mAh·g^{-1} at 0.5C and demonstrated a 72% capacity retention after 100 cycles. Furthermore, 3D-electrode architectures were fabricated, and are applicable to power hearing aids. Structured NMC333 film electrodes (50–80 μm thick) were prepared by laser printing using a pulsed laser (λ = 355 nm) at a fluence of 50 to 100 mJ·cm^{-2}. Such films exhibit a stable discharge capacity at a low C-rate (0.1C), but for a current density of 1C, the electrode reached 40% of the initial capacity [113]. Recently, Abe et al. elucidated the deterioration mechanism of pristine ZrO$_2$-coated NMC333 thin films prepared by PLD on a (110)STO substrate maintained at two temperatures of 25 and 700 °C; the oxygen pressure ranged from 3.3 to 10 Pa [114]. Characterization was performed by cyclic voltammetry, in situ X-ray diffraction, and in situ neutron reflectometry. As a result, ZrO$_2$ coating suppressed the low activity of the spinel phase formed at the surface and hence improved the cycleability of the thin film electrode.

3.7. Li-Rich Layered Oxides

The growth of Li-rich layered oxide $Li_{1.2}Mn_{0.55}Ni_{0.15}Co_{0.1}O_2$ on SRO/STO(100) and SRO/STO(111) substrate (where SRO and STO are $SrRuO_3$ and $SrTiO_3$, respectively) heated at $T_s = 600$ °C was reported by Bendersky et al. [115]. The transmission electron microscopy (TEM) images recorded using a high-angle annular dark field (HAADF) mode showed a predominant $Li_2(Mn,Ni,Co)O_3$ monoclinic phase. Johnston-Peck et al. confirmed the growth of monoclinic Li-rich thin films ($C2/m$ space group) using a target with the composition of $Li_{1.2}Mn_{0.55}Ni_{0.15}Co_{0.1}O_2$ and displayed the CV response with a non-aqueous electrolyte at a scan rate of 0.1 mV·s^{-1} [116]. Yan et al. intended to develop a PLD thin film electrode using an Li-rich layer structured oxide with a composition of $Li_{1.2}Mn_{0.54}Ni_{0.13}Co_{0.13}O_2$ (or written as $0.55Li_2MnO_3 \cdot 0.45LiNi_{1/3}Co_{1/3}Mn_{1/3}O_2$) [117]. Films deposited at $T_s = 650$ °C under $P_{O_2} = 46$ Pa and annealed at 800 °C showed the best electrochemical performance with an initial specific discharge capacity of 70 µAh·cm^{-2}·µm^{-1}. However, the differential capacity curves, dQ/dV, indicated that the layered structure gradually changed to the spinel phase during the charge–discharge cycling.

3.8. Li$_2$MO$_3$ (M = Mn, Ru)

Pulsed-laser deposited Li_2MnO_3 thin films at various thicknesses ($12 < \delta < 48$ nm) were grown using a KrF excimer laser on $Nb:SrTiO_3$(111) substrates from an Li-rich $Li_{3.2}MnO_3$ target. With synthetic conditions ($T_s = 650$ °C, $P_{O_2} = 75$ Pa, and $\Phi = 0.8$–1.1 J·cm^{-2}) and a laser frequency of $f = 1$ to 5 Hz, PLD Li_2MnO_3(111) films exhibited a single-phase with the $C2/m$ symmetry. The results of ICP analysis gave a composition of $Li_{1.90}Mn^{IV}O_{2.95}$, which indicates lithium and oxygen vacancies. The highest discharge capacity of 300 mAh·g^{-1} was delivered after 50 cycles by a 12.6-nm thick film [118]. PLD epitaxial Li_2RuO_3 thin films were successfully prepared with a (010) and (001) orientation on STO(110) and (111) substrates heated at $T_s = 500$ to 550 °C under $P_{O_2} = 3.3$ Pa. The initial charge–discharge capacities calculated using a theoretical density of 5.15 g·cm^{-3} were 120 and 105 mAh·g^{-1} at a 0.5C rate for the (010) and (001) orientation, respectively [119,120].

3.9. LiMn$_2$O$_4$ (LMO)

Since the early report in 1996 [47], numerous studies have been devoted to $LiMn_2O_4$ (LMO) thin films grown by laser ablation. A spinel structure ($Fd3m$ space group) was successfully prepared using different PLD conditions and applied as a positive electrode in thin-film lithium microbatteries (see, for example, [121]). Most prior works focused on the fundamental properties of PLD-grown LMO cathode films, aiming to deposit the highly structured and porous morphology required for a good operating electrode [7,122–136]. Morcrette et al. analyzed PLD film composition as a function of T_s and P_{O_2} using the Rutherford backscattering method and nuclear reaction analysis [122]. A stoichiometric LMO film was obtained at $T_s = 500$ °C and $P_{O_2} = 20$ Pa, while the Li/Mn ratio decreased with T_s and increased with P_{O_2}. The film of $Li_{0.6}Mn_2O_3$ (20% oxygen loss) was obtained under vacuum. The $Mn_3O_4 + MnO$ and $Mn_3O_4 + LiMn_2O_4$ mixed phases were successively grown in the range of $0.01 \leq P_{O_2} \leq 1$ Pa. Julien et al. defined the conditions of the disposition of LMO films grown on Si substrates. Impurity-free well-crystallized samples with a crystallite size of 300 nm were obtained at T_s as low as 300 °C and $P_{O_2} = 10$ Pa using an Li-enriched target (15% Li_2O excess) to avoid Li deficiency in the film [123]. The electrochemical features of the Li cell showed a specific capacity as high as 120 mC·cm^{-2}·µm^{-1} in the voltage range of 3.0 to 4.2 V vs. Li$^+$/Li, which was attributed to the high degree of film crystallinity [7,123]. Simmen et al. studied the relationship between $Li_{1+x}Mn_2O_{4-\delta}$ thin films and Li excess in the target and concluded that a film deposited from a composite target of $Li_{1.03}Mn_2O_{4-\delta}$ + 7.5 mol% Li_2O was the best, exhibiting a discharge capacity of 42 µAh·cm^{-2}·µm^{-1} [127]. Various substrates were successfully used for the epitaxial growth of $LiMn_2O_4$ (LMO) spinel thin films, such as Pt, Si, Au, MgO, Al_2O_3, and $SrTiO_3$. Gao et al. [128] reported a detailed mechanism of the epitaxial LMO film/substrate (current collector) interface formation. A coherent hetero-interface was formed with the substrate but a tetragonal Jahn–Teller distortion was observed, induced by oxygens' non-stoichiometry and the lattice misfit strain. PLD epitaxial LMO thin films were deposited on oriented Nb:SrTIO$_3$ substrates maintained at 950 °C from a sintered target with 100 wt.% excess Li$_2$O with different surface morphologies and orientations, such as (100)-oriented pyramidal, (110)-oriented rooftop, or (111)-oriented flat structure. The pyramidal-type LMO was cycled at a 3.3C rate, demonstrating a specific capacity of 90 mAh·g^{-1} after 1000 cycles [129]. Using oriented substrates, i.e., (111)Nb:SrTiO$_3$ (STO) and (001)Al$_2$O$_3$ single crystal, LMO films were grown with the (111) orientation under the following synthesis conditions: Li-enriched target (Li/Mn = 0.6), T_s = 650 °C, and P_{O_2} = 30 Pa [130]. Electrochemical tests emphasized the interactions between the films and substrate, and showed a plateau voltage at 3.6 to 3.8 V for LMO/STO and 3.8 to 4.1 V for LMO/alumina. Canulescu et al. [131] investigated the mechanisms of laser-plume expansion during the PLD of LMO films under P_{O_2} ranging between 10^{-4} and 20 Pa. The Li deficiency occurs as a result of the different behavior of the species at elevated T_s. Hussain et al. [132,133] obtained highly oriented LiMn$_2$O$_4$ thin films on oriented Si substrates heated in the range of 100 $\leq T_s \leq$ 600 °C under P_{O_2} = 10 Pa and with a laser fluence of Φ =10 J·cm^{-2}. Grains with a spherical shape (around 230 nm in diameter) changed to a flake-like structure at T_s = 600 °C (Figure 8a). The grain size varied almost linearly with the substrate temperature (Figure 8b).

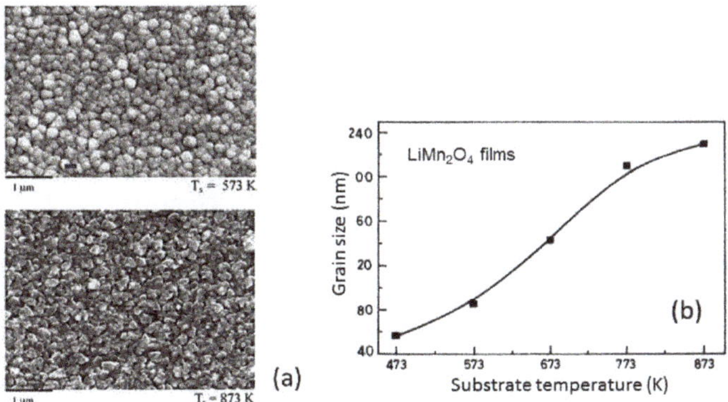

Figure 8. (**a**) SEM images of LiMn$_2$O$_4$ films deposited at T_s = 300 °C and T_s = 600 °C. (**b**) The grain size of PLD thin films as a function of T_s. (Reproduced with permission from [133]. Copyright 2007 Springer).

By applying an elevated-temperature PLD technique, Tang et al. [134] studied the influence of the substrate temperature (T_s) and the oxygen partial pressure (P_{O_2}) on LMO film crystallinity. LMO thin films deposited on Si (001)/0.2 µm-SiO$_2$ substrates at 575 °C under 13 Pa oxygen had a flat and smooth surface and exhibited mainly a (111) out-of-plane preferred texture (Figure 9a). Such films of the 300 nm thickness showed a very dense cross-section (density ~4.3 g·cm^{-3}) (see Figure 9b) [135]. The effect of stoichiometric deviations on the electrochemical performance of an LMO thin-film cathode was investigated by Morcrette et al. [136], while the kinetics of Li$^+$ ions in the LMO thin-film framework were documented by Yamada and coworkers [137]. A high-activation barrier of 50 kJ·mol^{-1} for Li-ion transfer was identified at the electrode/electrolyte interface for films deposited at T_s = 700 °C and P_{O_2} = 27 Pa. Albrecht et al. [138] reported the minimum crystallization temperature of spinel LMO thin films in a narrow annealing temperature range of around 700 °C. Electrochemical tests carried out with the galvanostatic cycling with the potential limits (GCPL) method proved that Li-ions are (de-)intercalated in different tetrahedral sites for which the processes occur at potentials that are slightly shifted by $U \approx$ 100 mV, which is similar to the previous results by Julien et al. [7].

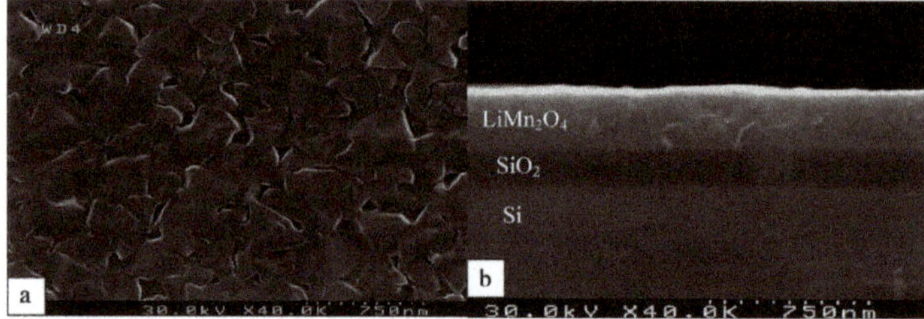

Figure 9. (**a**) Surface morphology and (**b**) cross-sectional picture of PLD-grown LMO thin film deposited on an Si(001) substrate covered by a 0.2 µm SiO_2 layer at 575 °C under a 13 Pa oxygen partial pressure. (Reproduced with permission from [135]. Copyright 2008 Elsevier).

Studies of the structure and electrochemical reactivity of heteroepitaxial $LiMn_2O_4/La_{0.5}Sr_{0.5}CoO_3$ (LMO/LSCO) bilayer thin films deposited on crystalline $SrTiO_3$ substrates show that LSCO reduced the lattice misfit strain with the substrate and favored a lower LMO surface roughness. However, a decrease of the electrical conductivity occurred during the electrochemical test (after first cycle) due to the lattice oxygen loss at the outermost layer (40 nm) [139]. Tang et al. reported a comparative investigation of the structures, morphologies, and properties of Li insertion for LMO films with different crystallizations. At T_s = 400 °C, LMO films consisted of nanocrystallites < 100 nm in size with rough surfaces that exhibited a discharge capacity of 61 $\mu Ah\cdot cm^2\cdot\mu m^{-1}$ with a capacity loss of 0.032% per cycle up to 500 cycles, while for T_s = 600 °C and P_{O_2} = 10 Pa, highly crystallized films showed an initial discharge capacity of 54.3 $\mu Ah\cdot cm^2\cdot\mu m^{-1}$ [134]. The intrinsic properties of PLD-grown LMO have been investigated by several techniques. Electrical measurements of LMO films showed that the conductivity is sensitive to T_s, as the activation energy that followed the Mott's rule increased with T_s up to E_a = 0.64 eV at T_s = 600 °C [133]. Singh et al. characterized the crystallinity and texture of LMO films deposited at T_s = 650 °C. Here, (111)-oriented films were grown on a doped Si substrate, while films deposited on a stainless-steel substrate exhibited a (001) orientation [140]. The thermo-power (or Seebeck coefficient) of PLD LMO films was reported to be 70 $\mu V\cdot K^{-1}$ [141].

PLD LMO films were subjected to an overcharge (5 V vs. Li^+/Li), which did not modify the structure and preserved the well-resolved voltage peaks at 4.1 and 4.2 V, while an overdischarge (2 V vs. Li^+/Li) led to a loss of capacity due to the structural disorder associated with the tetragonal transition, i.e., Jahn–Teller distortion [142]. Singh patented the fabrication of PLD $Li_{1-x}M_yMn_{2-2z}O_4$ films, where M is a doping element (M = Al, Ni, Co, Cr, Mg, etc.) and x, y, and z vary from 0.0 to 0.5 [143]. These defective spinel structures enhanced the oxygen content as compared to $LiMn_2O_4$ crystal. In particular, the oxygen-rich $Li_{1-\delta}Mn_{2-\delta}O_4$ films were superior cathode films, leading to excellent rechargeable battery performances. It is claimed that a high discharge rate of 25C produces only a 25% capacity loss and a specific capacity >150 $mAh\cdot g^{-1}$ remains after 300 cycles. Rao et al. reported the preparation of well-crystallized LMO films at a high substrate temperature of T_s = 700 °C and P_{O_2} = 13 Pa that delivered a capacity of 133 $mC\cdot cm^{-2}\cdot\mu m^{-1}$ at a very slow C/100 rate [144]. Several workers reported the evolution of the thin-film electrode/electrolyte interface, as the planar form of the film is the ideal geometry for such investigations [145–148]. Room temperature impedance measurements were carried out to identify the formation of the solid electrolyte interface (SEI) layer on a PLD LMO film cathode and the degradation mechanism during cycling in an aprotic electrolyte containing $LiPF_6$ salt. A reversible disproportionation reaction was suggested with the formation of the $Li_2Mn_2O_4$ and λ-MnO_2 phases at the surface [146]. Using epitaxial-film model electrodes, Hirayama studied the surface reaction and the formation of the SEI layer and the interfacial structural reconstruction during an initial battery process using in situ surface X-ray diffraction and reflectometry [147]. TEM images confirmed the surface reconstruction that occurred during the first charge, i.e., when a potential was applied. After 10 cycles, the SEI layer was observed on both the (111) and (110) surfaces and Mn dissolution appeared at the (110) surface [148]. Inaba et al. [149] investigated the surface morphology evolution of PLD LMO thin films grown on a PT substrate at T_s = 600 °C by electrochemical scanning tunneling microscopy (STM) with voltage cycling in the range of 3.5 to 4.25 V. The original LMO grains of 400 nm in size coexisted with small particles 120 to 250 nm in size, which appeared after 20 cycles and decreased to ~70 nm after 75 cycles through a kind of dissolution/precipitation process. LMO thin-film electrodes with a grain size of <100 nm deposited on a stainless steel substrate at T_s = 400 °C under a 26 Pa oxygen partial pressure displayed an excellent capacity of 62.4 $\mu Ah \cdot cm^{-2} \cdot \mu m^{-1}$ when cycled at a 20 $\mu A \cdot cm^{-2}$ current density in the voltage range of 3.0 to 4.5 V. A very low capacity fading was recorded for up to 500 cycles at 55 °C. Li^+-ion diffusion coefficients evaluated from EIS measurements were around 2.7×10^{-12} $cm^2 \cdot s^{-1}$ for an electrode charged at 4.0 V and 2.4×10^{-11} $cm^2 \cdot s^{-1}$ for 4.2 V [150]. Xie et al. [151] investigated the Li^+-ion transport in LMO thin films (~100 nm thick) grown on Au substrates at 600 °C at a deposition rate of 0.14 $nm \cdot min^{-1}$. The chemical diffusion coefficients determined by the EIS, GITTm, and PITT methods were in the range of 10^{-14} to 10^{-11} $cm^2 \cdot s^{-1}$ in the voltage range of 3.9 to 4.2 V. Table 5 lists some typical results on the kinetics of Li^+ ions in pulsed-laser deposited LMO thin films.

Table 5. Diffusion coefficients of Li^+ ions in PLD LMO thin film frameworks.

Growth Conditions	T_s (°C)	Method	Diffusion Coefficient $(cm^{-2} \cdot s^{-1})$	Ref.
P_{O_2} =13 Pa	800	GITT	2.5×10^{-11}	[47]
P_{O_2} = 13 Pa, $P = 10^8$ $W \cdot cm^{-2}$	300	GITT	10^{-11}–10^{-12}	[7]
P_{O_2} = 30 Pa, $\Phi = 1.0$ $mJ \cdot cm^{-2}$	600	EIS/PITT/GITT	10^{-11}–10^{-14}	[151]
P_{O_2} = 26 Pa, $P_{pulse} = 160$ mJ	400	PITT	10^{-10}–10^{-12}	[152]
P_{O_2} = 20 Pa, $\Phi = 1.1$ $mJ \cdot cm^{-2}$	500	Li radiotracer	1.4×10^{-10} (300 °C)	[153]
P_{O_2} = 0.2 Pa, $\Phi = 3.5$ $mJ \cdot cm^{-2}$	350	Li radiotracer	1.8×10^{-12} (350 °C)	[154]
P_{O_2} = 10 Pa, $\Phi = 2.0$ $mJ \cdot cm^{-2}$	650	Chronoamperometry	8.4×10^{-10}	[155]

The electrochemical behavior of Li-rich spinel $Li_{1.1}Mn_{1.9}O_4$ thin films grown by PLD on an Au substrate was reported by several workers [156,157]. The best performance was reported at a discharge current density of the 36C-rate for LMO films deposited for 30 min in P_{O_2} = 30 Pa and T_s = 600 °C with an Nd:YAG laser (266 nm) adjusted to an energy fluence of 1 $J \cdot cm^{-2}$ [156]. Nanocrystalline LMO films with grains less than 100 nm were deposited on a stainless-steel substrate at T_s = 400 °C and P_{O_2} = 26 Pa using a PLD pulse power of 100 mJ at the frequency of 10 Hz. The film cycled over 100 cycles delivered a specific capacity of 118 $mAh \cdot g^{-1}$ at a current density of 100 $A \cdot cm^{-2}$ [157]. Using reflectometry measurements, Hirayama et al. [158] studied the structural modifications at the electrode/electrolyte interface of a lithium cell, in which the LMO electrodes were prepared as epitaxial films by the PLD method with different orientations. The respective orientation of the LMO film corresponded to that of the substrate plane, i.e., the (111), (110), and (100) planes of the $SrTiO_3$ substrate. No density change was observed for the (110) and (100) planes, whereas a defect layer was detected in the (111) plane. ZrO_2-modified $LiMn_2O_4$ thin films prepared via PLD consisting of amorphous ZrO_2 formed on the grain boundary and the outer layer of the LMO matrix [159]. The high capacity retention of 82% after 130 cycles of films of $xZrO_2$-$(1-x)LiMn_2O_4$ ($x = 0.025$) monitored at the 4C rate was attributed to the decrease of the charge transfer resistance (R_{ct}).

Epitaxial $LiMn_2O_4/La_{0.5}Sr_{0.5}CoO_3$ (LMO/LSCO) bilayer thin films with sub-nano flat interfaces were deposited on (111)-oriented STO substrates at T_s = 650 °C in P_{O_2} = 4 Pa. After the first charge–discharge cycle, the decrease of the electrical conductivity of the LSCO buffer layer due to lattice oxygen loss induced capacity fading [139]. The PLD growth of a multilayer LMO thin film electrode demonstrated the compensation of lithium loss during deposition [160]. Such a sample prepared in the PLD conditions (T_s = 650 °C, Φ = 530 $mJ \cdot cm^{-2}$, and P_{O_2} = 1.3 Pa) showed the typical two pairs of voltammetry peaks at 0.82 and 1.02 V vs. Ag/AgCl in an aqueous cell. Kim et al. [161] prepared a $Li_{0.17}La_{0.61}TiO_3/LiMn_2O_4$ (LLTO/LMO) hetero-epitaxial electrolyte/electrode by PLD with an energy fluence of Φ = 1.7 $J \cdot cm^{-2}$ in a P_{O_2} = 6.6 Pa atmosphere. The typical herostructure is composed of a 17.5-nm thick LMO, 7-nm thick interfacial layer, and 26.5-nm thick LLTO deposited on a (111)-oriented

STO substrate. Voltammograms of the first and second cycles displayed redox peaks around 3.8 V attributed to an oxygen-deficient LMO and around 4.0 and 4.2 V, which are the typical redox voltages of the LMO spinel. Suzuli et al. [162] prepared multi-layer epitaxial $LiMn_2O_4/SrRuO_3$ (LMO/SRO) thin film electrodes deposited for 30 min via PLD on (111)STO substrates heated at 650 °C using an $Li_{1.2}Mn_2O_4$ target in P_{O_2} = 6.6 Pa. The LMO(33 nm)/SRO(38 nm) film exhibited a discharge capacity of 125 mAh·g^{-1} with the typical plateau regions of LMO in the charge–discharge reaction. Yim et al. substituted Sn for Mn in PLD LMO thin films [163]. The $LiSn_{x/2}Mn_{2-x}O_4$ films were prepared on a Pt/Ti/SiO$_2$/Si(100) substrate in the conditions of T_s = 450 °C, P_{O_2} = 26.7 Pa, Φ = 4.6 J·cm^{-2}, 10 Hz pulse frequency, and 4 cm target-substrate distance. XPS and EXAFS measurements showed that Sn^{2+} cations replace Mn^{3+} ions, which resulted in an increase of the valence of Mn in the spinel lattice. A high specific capacity of ~120 mAh·g^{-1} and cycleability with a capacity retention >81% at the 4C rate after 90 cycles was attributed to the Mn-deficient structure. A multi-layer PLD process was utilized to deposit LMO films (90 nm thick) on Si-based substrates coated with Pt as the current collector [164]. A reversible capacity of 2.6 µAh·cm^{-2} (corresponding to a specific capacity of $\approx$28 µAh·cm^{-2}·µm^{-1} or 66 mAh·g^{-1} assuming a dense film with 4.3 g·cm^{-3}) was reached at an extremely high current density of 1889 µA·cm^{-2} (equivalent to the 348C rate) with a capacity retention of 86% over 3500 cycles. A significant non-diffusion-controlled contribution (pseudocapacitive-like) was evidenced by cyclic voltammetry; however, the two typical voltage plateaus in the GCD of LMO (around 4 V) indicates that the faradaic redox reaction is the main process. For an easy comparison, Table 6 lists the electrochemical properties of PLD-prepared LMO thin film electrodes.

Table 6. Electrochemical properties of PLD-prepared LMO thin film electrodes. J is the current density, δ is the film thickness, and ΔC_c is the capacity fading per cycle.

PLD Conditions $T_s/P_{O_2}/\Phi$	Specific Capacity	Electrochemical Parameters	Ref.
300 °C/13 Pa/0.5 J·cm^{-2}	120 mC·cm^{-2}·µm^{-1}	J = 5 µA·cm^{-2}; δ = 200 nm; ΔC_c = 0.14%	[7]
25 °C/0.2 Pa/3.5 mJ·cm^{-2}	1.6 µAh·cm^{-2}	J = 4.4 µA·cm^{-2}; δ = 100 nm; ΔC_c = 0.4%	[121]
500 °C/20 Pa/2 J	n/a	ΔC_c = 0.05%/cycle	[122]
500 °C/20 Pa/4.3 J·cm^{-2}	42 µAh·cm^{-2}·µm^{-1}	J = 13 µA·cm^{-2} (1C); δ = 0.3 µm; ΔC_c = 0.025%	[127]
13 Pa/2.3 J·cm^{-2}	130 mAh·g^{-1}	I =5 µA (3.3C); δ = 110 nm; ΔC_c = 0.14%	[129]
575 °C/13 Pa/160 mJ	111 mAh·g^{-1}	J = 50 µA·cm^{-2}; δ = 0.3 µm; ΔC_c = 1.5%	[135]
600 °C/30 Pa/1 J·cm^{-2}	96 mAh·g^{-1}	J = 6 µA·cm^{-2}; δ = 100 nm;	[151]
600 °C/30 Pa/1 J·cm^{-2}	170 mC cm^{-2}·µm^{-1}	I = 1 mA (180C); δ = 120 nm	[156]
400 °C/26 Pa/160 mJ	142 mAh·g^{-1}	J = 20 µA·cm^{-2}; δ = 0.4 µm; ΔC_c = 0.06%	[157]
450 °C/26.7 Pa/4.6 J·cm^{-2}	120 mAh·g^{-1}	J = 330 mA·g^{-1}; δ = 140 nm; ΔC_c = 0.2%	[163]
650 °C/2.6 Pa/457 mJ·cm^{-2}	26 µAh·cm^{-2}·µm^{-1}	J = 1.88 mA·cm^{-2} (348C); δ = 90 nm	[164]

3.10. LiNi$_{0.5}$Mn$_{1.5}$O$_4$ (LNM)

$LiNi_xMn_{2-x}O_4$ is a substituted oxide spinel that operates at high voltages >4.5 V upon Li extraction. Substituted spinel films of $Li_xMn_{2-y}M_yO_4$ where M = Ni, Co and $0 \leq y \leq 0.25$ as-prepared with a crystalline morphology (0.3 µm thick) showed the typical features of high-voltage electrodes without carbon additive or binder materials in the range of 2.0 to 5.8 V vs. Li$^+$/Li [165]. Cyclic voltammetry showed that: (i) PLD LiMn$_{1.9}$Ni$_{0.1}$O$_4$ films charged at 5.7 V do not show capacity fading; (ii) LiMn$_2$O$_4$ and LiMn$_{1.75}$Co$_{0.25}$O$_4$ films present a good stability to 5.6 and 5.4 V vs. Li$^+$/Li, respectively; and (iii) below 3 V the films exhibit the typical Jahn–Teller distortion. The compound, LiNi$_{0.5}$Mn$_{1.5}$O$_4$ (LNM), is more interesting because of the oxidation state of cations: Ni^{2+} can be oxidized twice (i.e., 2e$^-$ transfer) during charge while Mn^{4+} is electrochemically inactive. Xia et al. showed that laser-ablated LNM films 0.3 to 0.5 µm thick deposited on a stainless steel substrate heated at 600 °C under an oxygen partial pressure of 26 Pa exhibit excellent capacity retention (i.e., ~120 mAh·g^{-1} after 50 cycles) in the voltage range of 3 to 5 V vs. Li$^+$/Li [166]. Well-crystallized oxygen deficient LiMn$_{1.5}$Ni$_{0.5}$O$_{4-\delta}$ films deposited by PLD at a controlled fluence of Φ = 2 J·cm^{-2}, T_s = 600 °C, and P_{O_2} = 26 Pa for 40 min exhibited a stepwise voltage profile near 4.7 V and a small plateau in the 4 V region. These disordered spinel structures had a stable specific capacity of 55 $\mu A\ h\ cm^{-2}\cdot\mu m^{-1}$ in the voltage range of 3 to 5 V vs. Li^+/Li. The good rate capability was due to the high kinetics for Li diffusion in the range of 10^{-12} to 10^{-10} $cm^2\ s^{-1}$ measured by the potentiostatic intermittent titration technique (PITT). These values are comparable to that of layered LCO [167,168]. Epitaxial LNM films were grown on single-crystal oriented $SrTiO_3$ (STO) substrates from an Li-enriched target with Li/(Ni + Mn) = 0.6. The film orientation, i.e., (100)-, (110)-, and (111)-oriented, were the replica of those of the STO substrates [169]. Depending on the film orientation and thickness, the discharge profiles exhibited two to three plateaus around 3.9, 4.5, and 4.7 V vs. Li^+/Li, which were attributed to the Mn^{3+}/Mn^{4+}, Ni^{2+}/Ni^{3+}, and Ni^{3+}/Ni^{4+} redox couples, respectively. Note that the emergence of the Mn^{3+}/Mn^{4+} redox couple was due to the introduced oxygen vacancies [170].

Other 5-V class cathode thin films include $LiCoMnO_4$. PLD $LiCoMnO_4$ films prepared under standard conditions (i.e., $T_s = 500\ ^\circ C$, $P_{O_2} = 20–100\ Pa$, and $\Phi = 2\ J\cdot cm^{-2}$) had a composition of Li:Co:Mn = 0.99:0.98:1. These films were tested in the voltage range of 3.0 to 5.5 V vs. Li^+/Li in SSMB consisting of $Li/Li_3PO_4/LiCoMnO_4$ fabricated on $Pt/Cr/SiO_2$ substrates [171]. Cyclic voltammetry showed that the higher capacity in the 5-V region was obtained for the film grown under $P_{O_2} = 100\ Pa$ (Figure 10). A specific discharge capacity of 90 $mAh\cdot g^{-1}$ remained after 20 cycles. Epitaxial $Li_{0.92}Co_{0.65}Mn_{1.35}O_4$ film with a cubic spinel structure was grown on a $SrTiO_3(111)$ single-crystal substrate using a layer-by-layer technique, which consisted of repeating a $Li_{1.2}Mn_2O_4/Li_{1.4}CoO_2$ deposition process at $T_s = 650\ ^\circ C$ and $P_{O_2} = 6.6\ Pa$ using a KrF excimer laser ($\lambda = 248$ nm) [172]. For a film area of 0.7 mm^2, thickness of 33.4 nm, and density of 4.38 $g\ cm^{-3}$, the specific discharge capacity was 340 $mA\cdot g^{-1}$ at the second cycle. A capacity retention of 80% was observed after 20 cycles.

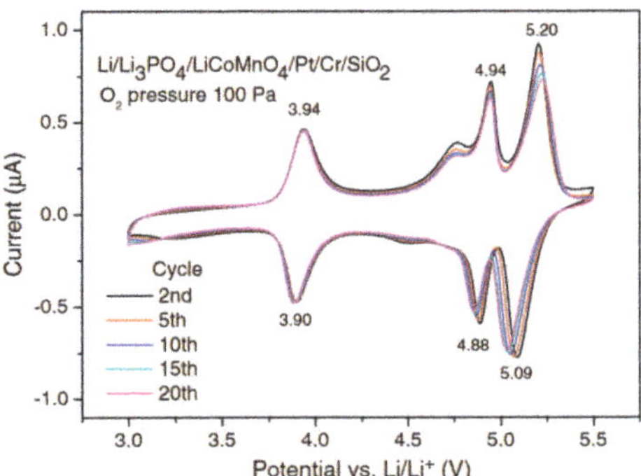

Figure 10. Cyclic voltammogram recorded at a 0.5 $mV\cdot s^{-1}$ scan rate of a $Li/Li_3PO_4/LiCoMnO_4$ thin film battery. The $LiCoMnO_4$ film cathode was grown under $P_{O_2} = 100$ Pa. (Reproduced with permission from [171]. Copyright 2014 Elsevier).

3.11. MnO_2

Due to its environmental compatibility and low cost, manganese oxides are promising candidate materials for supercapacitor applications using neutral aqueous solution as the electrolyte (i.e., 0.5 $mol\cdot L^{-1}$ K_2SO_4) [173]. Xia et al. prepared a dense Mn_3O_4 spinel thin film grown by PLD, which transformed to nanoporous MnO_x after electrochemical lithium insertion/extraction. After 2000 cycles, the MnO_x film deposited at $T_s = 600\ ^\circ C$ under $P_{O_2} = 26$ Pa demonstrated a specific capacitance of 193 $F\cdot g^{-1}$ at a current density of 5 $A\cdot g^{-1}$ [174]. The fundamental aspects of the redox reaction were investigated on amorphous MnO_x and crystalline Mn_2O_3 films prepared by PLD onto Si and (316)-stainless steel substrates [175]. Using standard conditions ($\Phi = 2–3\ J\cdot cm^{-2}$, $P_{O_2} = 13$ Pa, $T_s = 200–500\ ^\circ C$) Co-doped MnO_x films were grown from a hybrid Co_3O_4/Mn_3O_4 target. Non-doped and 3% Co-doped amorphous films (i.e., MnO_x and $Mn_{0.97}Co_{0.03}O_x$ film) exhibited a specific capacitance of 45 and 99 $F\cdot g^{-1}$ at a 5 $mV\cdot s^{-1}$ scan rate, respectively. The V_2O_5/Mn_3O_4 target was used to prepare PLD-grown V-doped MnO_x thin films with a

V content in the range of 3.3 to 10 atm. %. The specific capacitance of the crystalline PLD Mn_2O_3 film reached the value of 290 $F \cdot g^{-1}$ at a 1 $mV \cdot s^{-1}$ scan rate, while the 10 atm. % V-doped film ($Mn_{0.9}V_{0.1}O_2$) had a higher specific current value [176].

3.12. LiMPO$_4$ (M = Fe, Mn) Olivines

LiFePO$_4$ (LFP) thin-film electrodes have been successfully fabricated by pulsed-laser deposition [156,177–179]. It was shown that, due to the film thickness and carbon content, the electrochemical performances are very sensitive, i.e., electronic conductivity and Li-ion diffusion. Iriyama et al. reported the PLD growth of olivine structured LFP thin films and their electrochemical properties characterized by cyclic voltammetry and charge–discharge tests [180,181]. The typical olivine features were evidenced by CV measurements in the range of 2.0 and 5.0 V vs. Li$^+$/Li, i.e., a single couple of anodic and cathodic peaks at ~3.4 V. Song et al. synthesized PLD LFP films with a low carbon content (<1 wt.%) on stainless steel substrates utilizing an Ar atmosphere [182]. The 75-nm thick films showed reversible cycling of more than 80 $mAh \cdot g^{-1}$ after 60 cycles. Furthermore, 156-nm thick films grown using a target–substrate distance reduced to 5 cm had a layered surface texture and delivered more than 120 $mAh \cdot g^{-1}$ with a good capacity retention. LFP thin films with a needle-like morphology were prepared by an off-axis PLD technique [183]. The effect of the substrate on the structure and morphology was examined by Palomares et al. for PLD film deposited under argon gas kept at a pressure of 8 Pa [184]. Stainless steel was demonstrated to be the best substrate for the single-phase olivine (*Pnma* space group) with a temperature set at 500 °C.

LiFePO$_4$ deposited by pulsed-laser deposition proved to be effective as a thin film electrode. Tang et al. stated that a well-crystallized pure olivine phase was grown using optimized deposition parameters ($T_s = 500\,°C$, $P_{Ar} = 20$–30 Pa, pulse power of 120 mJ, pulse frequency of 10 Hz, $\lambda = 248$ nm) [185]. An electrochemical capacity of 38 $\mu Ah\ cm^{-2} \cdot \mu m^{-1}$ at the C/20 rate (36 $\mu Ah \cdot cm^{-2} \cdot \mu m^{-1}$ at a rate of C/4) was measured at 25 °C. High substrate temperatures ($500 \leq T_s \leq 700\,°C$) favored the presence of Fe^{3+} impurities, i.e., $Li_3Fe_2(PO_4)_3$ and $Fe_4(P_2O_7)_3$. In a second article, the same group analyzed the kinetics of Li$^+$ ions in PLD LFP films using CV, GITT, and EIS measurements [179]. CV data provided average D^* values of 10^{-14} $cm^2 \cdot s^{-1}$, while D^* deduced from both GITT and EIS techniques was in the range of 10^{-14} to 10^{-18} $cm^2 \cdot s^{-1}$. A maximum D^* value was observed at $x = 0.5$ for Li_xFePO_4. Lu et al. prepared different composite thin films, i.e., LiFePO$_4$–Ag and LiFePO$_4$–C, with the aim of enhancing the electronic conductivity [177,178]. It was found that films grown with 2 mol% carbon and annealed at 600 °C for 6 h had an improved coulombic efficiency. Well-crystallized olivine-type structure LFP films were obtained by PLD coupled with high temperature annealing of 650 °C. The first discharge capacity was 27 $mAh \cdot g^{-1}$ with a retention of only 49% after 100 cycles. The low reversible capacity and poor cycling performance was attributed to the existence of an Fe$_2O_3$ impurity produced by the high temperature treatment and poor intrinsic conductivity [186]. Sauvage et al. published several reports on the electrochemical properties of PLD LFP thin films grown in different configurations [187–190]. First, it was shown that well-crystallized and homogeneous 300-nm thick LFP films deposited on Pt-capped Si substrates have intrinsic Li insertion properties evaluated both in aqueous and non-aqueous electrolytes, i.e., voltage plateau at 3.42 V vs. Li$^+$/Li [187]. Second, the influence of the film thickness was studied in the range of 12 to 600 nm [188]. Third, the effect of the texture on the electrochemical performance was analyzed for PLD films deposited on a polycrystalline α-Al$_2O_3$ substrate coated with a 20-nm thick Pt layer from an LiFePO$_4$ pellet as the target. The standard PLD conditions were used (i.e., ($\Phi = 2\ J \cdot cm^{-2}$, $P_{Ar} = 8$ Pa, $T_s = 600\,°C$) [189]. Finally, the electrochemical stability of LFP films was analyzed as a function of the exposition to the most common lithium salt and for different current collectors (i.e., Si, Pt, Ti, Al, and (304)-stainless steel) [190]. A 270-nm thick film tested by CV at a 2 $mV \cdot s^{-1}$ scanning rate in 1 $mol \cdot L^{-1}$ LiClO$_4$ in EC/DMC solution delivered a specific capacity of 1.52 $\mu Ah\ cm^{-2}$ after 150 cycles. Recently, Raveendran et al. reported the properties of FeSe and LiFeO$_2$/FeSe bi-layers prepared by PLD as cathode materials [191]. Mangano-olivine LiMnPO$_4$ (LMP) thin films were fabricated on Pt-coated SiO$_2$ glass substrates using PLD parameters, e.g., $\Phi = 1.58\ J \cdot cm^{-2}$, $T_s = 400$–$700\,°C$, and $P_{Ar} = 2$–100 Pa [45]. LMP films (50-nm thick, 0.09 cm^2 area) were applied in Li/Li$_3$PO$_4$/LiMnPO$_4$ microbatteries for 500 cycles. From the CV measurement, a capacity of 28 mAh·g^{-1} at 20 mV min^{-1} was reported.

3.13. V$_2$O$_5$

Another candidate material for the cathodes of microbatteries is V$_2$O$_5$, in which about 1 mol of Li$^+$ ions can be inserted and extracted without the phase transformation of V$_2$O$_5$, leading to a theoretical specific capacity of 147 mAh·g^{-1}. Due to its stable layered structure and its ability to accommodate large amounts of Li ions, V$_2$O$_5$ has been widely studied for the development of electrochromic displays, color memory devices, and lithium-battery cathodes [192]. Extensive works have evidenced the advantages of PLD for the preparation of V$_2$O$_5$ films with a good reproducible stoichiometry similar to the target material [193–207]. The first work related to the growth of V$_2$O$_5$ thin film by PLD as an electrode for a thin-film lithium battery was reported by the National Renewable Energy Labs (USA) [193] followed by Julien's group [194,195]. A major advantage of laser ablation deposits is that it is possible to prepare thin layers of crystallized V$_2$O$_5$ under oxygen at a relatively low temperature of 200 °C [193]. The growth mechanism of PLD V$_2$O$_5$ thin films has been proposed by Ramana and coworkers [196]. It was reported that the grain size, surface roughness, and global morphology are highly sensitive to the nature and temperature of the substrate for films deposited in an oxygen partial pressure of P_{O_2} = 13 Pa. The functional influence of the growth temperature on the grain size for films deposited onto various substrates was also evidenced. Two main features should be pointed out: (i) The exponential variation of the grain size over the substrate temperature range of 25 to 500 °C; (ii) the variation is dependent on the substrate material, which is larger for the Si(00) wafer. McGraw et al. reported that pulsed-laser deposited V$_2$O$_5$ films can be grown on a number of low-cost substrates, including SnO$_2$-coated glass, on which highly textured (001) films are obtained at T_s = 500 °C under P_{O_2} in the range of 0.2 to 0.5 Pa [52,197,198]. PLD thin films of V$_2$O$_5$ were prepared for applications in lithium batteries using a ceramic V$_2$O$_5$ target and a KrF laser of a wavelength of 248 nm. Depending on the temperature of the substrates and the oxygen pressure during deposition, amorphous or crystallized layers are obtained. PLD-grown amorphous films exhibited a low capacity loss of ~2% over 100 discharge–charge cycles in the voltage range 4.1 to 1.8 V compared to 20% for crystalline film [45,199]. Thin layers of V$_2$O$_5$ were also prepared using a V$_2$O$_3$ target [200]. By making deposits at 200 °C with the same V$_2$O$_3$ target, amorphous layers were obtained in the absence of oxygen and layers crystallized in the presence of oxygen. Madhuri et al. [201] reported the successful crystallization of laser-ablated V$_2$O$_5$ thin films at T_s = 200 °C. These films were grown in the orthorhombic structure and exhibited a predominant (001) orientation. The growth of crystalline thin dense films without post-deposition annealing was claimed and the good electrochemical performance of PLD films was demonstrated. Iida et al. [202] addressed the electrochromic properties of V$_2$O$_5$ films deposited onto ITO glass as a function of the PLD parameters. The film recrystallization occurred in the range of 400 $\leq T_s \leq$ 500 °C and the best morphology was obtained for P_{O_2} = 13.3 Pa. McGraw et al. deposited thin films of V$_2$O$_5$ for applications in lithium batteries using a ceramic V$_2$O$_5$ target and a KrF laser, with a wavelength of 248 nm. Depending on the temperature of the substrates and the oxygen pressure during deposition, amorphous or crystallized layers were obtained [193,199]. Thin layers of V$_2$O$_5$ were also prepared using a V$_2$O$_3$ target. By making deposits at 200 °C with the same V$_2$O$_3$ target, amorphous layers were obtained in the absence of oxygen and layers crystallized in the presence of oxygen. Stoichiometric amorphous V$_2$O$_5$ films can be grown onto substrates maintained at low temperatures (T_s < 100 °C) using a sintered V$_2$O$_5$ target. Ramana et al. revealed that stoichiometric V$_2$O$_5$ films can be grown with a layered structure onto amorphous glass substrates at temperatures as low as 200 °C and an oxygen partial pressure of 100 mTorr [203]. The onset of crystallization occurred at 200 °C with an activation energy of 0.43 to n0.55 eV [204]. Correlations between the growth conditions, microstructure, and optical properties were investigated for V$_2$O$_5$ thin films deposited over a wide substrate temperature range of 30 to 500 °C by Rutherford backscattering spectrometry (RBS), X-ray photoelectron spectroscopy (XPS), scanning electron microscopy (SEM), and UV−vis−NIR spectral measurements. As shown in Figure 11, the film grain size follows a power law of the substrate temperature and the optical energy bandgap decreases from 2.47 to 2.12 eV with the increase of T_s from 30 to 500 °C [205]. Bowman and Gregg investigated the effect of the applied strain on the resistance of V_2O_5 thin films grown from both metallic vanadium and a ceramic V_2O_5 target using a laser fluence of ~3.0 and ~1.5 J cm^{-2}, respectively [208]. Deng et al. compared the growth of V_2O_5 films using femtosecond (f-PDL) and nanosecond (n-PDL) pulsed laser deposition using SEM, XRD, and Raman spectroscopy. Prior to annealing, f-PLD films showed a rougher texture and nano-crystalline character, while n-PLD films were much smoother and predominantly amorphous [209].

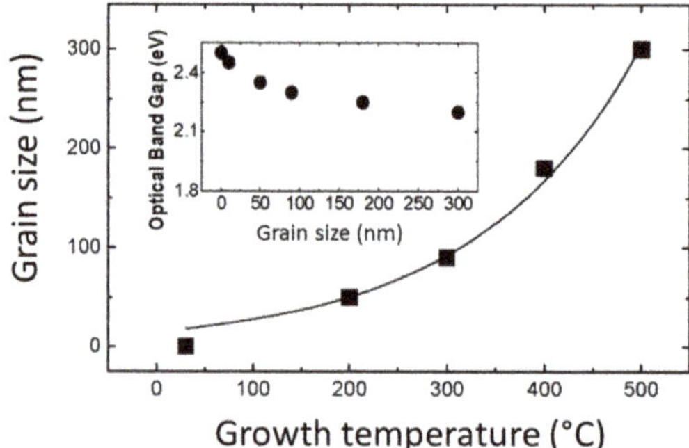

Figure 11. Variation of the grain size in V_2O_5 thin films as a function of the substrate temperature. (Reproduced with permission from [205]. Copyright 2005 American Chemical Society).

The PLD growth conditions were refined by an analysis of the surface properties for the production of high-quality V_2O_5 films. The investigations were carried out by AFM, SEM, FTIR, and XRD. AFM measurements showed a surface roughness of ~12 nm with a Gaussian-like height distribution of surface grains for films deposited at $T_s = 200$ °C under $P_{O_2} = 10$ Pa [210]. The local structure of 0.3-µm thick films grown on Si(100) substrates was characterized by Raman spectroscopy [211]. The influence of the deposition temperature on the microstructure was investigated by an examination of the rigid layer-like mode at 145 cm^{-1}, which showed a frequency shift with increasing T_s. The ability of the V_2O_5 thin film lattice to accommodate Li$^+$ ions was also investigated by Raman spectroscopy. The appearance of the δ- and γ-phases of Li$_x$V$_2$O$_5$ gave additional insight into the structural changes of lithiated films. From the photoluminescence spectra, Iida et al. evidenced a blue shift of the vanadyl V = O peak upon Li$^+$ insertion for different electric charge in the range of $0 \leq Q \leq 20$ mC of Li$^+$ [212]. From X-ray diffraction and Raman spectroscopy data, Shibuya et al. derived a $P_{O_2} - T_s$ phase diagram for V-O films grown on Si(100) substrates (Figure 12). The composition of V-O films was as follows: (i) A VO$_2$ monoclinic phase was formed at $T_s \geq 450$ °C and P_{O2} in the range of 5 to 20 mTorr; (ii) a V_2O_5 orthorhombic phase was obtained under oxidative conditions, i.e., at high P_{O_2}; (iii) a V_6O_{13} phase was grown under P_{O_2} between oxidative and reductive conditions; and (iv) metastable V_4O_9 and VO$_2$(B) phases were formed for lower T_s (≤400 °C) and lower P_{O_2} (≤30 mTorr) [213].

V_2O_5 thin films have been widely used as electrochromic electrodes but few reports are devoted to PLD-grown films. Fang et al. obtained thin films deposited on In$_2$O$_3$:SnO$_2$ (ITO)-coated glass and (111)Si wafer from a V_2O_5 target using an XeCl laser with a wavelength of 308 nm for applications in electrochromic devices [214,215]. Electrochromic tests over 60,000 cycles showed that a significant change in the optical density (bleached and colored states) was evaluated to be 0.13 at λ = 600 nm for as-prepared films at $T_s = 200$ °C. Crystallized c-axis oriented V_2O_5 films were obtained under oxygen and at a substrate temperature of 200 °C. The durability without long-term degradation of the electrochromic V_2O_5 films was tested over 8000 cycles in the voltage range of 1.2 to 1.4 V [216]. Ti-doped V_2O_5 thin films prepared by the pulsed laser ablation technique at $T_s = 200$ °C and $\Phi = 2$ J·cm^{-2} were studied as the electrode for an electrochromic display that exhibits a neutral brownish blue color.

The long-term durability was verified over 8000 cycles of a voltage cycled in the range from −1.0 to +1.0 V vs. SCE showing a charge of 35 mC·cm^{-2}. The good cycleability was attributed to the layered structure of PLD crystalline films with a parallel orientation to the substrate, suitable for Li$^+$-ions' transport [215]. PLD thin films of the system, WO$_3$-V$_2$O$_5$, were prepared with a laser fluence of 1 to 2 J·cm^{-2} on SnO$_2$/F-coated glass substrates at T_s = 25 °C under P_{O_2} = 0.1 mbar. Such films with low V contents cycled in the protonic medium. The true color neutrality is the main advantage of V-based WO$_3$ thin films; however, the cell capacity and coloration efficiency decrease with an increase of the V content [217]. The orthorhombic V$_2$O$_5$ phase is also applied as electrodes for sensors. Huotari reported that pure PLD films were obtained at Φ = 2.6 J cm^{-2}, T_s = 400 °C, and P_{O_2} = 1.0 Pa with a post-annealing treatment at 400 °C for 1 h in normal ambient conditions [218]. The efficient response to NH$_3$ at part-per-billion levels. indicates these films use as possible sensing materials for ammonia gas [219].

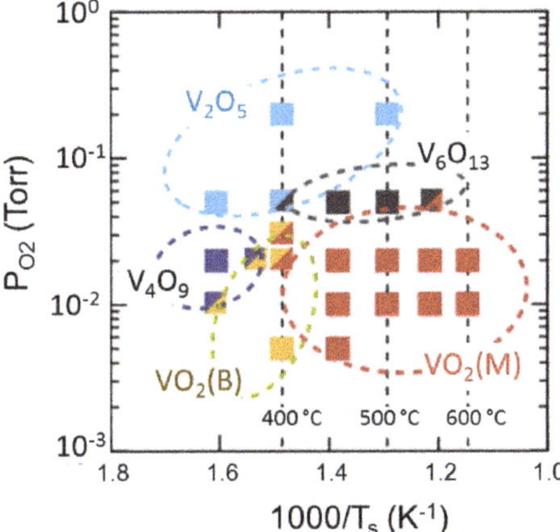

Figure 12. P_{O_2} vs. 1000/T_s phase diagram for films of vanadium oxides laser-pulse deposited on silicon substrates. (Reproduced with permission from [213]. Copyright 2015 AIP Publishing).

The electrochemical properties of V$_2$O$_5$ thin-film cathode material have been widely studied in cells with aprotic electrolytes (typically LiClO$_4$ dissolved in propylene carbonate). The electrochemical charge–discharge profiles of PLD V$_2$O$_5$ films were also found to be dependent on T_s, exhibiting a marked difference for V$_2$O$_5$ films grown at T_s < 200 °C when compared to those grown at $T_s \geq$ 200 °C. The effect of the substrate temperature and hence the microstructure on the kinetics of the lithium intercalation process in V$_2$O$_5$ films is remarkable. The applicability of the grown PLD V$_2$O$_5$ films in lithium microbatteries indicates that PLD V$_2$O$_5$ films in the temperature range of 200 to 400 °C offer better electrochemical performance than films grown at other temperatures due to their excellent structural quality and stability [25,220]. As an experimental fact, pulsed laser deposited V$_2$O$_5$ thin films exhibit a higher initial voltage than the crystalline material, i.e., ~4.1 vs. ~3.5 V (Li$^+$/Li). For instance, V$_2$O$_5$ thin-film cathodes, deposited from a V$_6$O$_{13}$ target at a fluence of ~12 J cm^{-2} on SnO$_2$-coated glass at T_s = 200 °C, were efficient for Li$^+$-ion incorporation. In (h00)-textured films, the specific capacity reached values between 50% and 80% of the theoretical value. On the other hand, amorphous films display a stable capacity corresponding to 1.2 F mol^{-1} in the voltage range of 4.1 to 1.5 V. Prior textured V$_2$O$_5$ films discharged beyond the threshold to 2.0 V vs. Li$^+$/Li showed an immediate and continuous capacity fading and a quasi-total amorphization after 10 cycles [193,197]. The chemical diffusion coefficient of Li$^+$ ions, D^*, measured by PITT was found to be in the range of 1.7 × 10^{-12} to 5.8 × 10^{-15} cm^2·s^{-1} in crystalline V$_2$O$_5$ films, which compares well to the value found in Li$_x$V$_2$O$_5$ phases, whereas D^* displayed a smooth and continuous decrease as the Li content increased in amorphous films [198].

In an attempt to apply PLD V_2O_5 films in SSMB, a thin-film microbattery was constructed using a glassy $Li_{1.4}B_{2.5}S_{0.1}O_{4.9}$ electrolyte film with an ionic conductivity of 5×10^{-6} S·cm^{-1} at 25 °C and an Li anode film. This $Li/Li_{1.4}B_{2.5}S_{0.1}O_{4.9}/V_2O_5$ cell delivered a capacity of ~400 mC·cm^{-2}·µm^{-1} at a current density of 15 µA·cm^{-2} [221]. $Ag_{0.3}V_2O_5$ and LiPON thin films with a smooth surface were grown by PLD in an N_2 and O_2 atmosphere, respectively. The $Li/LiPON/Ag_{0.3}V_2O_5$ SSMB displayed good cycleability at a current density of 7 µA·cm^{-2} in the voltage window of 1.0 to 3.5 V. The specific capacity was maintained at 40 µAh·cm^{-2}·µm^{-1} after 100 cycles [222]. Recently, amorphous vanadium oxide a-VO_x PLD films (650 nm thick) were grown on stainless steel substrates from a V_2O_5 PLD-target under P_{O_2} in the range of 0 to 30 Pa. Films prepared under P_{O_2} = 13 Pa had a smooth surface and bore an O/V atomic ratio of 2.13 with a higher atomic percentage of V^{5+} than that of V^{4+}. Electrochemical tests carried out in Li cells with 1 mol·L^{-1} $LiPF_6$ in ethylene carbonate (EC) and diethyl carbonate (DEC) (1:1 by volume) as the electrolyte showed a reversible specific capacity as high as 300 mAh·g^{-1} at the C/10 current rate and a capacity retention of 90% after 100 cycles [223]. Such studies were initiated by Zhang et al. in 1997 to obtain VO_x films PLD grown at 200 °C and exhibiting a specific capacity of 340 mAh·g^{-1} at a current density of 0.1 mA·cm^{-2} and a capacity loss <2% at the end of 100 cycles [201]. A summary of the electrochemical properties of PLD-grown vanadium oxide thin film electrodes is given in Table 7.

Table 7. Electrochemical properties of PLD-prepared vanadium oxide thin film electrodes. J is the current density, δ is the film thickness, and ΔC_c is the capacity fading per cycle.

PLD Conditions $T_s/P_{O_2}/\Phi$	Specific Capacity	Electrochemical Parameters	Ref.
200 °C/2.6 Pa/12 J·cm^{-2}	200 mAh·g^{-1}	Capacity loss of <2% over 200 cycles	[193]
300 °C/133 Pa/0.1 J·cm^{-2}	235 mC cm^{-2}·µm^{-1}	J = 5 µA·cm^{-2}; ΔC_c = 1.1%	[195]
100 °C/6 Pa/200 mJ	240 mAh·g^{-1}	0.1C current rate; δ = 650 nm; ΔC_c = 0.1%	[223]
200 °C	340 mAh·g^{-1}	Capacity loss <2% after 100 cycles	[201]
200 °C/10 Pa/2 J·cm^{-2}	340 mC cm^{-2}·µm^{-1}	δ = 100 nm; 8000 cycles (electrochromic)	[216]

3.14. V_6O_{13}

With the ability of vanadium cations (two V^{4+} every V^{5+}) to be reduced, the mixed-valence vanadium oxide, V_6O_{13}, the structure of which is formed by alternated single and double layers of VO_6 units, can insert reversibly about 6 mol of Li, giving a specific capacity of 311 mAh·g^{-1}. It makes this compound a good candidate for the cathode material of rechargeable batteries [224]. V_6O_{13} films were fabricated by the PLD technique using a pulsed KrF excimer laser (λ = 248 nm, 20 ns pulse duration, 10 Hz frequency, and 4 J·cm^{-2} laser fluence). A (100)-oriented Si substrate was maintained at a temperature of 500 °C. During the PLD process, the formation of crystalline V_6O_{13} films (dark-bluish color) and the vanadium oxidation state (+2.166) was monitored by controlling the processing temperature and O_2 partial pressure. The (002)-oriented V_6O_{13} thin films (50 nm thick) were obtained after a post annealing at 400 °C under an oxygen partial pressure of 100 Pa [221]. The discharge profiles for $Li//V_6O_{13}$ thin-film cells were recorded in the voltage range of 3.3 to 2.5 V at a current density of 5 µA cm^{-2} (Figure 13). A film as-grown at T_s = 250 °C exhibited a steady discharge curve with an insertion uptake of 6Li per V_6O_{13} formula unit, whereas the cell voltage decay was faster for a film deposited at T_s = 25 °C. However, the films deposited at T_s = 250 °C and annealed at 300 °C in an Ar atmosphere displayed a stepped discharge profile with the appearance of a voltage plateau at ca. 3.02 and 2.85 V vs. Li^+/Li.

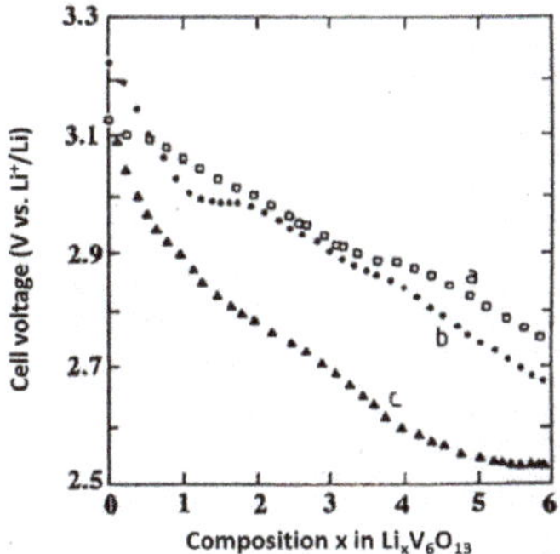

Figure 13. Discharge profiles vs. lithium uptake for Li//V_6O_{13} thin-film microbatteries. Active cathode films were grown with: (**a**) T_s = 250 °C, as-deposited; (**b**) T_s = 250 °C, annealed at 300 °C in Ar; (**c**) T_s = 25 °C, annealed at 300 °C in Ar.

3.15. FeF_2

Iron fluoride is a conversion-type cathode material with a high theoretical specific capacity of 571 mAh·g^{-1}. Several groups reported electronic additive-free FeF_2 films grown by the PLD technique at low temperatures [225–228]. The electrochemical properties of FeF_x films were reported to be dependent on the substrate temperature. Using an FeF_3 target, crystallized-like FeF_2 film ($P4_2/mnm$ space group) was obtained at T_s = 600 °C, while a mixed FeF_3-FeF_2 phase was grown at T_s = 25 °C and single FeF_3 phase was prepared at T_s = −50 °C [226]. FeF_x deposited on stainless steel substrates under vacuum (5 × 10^{-5} Pa) exhibited a capacity of ~600 mAh·g^{-1} at a current density of 0.56 μA·cm^{-2}. Santos-Ortiz et al. reported the PLD growth of polycrystalline FeF_2 thin films on oxide-etched Si(100) and glass substrates using standard conditions (T_s = 400 °C, Φ = 8 J·cm^{-2}, growth rate of ~6 nm·min^{-1}) [227]. A 50-nm thick PLD FeF_2 film on stainless steel substrates held at 400 °C showed an initial specific discharge capacity of 167 mAh·g^{-1} when cycled 200 times in the potential range of 1 to 4 V vs. Li$^+$/Li at the 1C current rate [228].

3.16. MoO_3

MoO_3 is an attractive cathode material for microbattery technology from several standpoints: (i) The orthorhombic a-phase is a layered structure favorable for Li insertion between slabs; (ii) Mo has the highest +6 oxidation state, making the high structural stability; (iii) the lattice can be reversibly inserted up to 1.5Li per mole of oxide, yielding a specific capacity of 280 mAh·g^{-1}; and (iv) the capacity of the dense film can reach a value of ≈130 μAh·cm^{-2}·μm^{-1}, almost twice the value for $LiCoO_2$ [229]. In addition to the use as cathode batteries, MoO_3 is a material applied in electrochromics, gas sensors, and electro-optics. For certain applications, high-quality films grown by PLD are required.

Currently, PLD MoO_3 thin films are grown using a KrF excimer laser (λ = 248 nm) with a fluence of 2 J cm^{-2} (energy of 300 mJ per pulse) and deposited on various substrates heated in the range of $25 \leq T_s \leq 500$ °C under an atmosphere of O_2 flow maintained at a pressure of $0.1 \leq P_{O_2} \leq 20$ Pa. In the prior report, Julien et al. showed that the structure analyzed by optical spectroscopy strongly depends on T_s: For T_s < 150 °C, an amorphous phase is formed, the β-MoO_3 phase grows at T_s ≈ 200 °C, and the layered α-MoO_3 phase appears at T_s = 300 °C [230–233]. Al-Kuhaili et al. reported the growth of polycrystalline MoO_3 films on unheated substrates using both XeF and KrF excimer lasers. By tuning the annealing temperature in the range of 300 to 500 °C, both the grain size and surface roughness increased. Films formed using the XeF laser (λ = 351 nm) and annealed at 400 °C have the best stoichiometry of $MoO_{2.95}$ [233]. Analyzing the growth mechanism, Ramana and Julien concluded that the thermochemical reaction during ablation strongly influences the structural characteristics of PLD MoO_3 films. Above T_s = 400 °C, the formation of compositional defects induces structural disorder, i.e., α-β-MoO_{3-x} phase mixture [234,235].

The applicability of PLD films to an Li microbattery was demonstrated by the best electrochemical features: A discharge capacity of 90 μAh cm^{-2} μm^{-1} was obtained for T_s = 400 °C, while only 53 μAh·cm^{-2}·μm^{-1} was delivered for T_s = 200 °C [236]. Puppala et al. investigated the microstructure and morphology of PLD MoO$_{3-x}$ thin films' growth for catalytic applications using a femtosecond laser (f-PLD) and a nanosecond excimer-laser (n-PLD). Substantially textured films with a partially crystalline phase prior to annealing were obtained by the f-PDL laser, while the n-PLD-grown MoO$_{3-x}$ films were predominantly amorphous with a smooth surface [237]. Sunu et al. claimed that as-deposited PLD films (T_s = 400 °C, Φ = 4–5 J·cm^{-2}, repetition rate of 15 to 20 Hz, and P_{O_2} = 500 Pa) are suboxide-like, i.e., mixture of η-Mo$_4$O$_{11}$ and χ-Mo$_4$O$_{11}$, which transformed to MoO$_3$ after annealing at 500 °C in air for 5 h [238]. Several works reported the PLD growth of films (MoO$_3$)$_{1-x}$(V$_2$O$_5$)$_x$ with $0.0 \leq x \leq 0.3$ prepared at room temperature under an oxygen pressure of 13.3 Pa. The effect of the V$_2$O$_5$ content on the coloring switching properties for thermochromic, gasochromic, photochromic, and electrochromic applications was investigated [239,240]. Contrary to pure MoO$_3$, the electrochromism of MoO$_3$-V$_2$O$_5$ films showed that the Mo oxidation state (+6) did not change considerably upon Li$^+$ insertion, while V^{5+} was reduced considerably to V^{4+} [239]. A similar improvement of the gas-sensing properties, i.e., the shortest response time and highest transmittance change, was observed for V$_2$O$_5$-doped MoO$_3$ films under an H$_2$ atmosphere [240].

3.17. WO$_3$

Tungsten oxide (WO$_3$) belongs to the class of "chromogenic" materials, i.e., materials exhibiting coloration effects through electro-, photo-, gas-, laser-, and thermochromism processes, which requires the high homogeneity provided by the PLD technique. Preliminary studies of the growth of WO$_3$ thin films by PLD were first attempted by Haro-Poniaowski et al. [233] in 1998. Later, Rougier et al. reported the PLD conditions for the growth of efficient WO$_3$ films as electrochromics (EC) components [241]. The microstructure of films deposited on SnO$_2$:F coated glass substrate is strongly sensitive to both the oxygen pressure and substrate temperature: (i) Crystallized films are formed for T_s = 400 °C and P_{O_2} = 10 Pa; (ii) amorphous films are obtained for P_{O_2} = 1 Pa at any T_s; (iii) for T_s = 25 °C and P_{O2} = 1 Pa, WO$_3$ films are blue colored and conductive; and (iv) colorless insulator films are grown for T_s = 25 °C and P_{O_2} = 10 Pa, which display the best electrochromic properties. Qiu and Lu showed that oxygen deficient WO$_{3-\delta}$ films with a deviated monoclinic structure were produced using PLD parameters as 2.5 J·cm^{-2}, P_{O_2} = 26 Pa, and a target-Si(100) substrate distance of d = 5 cm [242]. Ramana et al. investigated the structural transformations of PLD WO$_3$ as a function of the annealing treatment. Using standard conditions (Φ = 2 J·cm^{-2}, T_s = 300 °C, P_{O_2} = 13.3 Pa), films deposited on glass substrates (200–500 nm thick) showed an atomic ratio of O/W $\approx$ 2.96 $\pm$ 0.05. The monoclinic phase of the as-prepared film transformed to an orthorhombic phase at 350 °C and to a hexagonal phase at 500 °C [243,244]. By varying the substrate temperature in the range of 150 to 800 °C and the oxygen pressure from 1 to 40 Pa, Mitsugi et al. obtained WO$_3$ films with a different microstructure: Amorphous, crystallized tetragonal, and triclinic phases [245]. Hussain et al. obtained amorphous, polycrystalline, and nanocrystalline WO$_3$ phases, and iso-epitaxial WO$_3$(00l) thin films deposited on single-crystal SrTiO$_3$ substrates at 600 °C and under P_{O_2} = 18 Pa [246]. Suda et al. deposited PLD WO$_3$ thin films on flexible ITO substrates. They showed that films, prepared at T_s < 300 °C, are amorphous and polycrystalline phases were obtained at T_s > 400 °C, while the crystallinity of the film on glass substrates was not dependent on P_{O_2} [247]. Films deposited at 400 °C were porous with a nanocrystalline triclinic structure and showed the best cycleability [216,248,249].

The suitability of PLF WO$_3$ films for EC applications was investigated as a function of the partial oxygen pressure during deposition. Studies of the texture and morphology of PLD 30-nm thick WO$_3$ films deposited on Si(100) and SrTiO$_3$(100) substrates under an O$_2$ background of 2.5 Pa showed that: (i) The laser fluence (in the range of 5 to 15 J·cm^{-2}) strongly influences the texture, (ii) the films grown on STO are biaxially textured with a smooth surface, and (iii) films deposited on Si are granular [250]. The fabrication of WO$_3$ thin films with color neutrality for applications as EC materials was realized by the deposition of films containing 20% of vanadium onto SnO_2:F coated glasses at $T_s = 20\,°C$ under $P_{O_2} = 10\,Pa$. The blue color in the reduced state ($-0.4\,V$) of the W-O-V films lost intensity and turned grey-blue (transmittance of 50%) as the V concentration increased [251]. Highly transparent WO_3 films exhibiting strong coloration and fast and full bleaching were prepared under PLD conditions ($\Phi = 1\,J\cdot cm^{-2}$, $T_s = 250\,°C$, $P_{O_2} = 16\,Pa$, and $d = 40\,mm$) [252]. WO_3 films were also prepared using similar PLD parameters for applications in gas sensors [253–255].

4. Solid Electrolyte PLD Films

For the development of solid-state thin film batteries, thin films of solid electrolytes with excellent performances, i.e., high ionic conductivity (σ_i), good stability against the lithium anode, large electrochemical window (ΔV), and poor electronic conductivity (σ_e), are currently required. To fulfill these requirements, the thin films of oxide-, phosphate-, or sulphide-based solid electrolytes were grown by the PLD technique [256–258]. The facile manufacture of such thin films is due to the easy control of the PLD chamber's atmosphere. Table 8 lists some typical solid electrolyte thin films prepared by PLD [41,259–264].

Table 8. Electrical properties of PLD-grown solid-electrolyte thin films.

Electrolyte	Ionic Conductivity (S cm^{-1})	Electronic Conductivity (S cm^{-1})	Ref.
$Li_{3.3}PO_{3.9}N_{0.17}$ (LiPON)	1.6×10^{-6}	$>10^{-14}$	[259]
$Li_{3.4}V_{0.6}Si_{0.4}O_4$ (LVSO)	2.5×10^{-7}	7.4×10^{-13}	[41]
$Li_{3.25}Ge_{0.25}P_{0.75}S_4$ (thio-LISICON)	2.2×10^{-3}	1.5×10^{-7}	[260]
$Li_{0.5}La_{0.5}TiO_3$ (LLTO)	2.2×10^{-5}	3.5×10^{-11}	[261]
Li_2O-Si_2O (LSO)	2.2×10^{-4} (200 °C)	–	[262]
β-$LiAlSiO_4$	4.0×10^{-7} (225 °C)	–	[263]
$Li_6BaLa_2Ta_2O_{12}$ (garnet-type)	2.0×10^{-6} (25 °C)	2.9×10^{-13}	[264]

4.1. LiPON

In the early 1990s, Bates et al. prepared Li_3PO_4 thin films using a sputter-deposition technique in the presence of N_2 gas that resulted in a nitrogen-doped lithium phosphate (called LiPON) of a typical chemical composition, $Li_{3.3}PO_{3.9}N_{0.17}$ to $Li_{2.9}PO_{2.9}N_{0.7}$. The structure consists of doubly and triply coordinated nitrogen atoms, which form cross-links between the phosphate chains [20]. LiPON displays a high chemical stability and an ionic conductivity of $2 \times 10^{-6}\,S\cdot cm^{-1}$ at 25 °C [265]. The growth of LiPON thin films by pulsed-laser deposition is also realized in nitrogen partial pressure with a moderate laser power influence [259,266].

Zhao et al. reported the growth LiPON thin films on three different substrates (i.e., Si wafer, Au-coated Si, and Al-coated glass plate) by reactive PLD in an N_2 gas atmosphere in the range of 50 to 200 mTorr using a Li_3PO_4 target. The target was ablated by the beam of a Nd:YAG laser at the fluence of 5 to 20 mJ·cm^{-2}. The influence of the ambient N_2 pressure and the laser fluence on the ionic conductivity was systemically examined and the best result of $1.6 \times 10^{-6}\,S\cdot cm^{-1}$ with an activation energy of 0.58 eV at 25 °C was obtained for a film prepared under 200 mTorr at $\Phi = 15\,J\cdot cm^{-2}$. The mechanism of the nitridation of Li_3PO_4 was carried out by XPS measurements, showing that σ_i increases with the N/P ratio [259]. West et al. showed that a 17-nm thick layer of LiPON deposited at the solid electrolyte–electrode interface decreased the charge-transfer resistance from 4470 to 760 cm^{-2} in a Li/LiPON/LNM cell. The PLD amorphous films with $\sigma_i = 1.5 \times 10^{-8}\,S\cdot cm^{-1}$ at 25 °C were deposited from a crystalline Li_2PO_2N target under the flow of N_2 gas at $P_{N_2} = 1\,Pa$ [267].

4.2. $Li_xLa_{2/3+y}TiO_{3-d}$ (LLTO)

Solid electrolytes, such as lithium lanthanum titanium oxides, $Li_xLa_{2/3+y}TiO_{3-\delta}$ (LLTO), based on a perovskite-like structure can accept vacancies at the Li (or La) and oxygen sites and show properties depending on the composition, with an electronic conductivity when Ti^{3+} cations (instead of Ti^{4+}) are present and an ionic conductivity for Li-rich material. The typical growth of LLTO thin films fabricated by the laser ablation technique is obtained at deposition temperatures in the range of 600 to 800 °C under a controlled oxygen pressure from 0.1 to 100 Pa [268]. LLTO films, such as $Li_{3x}La_{(2/3)-x}TiO_3$, exhibit a high ionic conductivity of up to 10^{-5} S·cm^{-1} when deposited with pulsed laser deposition [269–271]. $Li_{0.5}La_{0.5}TiO_3$ (LLTO) PLD thin films, prepared at 400 to 600 °C, are amorphous and show an ionic conductivity of ~2×10^{-5} S·cm^{-1} at room temperature. Contrary to crystalline films, the amorphous LLTO exhibits good stability in contact with lithium metal anodes. Half-cells based on $LiCoO_2$ films covered with LLTO films deposited by pulsed laser deposition could be cycled for hundreds of cycles [269]. Furusawa et al. prepared amorphous LLTO films at T_s = 25 °C with a uniform thickness (0.46–0.63 μm) using a laser energy of 180 mJ per pulse at 10 Hz [270]. The authors stated a controlled pressure of ~10^{-6} Torr but did not mention the presence of O_2 gas. The highest σ_i of 1.2×10^{-3} S·cm^{-1} (E_a = 0.35 eV) obtained for $Li_{0.5}La_{0.5}TiO_3$ films deposited on an Ag substrate was due to the absence of grain boundaries. Maqueda optimized the PLD growth parameters to prepare $La_{0.57}Li_{0.29}TiO_3$ dense films at T_s = 700 °C under P_{O_2} = 15 Pa with smooth surfaces [271]. The obtained nano-crystalline films exhibited domains, which are cubic and tetragonal modifications of the perovskite phase. Transport measurements showed an ionic conductivity of 8.2×10^{-4} S·cm^{-1} at 25 °C with E_a = 0.34 eV. Epitaxial $Li_{0.33}La_{0.56}TiO_3$ solid electrolyte thin films were grown on $NdGaO_3(110)$ by PLD at T_s higher than 900 °C under P_{O_2} = 5 Pa [272]. These films showed a conductivity σ_i of 3.5×10^{-5} S·cm^{-1} at 25 °C with E_a = 0.35 eV (Figure 14).

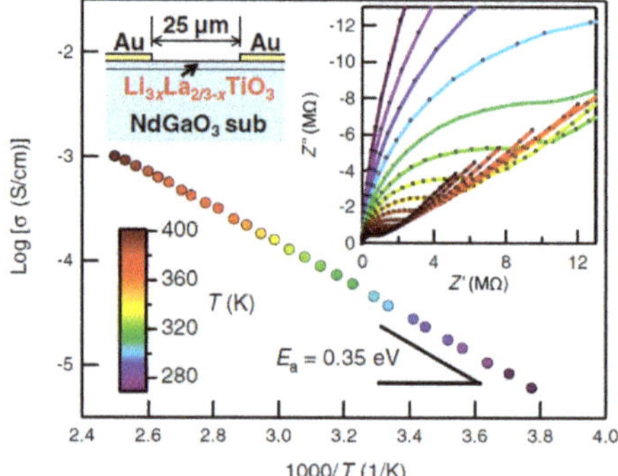

Figure 14. Temperature of the in-plane ionic for conductivity epitaxial $Li_{0.33}La_{0.56}TiO_3$ solid electrolyte thin films (36-nm tick) deposited on $NdGaO_3(110)$ by PLD at T_s higher than 900 °C under P_{O_2} = 5 Pa. (Reproduced with permission from [272]. Copyright 2012 Elsevier).

The influence of different substrates and excess lithium in the target on the microstructure and ionic conductivity of PLD LLTO thin films was examined by Aguesse et al. [273] Despite a large lattice mismatch of up to +8.8% with the substrate, the epitaxial growth of LLTO is possible on different (001) oriented $LaAlO_3$, $SrTiO_3$, and MgO substrates using a sintered $Li_{0.37}La_{0.54}TiO_3$ target and PLD parameters, such as T_s = 750–880 °C, P_{O_2} = 4–20 Pa, and laser fluence of 1.07 J cm^{-2}. An ionic conductivity as high as 19.2×10^{-3} mS·cm^{-1} at 25 °C was obtained for 170-nm thick LLTO films grown on an STO substrate from an ablated 10 mol% lithium excess target. PLD LLTO films with a σ_i of 3×10^{-4} S·cm^{-1} and σ_e of 5×10^{-11} S·cm^{-1} were obtained by controlling the background P_{O_2} and T_s. Amorphous LLTO films were utilized in SSMB cycled up to 4.8 V vs. Li^+/Li with high voltage $LiNi_{0.5}Mn_{1.5}O_4$ spinel cathode thin films [274].

Another class of LLTO electrolytes consists of Ti-based solid electrolytes with a garnet-like structure, first reported by Weppner et al. [275]. $Li_6BaLa_2Ta_2O_{12}$ thin films were deposited on an MgO(100) substrate by the ablation of a target with a 5 mol% Li_2O excess. In standard PLD conditions

(T_s = 550 °C, P_{O_2} = 5 Pa, laser fluence of Φ = 2 J·cm^{-2}, 40,000 laser pulses), an ionic conductivity of 2×10^{-6} S·cm^{-1} at 25 °C with an activation energy of 0.42 eV was obtained. This is comparable with the σ_i of LiPON. The electronic conductivity varied from 2.87×10^{-13} to 3.47×10^{-10} S·cm^{-1} in the range of the polarization voltage from 2.8 to 4.3 V [273]. Saccoccio et al. fabricated garnet $Li_{6.4}La_3Zr_{1.4}Ta_{0.6}O_{12}$ films via PLD and studied the impact of PLD parameters (fluence of 1 to 4 J·cm^{-2}, T_s in the range of 50 to 700 °C, and a post-annealing process) on the structural and transport properties. The ionic conductivity was measured by impedance spectroscopy. It was concluded that σ_i is not dependent on T_s but is strongly affected by the laser fluence [276]. $Li_7La_3Zr_2O_{12}$ (LLZO) garnet-like thin films were deposited on Si_3N_4/Si substrates at temperatures in the range of $50 \leq T_s \leq 750$ °C under a fixed background of P_{O_2} = 1.3 Pa with a KrF excimer laser set at 0.6 J·cm^{-2} [277]. The best material, which exhibited an ionic conductivity of 6.3×10^{-3} S cm^{-1} at 400 °C (E_a = 0.6 eV), was obtained at T_s = 300 °C. The review of garnet-like solid electrolyte thin films grown via PLD is summarized in Table 9.

Table 9. Literature review of garnet-like solid electrolyte thin films grown via PLD.

Material	Substrate	σ_i at 25 °C (S cm^{-1})	E_a (eV)	Ref.
$Li_6BaLa_2Ta_2O_{12}$	MgO(100)	2×10^{-6} at 25 °C	0.42	[278]
$Li_7La_3Zr_2O_{12}$	Si_3N_4/Si	6.3×10^{-3} at 400 °C	0.60	[279]
$Li_7La_3Zr_2O_{12}$	$SrTiO_3$, sapphire	7.4×10^{-7} at 25 °C	0.32	[280]
$Li_7La_3Zr_2O_{12}$	Si, SiO_2, MgO	1.6×10^{-6} at 25 °C	0.35	[281]
Al-doped $Li_7La_3Zr_2O_{12}$	$Gd_3Ga_3O_{12}$	2.5×10^{-6} at 25 °C	0.52	[275]
Al-doped $Li_7La_3Zr_2O_{12}$	MgO	8.3×10^{-4} at 300 °C	0.60	[264]

4.3. P- and Si-Based Electrolytes

Several phosphorus- or silicon-based oxides and sulfides are solid electrolytes for lithium batteries, such as Li_3PO_4, Li_4SiO_4-Li_3PO_4, Li_2S-P_2S_5 glass ceramic, $Li_{2+2x}Zn_{1-x}GeO_4$ (LiSICON), $Li_{3.25}Ge_{0.25}P_{0.75}S_4$ (thio-LiSICON), etc., that can be prepared as thin films. Kuwata et al. prepared high quality Li_3PO_4 thin films by PLD for applications in $Li/Li_3PO_4/LiCoO_2$ all-solid-state thin-film batteries. The Li_3PO_4 film exhibited an ionic conductivity of 4×10^{-7} S·cm^{-1} at 25 °C and an activation energy of 0.58 eV. This solid electrolyte showed an electrochemical stability in the potential range of 0.0 to 4.7 V vs. Li^+/Li and was applied in $Li/Li_3PO_4/LiCoO_2$ cells [14,282]. Amorphous PLD thin films of LiSICON display higher conductivities than that of Li_4SiO_4 and Li_3PO_4 films. The solid electrolyte $0.5Li_4SiO_4$–$0.5Li_3PO_4$ dense films deposited on an Si wafer at Φ = 2–6 J·cm^{-2} under an argon gas of P_{Ar} = 0.01–5 Pa had an ionic conductivity of 1.6×10^{-6} S·cm^{-1} at 25 °C and an activation energy of 52 kJ·mol^{-1} [283]. Nakagawa et al. determined that PLD Li_2SiO_3 films stable to CO_2 have an ionic conductivity of 2.5×10^{-8} S·cm^{-1} at 25 °C lower than that of Li_2SiO_3 films (i.e., 4.1×10^{-7} S·cm^{-1}), which are unstable to CO_2 [284]. PLD thin films of lithium meta-silicate (LSO) deposited at a growth rate of 0.17 Å per pulse on various substrates (i.e., SiO_2, quartz, sapphire, Al_2O_3 ceramic, and MgO) from an Li_2SiO_3 sintered tablet were grown in the amorphous state. The ionic conductivity slightly depends on the substrate species with the best results (σ_i = 4.5×10^{-4} S·cm^{-1} at 300 °C, E_a = 0.88 eV) found for an 80-nm thick film deposited on SiO_2 glass [262,285]. The PLD conditions for the growth of thio-LiSICON $Li_{3.25}Ge_{0.25}P_{0.75}S_4$ solid electrolyte thin films were carefully chosen (especially the Li content of 3.2 in the target, which maintains the number of Li vacancies) to obtain a high σ_i value of 1.7×10^{-4} S·cm^{-1} at 25 °C [265]. PLD $80Li_2S$-$20P_2O_5$ thin film prepared under P_{Ar} = 5 Pa exhibited an ionic conductivity and activation energy of 7.9×10^{-5} S·cm^{-1} and 43 kJ mol^{-1} at 25 °C, respectively. Heat treatment increased the σ_I to 2.8×10^{-4} S·cm^{-1} [286]. To avoid the formation of a Li-deficient phase, such as $Li_4P_2S_6$, an Li_2S-enriched Li_3PS_4 target was used to grow PLD solid-electrolyte thin films. Using an $Li_{3.42}PS_{4.21}$ target, PLD Li_3PS_4 films exhibited a higher ionic conductivity of 5.3×10^{-4} S·cm^{-1} at 20 °C [287].

4.4. PLD Electrolyte as Buffer Layers

Solid-state electrolyte (SSE) thin films have been used as a conductive buffer layer for the reduction of high resistance at the electrode/SSE interface of high-power all-solid-state lithium batteries. Coating the Li_3PO_4 thin films on electrode materials by the PLD method was found to be efficient for this purpose. Konishi et al. [288] reported the effect of surface Li_3PO_4 coating on $LiNi_{0.5}Mn_{1.5}O_4$ epitaxial thin film electrodes. Amorphous Li_3PO_4 film (1–4 nm thick) was deposited at 25 °C with a laser energy of 150 mJ under P_{O_2} = 3.3 Pa. It was also pointed out that such a coating reduces the Mn dissolution in the non-aqueous electrolyte. Yubuchi et al. [289] fabricated the same coated electrode with Φ = 2 J cm^{-2} but under a lower oxygen gas pressure of 0.01 Pa. With a 100-nm thick Li_3PO_4 deposit, the total resistance of the Li cell decreased from 15 to 350 Ω. A PLD protective coating of $80Li_2S–20P_2S_5$ solid electrolytes on $LiCoO_2$ particles was performed at room temperature under Ar gas at P_{Ar} = 5 Pa with a fluence of ca. 2 J·cm^{-2} (200 mJ per pulse). After SSE deposition for 120 min, the deposited film was ~150-nm thick, corresponding to 3 wt.% $LiCoO_2$. Annealing the SSE deposit at 200 °C increased the capacity of the all-solid-state cell [42]. Another example of the buffer function of the $Li_2S–P_2S_5$ solid electrolyte is given by the PLD coating of NiS-carbon fiber composite electrodes. The high ionic conductivity of $80Li_2S–20P_2S_5$ film deposited on an Si wafer was 7.9×10^{-5} S·cm^{-1} at 25 °C [290]. A capacity of 300 mAh·g^{-1} was delivered after 50 cycles at a current density of 3.8 mA·cm^{-2} (1C-rate). This SSE coating favors the lithium ion and electron conduction paths in the NiS framework. Ito et al. [291] successfully deposited $Li_2S–GeS_2$ thin films as the buffer electrolyte ($\sigma_i = 1.8 \times 10^{-4}$ S·cm^{-1}) on $LiCoO_2$ particles by the PLD technique. The amorphous $78Li_2S–22GeS_2$ solid electrolyte thin films prepared using standard PLD conditions exhibited an ionic conductivity of 1.8×10^{-4} S·cm^{-1} at 25 °C. These SSE films were applied to form an electrode–electrolyte buffer interface with $LiCoO_2$ [291]. The coating of a $LiNbO_3$ SSE buffer coated onto the $LiMn_2O_4$ cathode resulted in an enhancement of the high rate capability and cycling stability of the electrode [292]. A similar process ensured a high thermal stability for the $LiNi_{0.8}Co_{0.15}Al_{0.05}O_2$ electrode operating over 500 charge–discharge cycles at 150 °C [293].

4.5. Li₂O-V₂O₅-Si₂O (LVSO)

4.5. Li_2O-V_2O_5-Si_2O (LVSO)

PLD films of $Li_{2.2}V_{0.54}Si_{0.46}O_{3.4}$ are amorphous solid-state electrolytes of the system, Li_2O-V_2O_5-Si_2O (LVSO), which exhibits a conductivity of ~2.5×10^{-7} S·cm^{-1} at 25 °C [44]. Li_4SiO_4 thin films were successfully deposited by PLD using both an Nd:YAG laser (λ = 266 nm) and ArF excimer laser (λ = 193 nm) at the fluence of 2.5 J·cm^{-2} in the flow of O_2 gas at P_{O_2} = 0.2 Pa. Having a conductivity of 4.1×10^{-7} S·cm^{-1} at 25 °C, thermally activated with E_a = 0.52 eV, these films were applied in SSMBs [44]. Zhao et al. prepared PLD Li–V–Si–O thin films' electrolytes on an Si wafer and Al-coated glass as substrates placed 4 cm from the target. The film deposition was carried out at a fluence of 1.2 J cm^{-2} under an ambient of P_{O_2} = 6 Pa [294]. For T_s = 300 °C, the Li–V–Si–O film exhibited $\sigma_i = 3.98 \times 10^{-7}$ S·cm^{-1} at 25 °C and E_a = 0.55 eV. Workers at Kawamura's lab reported the PLD growth of several LVSO solid electrolytes. The $Li_{2.2}V_{0.54}Si_{0.46}O_{3.4}$ film deposited with a continuous flow of O_2 gas maintained at P_{O_2} = 0.2 Pa displayed an ionic conductivity of 2.5×10^{-7} S·cm^{-1} at 25 °C with an activation energy of 0.54 eV [41]. PLD amorphous $0.6(Li_4SiO_4)–0.4(Li_3VO_4)$ films deposited on Si(111) or fused silica plate exhibited an ionic conductivity of 10^{-7} S cm^{-1}·at 25 °C, which is one order higher than the value for PLD Li_2TiO_3 film [295]. All-solid-state thin film batteries were fabricated using both LCO and LMO PLD film cathodes and amorphous LVSO solid electrolytes as shown in Figure 15 [121].

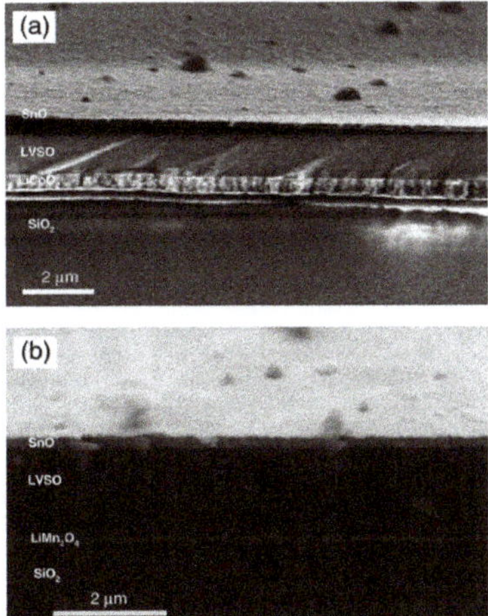

Figure 15. SEM cross-sectional images of solid-state thin film lithium batteries, (**a**) SnO/LVSO/LCO and (**b**) SnO/LVSO/LMO. (Reproduced with permission from [121]. Copyright 2006 Elsevier).

4.6. LiNbO₃

Because of its high room-temperature ionic conductivity and low electronic conductivity (10^{-5} and 10^{-11} S·cm^{-1}, respectively), LiNbO$_3$ (*R3c* crystal structure) is considered as a good SSE for electrode coating [296]. LiNbO$_3$ was applied as a buffer layer between an LCO cathode and thio-LISICON electrolyte (Li$_{3.25}$Ge$_{0.25}$P$_{0.75}$S$_4$). The resultant electrochemical cell showed low interfacial resistance and a high-rate capability [297]. A high quality was obtained for PLD LiNbO$_3$ thin films deposited at 730 °C on sapphire substrates by using a relatively high oxygen partial pressure of P_{O_2} = 133 Pa and a laser fluence of 3 to 5 J cm^{-2} [298]. Contrastingly, Perea et al. prepared PLD LiNbO$_3$ films using a lower laser fluence (0.8 to 1.6 J·cm^{-2}) in a residual pressure of $\approx 4 \times 10^{-4}$ Pa [299].

5. Negative Electrode PLD Films

5.1. TiO₂

Due to the theoretical capacity of ~335 mAh·g^{-1} of titanium dioxide (comparable to ~372 mAh·g^{-1} for graphite and the small volume expansion (~4% for anatase)) significant interest has been devoted to the applied anode material in Li-ion batteries. The tetragonal anatase polymorph of TiO$_2$ is a good anode candidate due to its insertion potential of around 1.5 V vs. Li$^+$/Li [300]. Several works of the literature report the growth of TiO$_2$ thin films with either a rutile or anatase structure fabricated by the PLD technique [301,302]. The growth conditions were studied on TiO$_2$ films deposited by PLD using an Nd:YAG laser (532 nm wavelength beam) and a rutile-type TiO$_2$ target. The effects of the substrate temperature (T_s) and oxygen partial pressure (P_{O_2}) were investigated by Raman spectroscopy [13]. The parameters of T_s = 300 °C and P_{O_2} = 50 mTorr were optimized to obtain crystalline TiO$_2$ films with a preferential (110) orientation. Kim et al. discussed the effects of the target morphology and target density on the size and distribution density of crystalline in PLD rutile-type TiO$_2$ films deposited on (100)-oriented Si wafers maintained at 700 °C in a chamber with an oxygen partial pressure of 1.33 Pa [303]. A nearly particulate-free film was obtained from a dense target and the laser shots were adjusted for clear ripple patterns from the target surface. The optical bandgap energies of TiO$_2$ PLD films grown on an α-Al$_2$O$_3$ (0001) substrate with an anatase and rutile structure were evaluated to be 3.22 and 3.03 eV, respectively [304]. Inoue et al. reported that films deposited at T_s = 150 °C have an anatase structure, while T_s = 300 °C provides rutile-type TiO$_2$ films [305]. Choi et al. [306] prepared anatase TiO$_2$ thin films with nanograins of 11 to 28 nm using a TiC target with T_s = 500 °C under 4 Pa O$_2$ gas.

5.2. Li₄Ti₅O₁₂ (LTO)

$Li_4Ti_5O_{12}$ (LTO) cubic structure ($Li[Li_{1/3}Ti_{5/3}]O_4$ in spinel notation), considered as a "zero-strain" anode material, exhibits the advantage of very minor volumetric changes (<0.2%) upon cycling. This electrode displays a large voltage plateau at ~1.5 V vs. Li^+/Li and a theoretical specific capacity of 175 mAh·g^{-1} [307]. The first PLD growth of LTO thin films deposited onto $Pt/Ti/SiO_2/Si$ substrates using a KrF excimer laser beam (248 nm, 250 mJ) were reported by Deng et al. [308]. Films annealed at 800 °C (410 nm thick) exhibited a cubic structure with a lattice constant 8.375 Å larger than that of the LTO crystal (8.359 Å). The SEM cross-section image (Figure 16a) revealed the porous morphology induced by the high temperature treatment. The discharge specific capacity was the largest for films annealed at 700 °C due to the optimized adhesion strength between the film and substrate (Figure 16b). The anode films discharged at a current density of 10 μA·cm^{-2} (0.58C rate) showed excellent cycleability; the discharge capacity remained as 149 mAh·g^{-1} after 50 cycles. $Li_4Ti_5O_{12}$ films (545 nm thick) deposited on conducting fluorine-doped tin oxide (LTO/FTO) with a crystallite size of 50 to 80 nm were investigated as electrochromic active material with the highest contrast at a wavelength of 705 nm (transmittance change of ~48%) [309]. Epitaxial LTO thin-film grown on $SrTiO_3$ single crystal from an Li-rich target, $Li_{5.2}Ti_5O_{12}$, have a structural orientation identical to the substrate and are impurity-free when deposited at T_s = 700 °C. The electrochemical features of LTO film anodes (20 nm thick) exhibited discharge capacities of ~200 and ~250 mAh·g^{-1} for the (100)- and (111)-orientation, respectively [310]. Kim et al. prepared nano-sized epitaxial LTO(110) deposited on $Nb:SrTiO_3(110)$ substrate. These films (~28 nm thick) were tested by cyclic voltammetry at a scan rate of 1 mV·s^{-1} and exhibited redox peaks at 1.53 and 1.60 V, corresponding to the insertion and extraction of Li^+ ions. As-deposited films at a substrate temperature of 700 °C in a 6.6 Pa oxygen partial pressure exhibited a high initial capacity (~200 mAh·g^{-1}) but poor stability [311]. Kumatani et al. investigated the PLD growth process of epitaxial LTO films deposited on an $MgAl_2O_4$ (111) substrate. With T_s = 800 °C and P_{O_2} = 1 × 10^{-3} Torr, LTO films had excellent crystallinity and a low resistivity of 3.3 × 10^{-4} Ω cm. at 25 °C. At lower P_{O_2}, the PLD $LiTi_2O_4$ film was formed, while at higher P_{O_2}, Ti was segregated as TiO_2 rutile and $Li_{0.74}Ti_3O_6$ [312].

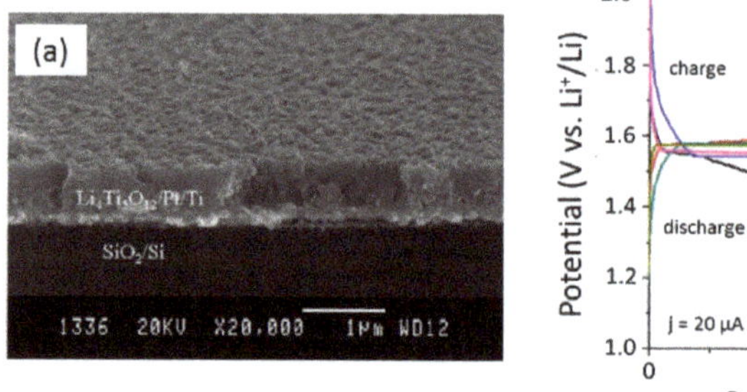

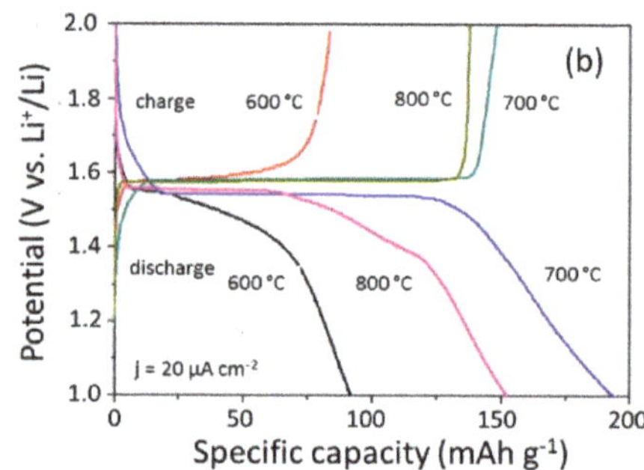

Figure 16. (a) SEM cross-section image of LTO film (410 nm thick) heat treated at 800 °C. (b) Charge–discharge profiles recorded at 20 μA cm^{-2} (i.e., ~1.15 C) current density in the voltage range of 1 to 2 V vs. Li^+/Li of PLD films heated at various temperatures. (Reproduced with permission from [308]. Copyright 2009 Elsevier).

Studies of the electrochemical performance and kinetic behavior of PLD LTO films deposited on $Pt/Ti/SiO_2/Si$ substrates were reported by Deng et al. [313]. Using an Li-rich target (i.e., excess 5 wt.% Li_2O), the films annealed at 700 °C for 2 h in air were well-crystallized items with densely packed grains. The galvanic charge–discharge plateau was observed around 1.56 V and an initial specific capacity of 159 mAh g^{-1} was delivered with a retention of 93.7% after 20 cycles. The diffusion coefficient of Li^+ ions in such an LTO framework was in the range of 10^{-15} to 10^{-12} cm^2·s^{-1}. The energy barrier of the diffusion of lithium ions was estimated to be E_a = 0.11 eV in LTO (111)-oriented PLD thin films (190 nm thick) grown on a spinel $MgAl_2O_4$ (111) substrate [314].

Zhao et al. reported the optical properties of epitaxially grown LTO films on (001)-oriented $MgAl_2O_4$ substrate. The optical bandgap of 3.14 eV was measured for 86 nm thick films (surface roughness of 4.61 nm) [315]. Schichtel et al. fabricated all-solid-state microbatteries with LTO as the positive electrode. PLD films were obtained on various substrates at T_s = 650 °C under a 0.3 Pa pure oxygen atmosphere using a commercially available LTO powder. As-prepared films (650 nm thick) revealed columnar growth that allowed a coulombic efficiency >97% after the second cycle and a discharge capacity of 33 $\mu Ah \cdot cm^{-2}$ at a 3.5 $\mu A\ cm^{-2}$ current density [43]. Pfenninger et al. demonstrated that LTO thin films deposited by PLD on an MgO substrate kept at 500 °C using a dense $Li_{7.1}Ti_5O_{12}$ target sintered at 1000 °C for 12 h are compatible with the $Li_{6.25}Al_{0.25}La_3Zr_2O_{12}$ electrolyte pellet. Such films display a stable structure and cycleability almost close to 175 $mAh \cdot g^{-1}$. The typical voltage plateau at 1.57 V (oxidation) and 1.53 V (reduction) was observed at a rate of 2.5 $mA \cdot g^{-1}$ [316]. Among the $Li_{1+x}Ti_{1-x}O_4$ ternary system, $LiTi_2O_4$ thin films were grown by the PLD route in the temperature range of 400 to 800 °C using a target with a higher Li/Ti ratio of 0.8 [317]. Chopdekar et al. grew epitaxial PLD $LiTi_2O_4$ thin films on various crystalline-oriented substrates, such as single crystalline substrates of $MgAl_2O_4$, MgO, and $SrTiO_3$ [318]. The authors state the PLD conditions with T_s held at 450 to 600 °C in a vacuum of better than 5×10^{-6} Torr without any mention of the oxygen partial pressure, while Kumatani determined that stoichiometric $LiTi_2O_4$ thin films were obtained at a P_{O_2} of 5×10^{-6} Torr with T_s = 800 °C [312]. Recently, PLD LTO films grown on Nd-doped oriented STO substrates at T_s = 700 °C under P_{O_2} = 20 Pa showed high discharge capacities of 280 to 310 $mAh \cdot g^{-1}$. The best rate performance of 30C was obtained for the (100)-oriented $Li_4Ti_5O_{12}$ films [319].

5.3. LiNiVO₄

Amorphous $LiNiVO_4$ thin-film anodes for microbatteries were grown by pulsed laser deposition using a sintered $Li_{1.2}NiVO_4$ target. The film grown at T_s = 25 °C and P_{O_2} = 8 mTorr showed the best electrochemical performance with a retainable capacity as high as 410 $\mu Ah \cdot cm^{-2} \cdot \mu m^{-1}$ after 50 cycles [320].

5.4. TiNb₂O₇

An alternative to LTO, titanium-niobium oxide, $TiNb_2O_7$ (TNO), is considered a promising anode material for long life Li-ion batteries, due to its high Li^+ ion transport, average voltage of 1.66 V, and theoretical capacity of ~387 $mAh \cdot g^{-1}$ [321]. Fabrication of PLD $TiNb_2O_7$ thin films as anode electrodes for Li-ion micro-batteries was demonstrated by the ablation of a Nb_2O_5 + TiO_2 mixture as a target at a laser fluence of 4.6 $J \cdot cm^{-2}$. Pure monoclinic TNO films were deposited on $Pt/TiO_2/SiO_2/Si(100)$ substrates at 750 °C under an O_2 gas of P_{O_2} = 6–13 Pa. The 380-nm thick films grown at P_{O_2} = 13 Pa delivered an initial specific capacity of 142 $\mu Ah\ cm^{-2}\ \mu m^{-1}$ at a current density of 50 $\mu A \cdot cm^{-2}$ with a 58% capacity retention after 25 cycles [322]. Recently, the same co-workers reported a high specific discharge capacity of 226 $\mu Ah \cdot cm^{-2} \cdot \mu m^{-1}$ (~460 $mAh \cdot g^{-1}$) at a current density of 17 $\mu A \cdot cm^{-2}$ for amorphous TNO films grown by PLD. Li^+ diffusion coefficient of $\approx 10^{-13}\ cm^2 \cdot s^{-1}$ and an electronic conductivity of $\approx 10^{-9}\ S \cdot cm^{-1}$ were also reported [323].

5.5. Silicon

Pulsed-laser deposited silicon thin films have been widely studied for applications in opto-electronics. With a large theoretical capacity (4200 $mAh \cdot g^{-1}$), silicon is also considered as a promising anode material for the replacement of graphite anode (LiC_6, 372 $mAh \cdot g^{-1}$) for Li-ion batteries [324]. Despite the huge volume expansion of >300% during lithiation up to $Li_{22}Si_5$, it is possible to obtain anodes with Si thin films grown by physical vapor deposition (PVD), reaching a cycling life of up to 3000 cycles due to the limited volume change in the 2D film [325]. For example, a film deposited on Ni foil maintained a capacity of 3000 $mAh \cdot g^{-1}$ at a 12C rate over 1000 cycles [326]. PLD-grown Mg_2Si thin film (30–380 nm thick) exhibited electrochemical activity with a stable cycling behavior in the voltage range of 0.1 to 1.0 V vs. Li$^+$/Li; however, the initial irreversible capacity loss increased with the film thickness. The superior capacity of the 30-nm thick film was attributed to the formation of Li-Si alloys at the Si-rich surface [327]. Park et al. prepared PLD amorphous Si (a-Si) thin films on a stainless-steel substrate at temperature of 500 °C under an Ar gas pressure P_{Ar} = 6.5 mPa. Furthermore, 1.5-μm thick a-Si films were obtained at the growth rate of 25 nm·min^{-1}. Electrochemical tests carried out in the voltage range of 0.005 to 1.5 V showed a first discharge capacity of 9 to 0.7 μAh·cm^{-2} with a 54.4% coulombic efficiency. Although, after 70 cycles, the 1-μm thick Si film exhibited a good cyclic performance [328]. Xia et al. reported the growth of a-Si using the standard conditions (T_s = 25 °C, P = 1.3 mPa, fluence of 150 to 160 mJ per pulse, deposition time of 30 min). Electrochemical tests showed that 120-nm thick a-Si films exhibited an initial charge capacity of ~64 μAh·cm^{-2} at a current density of 100 μA·cm^{-2}, a discharge capacity of ~50 μAh cm^{-2} was maintained after 40 cycles, and the diffusion coefficient of Li ions determined from the cyclic voltammograms was ~10^{-13} cm^2·s^{-1} [329]. Some Si-based composite thin films prepared by PLD combine the advantages of both components. The most popular are the carbon-based composites [330–333]. Chou et al. obtained a flexible anode material by the deposition of Si film onto single-wall carbon nanotubes (SWCNTs) using standard PLD conditions (λ = 248 nm, T_s ≈ 30 °C, Φ = 3 J·cm^{-2}, P_{Ar} = 13 Pa, target–substrate distance of 50 mm). After 50 cycles, this Si/SWCNT nanocomposite delivered a specific capacity of 163 mAh·g^{-1} at a 25 mA·g^{-1} current density, which is 60% higher than for CNT [330]. The ultrathin film of Si grown by PLD was deposited on multilayer graphene (MLG) by CVD on an Ni foam substrate. The specific capacity of this binder-free Si/MLG anode appeared to be stable at ca. 1400 mAh·g^{-1} [331]. Silicon nitride SiN$_{0.92}$ thin films were prepared by PLD and investigated as a negative electrode in lithium batteries. A 200-nm thick film grown on buffed stainless-steel substrates kept at 25 °C from an Si$_3$N$_4$ pellet as the target delivered a high specific capacity of 1300 mAh·g^{-1} after 100 cycles [334].

5.6. Graphene

Most of the commercial lithium batteries have a carbon anode. Graphene is the most conductive form of carbon, and as such, it is considered as a promising electrode, especially when it is doped with nitrogen [229]. A recent review was devoted to PLD-graphene synthesized from a solid carbon source [335]. Since nitrogen modifies the intrinsic properties of graphene, it is important to control its concentration. PLD, which allows for a one-step synthesis of N-doped carbon films, is particularly suited to this purpose. The first N-doped amorphous carbon film (a-C:N) synthesized by PLD dates from 2013 [336]. It contained 2 at. % of nitrogen. More recently, using the same approach, the upper nitrogen concentration in PLD a-C:N film was raised to 3.2 at.% [337]. These films, however, were not used as electrodes. On the other hand, a N-doped graphene (NG) electrode prepared by PLD coupled with in-situ thermal annealing (PLD-TA) was achieved by Fortgang et al. [338]. More recently, Bourquard et al. used the PLD-TA process to form an N-doped graphene film by high temperature condensation of the laser-induced carbon plasma plume onto the Si electrode previously covered by an Ni catalytic film [339], using a protocol published by the same group [340] Carbon was ablated at 780 °C from the graphite target using a femtosecond laser (λ = 800 nm, pulse width of 60 ns, repetition rate of 1 kHz, and Φ = 5 J·cm^{-2}) at a distance of 36 mm from the graphite target, with P_N = 10 Pa in the vacuum chamber. The electrochemical properties were measured with the thus-obtained 40 nm-thick film with a nitrogen concentration of 1.75% as the working electrode and an active area of 0.07 cm^2, saturated calomel electrode as the reference electrode, and platinum as the counter electrode, in a 0.5 mol·L^{-1} 1,1′ ferrocene-dimethanol solution of 0.1 mol·L^{-1} NaClO$_4$. Aiming to detect H$_2$O$_2$ in 0.1 mol L^{-1} phosphate buffer saline (PBS) solution (pH 7.4), linear sweep voltammetry was used in the anodic range from 0 to 1000 mV with a scan rate of 100 mV·s^{-1}. The electrode showed excellent reversibility, 60 mV, close to the theoretical value, and a detection limit of 1 mM of hydrogen peroxide, which constitutes a major improvement of the electroanalytical oxidation of H$_2$O$_2$ in comparison with undoped graphite electrodes. Such results are recent, and to our knowledge, no such electrode has been tested yet as an anode for lithium batteries.

5.7. Other PLD Anodes

Significant efforts have been devoted to the design and development of new PLD-grown anode materials for SSMBs, yielding a high energy density (from 500 to 1500 mAh·g^{-1}) and electrochemical stability [341–354]. Table 10 summarizes the PLD-growth conditions and electrochemical properties of some new anodes proposed for lithium microbatteries. All transition-metal oxide M_xO_y materials are subjected to electrochemical lithiation via a conversion reaction, which implies a two-step process, i.e., first, fine metallic (M) nanoparticles embedded in an insulating matrix, such as Li_2O, are in situ formed during the first (irreversible) discharge, and secondly, an alloying reversible reaction (Li-M) occurs on subsequent cycles [341].

Recently, Wu et al. proposed a novel anode consisting of a Li_3P-VP nanocomposite fabricated by PLD [353] using a 5-Hz Nd:YAG laser (λ = 355 nm, Φ = 2 J·cm^{-2}) concentrated on the target surface with an incident angle of 45°, with P_{Ar} =10 Pa in the vacuum chamber. The stainless-steel substrate was placed 3 cm from the surface and kept at 400 °C. The weight of the film thus obtained (without the substrate) was 0.10 mg·cm^{-2}. The excessive lithium in this composite was used to stabilize the VP_2 structure after the first charge. Electrochemical tests were made with this film as the working electrode, while lithium metal sheets were used as counter and reference electrodes with 1 mol·L^{-1} $LiPF_6$ in EC:DMC (1:1 in volume) electrolyte. When cycled in the range of 0.01 and 4 V vs. Li$^+$/Li at a current density of 5 μA·cm^{-2}, a capacity of 1040 mAh·g^{-1} was delivered at the second discharge, 987 mAh·g^{-1} at the 50th cycle.

Table 10. Growth conditions and electrochemical properties of new anode materials.

Anode Material	T_s (°C)	Ambient (Pa)	Laser Energy (J cm^{-2})	Electrochemical Properties	Ref.
C-monocolumn	25	(N$_2$) 6.7	0.08	315 mAh·g^{-1} at 0.1 A·g^{-1} (50th cycle)	[342]
NiFe$_2$O$_4$	650	(O$_2$) 2.0	2.5	370 mAh·g^{-1} at 25 μA·cm^{-2}	[343]
Co$_2$O$_3$	700	(O$_2$) 100	2.0	750 mAh·g^{-1} at 2C (350th cycle)	[344]
Cu$_2$Sb	25	(Ar) 1.3	3–4	>135 mAh·g^{-1} at 35 μA·cm^{-2} (100th cycle)	[345]
Fe$_2$O$_3$	25	(O$_2$) 0.3	0.5	905 mAh·g^{-1} at 0.1 A·g^{-1} (200th cycle)	[346]
Co$_3$O$_4$	600	(O$_2$) 40	2	600 mAh·g^{-1} at 10 μA·g^{-1} (2th cycle)	[347]
Co$_3$O$_4$/Co(OH)$_2$	250	(O$_2$/H$_2$) 30	–	360 mAh·g^{-1} at 32 A·g^{-1} (1000th cycle)	[348]
In$_2$O$_3$	200	(O$_2$) 5	2	500 mAh·g^{-1} at 10 μA·g^{-1} (30th cycle)	[349]
Sb$_2$Se$_3$	200	(Ar) 5	2	530 mAh·g^{-1} at 5 μA·g^{-1} (100th cycle)	[350]
FeOF	25	(Ar) 5	2	110 mAh·g^{-1} at 5 μA·g^{-1} (100th cycle)	[351]
SrLi$_2$Ti$_6$O$_{14}$	700	(O$_2$) 2	2	130 mAh·g^{-1} at C/20 (40th cycle)	[352]
MnO	500	(Ar) 20	5	425 mAh·g^{-1} at C/8 (25th cycle)	[352]
VP$_2$	400	(Ar) 10	2	987 mAh·g^{-1} at 5 μA·cm^{-2} (50th cycle)	[354]

6. Discussion

There is general agreement on the beneficial results of the pulsed-laser deposition of thin films used as active materials in lithium microbatteries. This is a technique prone to fabricate thin film rapidly, from a small amount of target material, while maintaining the ideal stoichiometry by the control of different growth parameters. We observed (Table 2) that there is a strong trend to develop microcell technologies using $LiCoO_2$ film (typical thickness of 4-μm) as the cathode, LiPON thin film (typical thickness of 1 μm) as the solid electrolyte, and Li thin film anode, which may have advantages in terms of the following key requirements: High energy density, high voltage, sustainability, and easy fabrication. For such microbatteries, let us recall the energy units used by authors. For a comparison of the volumetric specific energy, one generally refers to that of the cathode material (i.e., the limiting electrode); thus, considering a dense LCO film (d = 4.3 g·cm^{-3}), a gravimetric specific energy of 100 mAh·g^{-1} is equivalent to 43 μAh·cm^{-2}·μm^{-1} or 154.8 mC·cm^{-2}·μm^{-1}.

Let us compare and discuss the growth conditions that allow the best electrochemical performance of each component, i.e., cathode, anode, and electrolyte. Regarding the growth of lithiated oxides used as cathode materials, an excess of lithium (at least ~15 wt.%) is mandatory for any material to compensate the loss of volatile lithium species during the ablation process. Given the different materials' available candidates for the cathode, LiCoO$_2$ appears to be the most electrochemically efficient. From Figure 17, comparing the Ragone plots of LiCoO$_2$ and LiMn$_2$O$_4$ thin films, each curve displays a series of discharge profiles for a lithium microbattery with the cathode thickness (in μm). As the capacity delivered by a cathode is proportional to the mass of the material, thicker films provide high energies, but often at the expense of the power performance [26]. The specific energy for a planar Li//LiCoO$_2$ cell with a thick cathode can reach 500 μWh·cm^{-2}·μm^{-1}. However, the rate capability depends strongly on the plane orientation of the film, which can be controlled by the nature of the substrate. Thus, the preferred orientation is the (003)-plane parallel [29]. However, a surface–electrolyte interface (SEI) is formed on epitaxial LiCoO$_2$ films with different orientations of (104), (110), and (003) that result in anisotropic reactions of intercalation activity [102]. It has also been demonstrated that the minimized strain energy in thick LCO films allows preferential (101) and (104) textures [32]. Nevertheless, the best well-crystallized LCO thin films were fabricated in the following PLD conditions: T_s = 500 °C, P_{O_2} = 13 Pa, Φ = 2 J·cm^{-2}, substrate–target distance of 30 to 40 mm, and laser beam-target incident angle of 45°. Interesting results reported in Ref. [53] showed impurity-free LCO films, highly (003)-oriented with a very small lattice expansion during charge (at 4.2 V), when grown on stainless steel substrates at relatively low temperatures ($T_s \approx$ 300 °C). In this case, the films had a texture between the amorphous and well-crystalline state with very small grains, which is suitable for short pathways for electrons and ions during the (de)intercalation reaction.

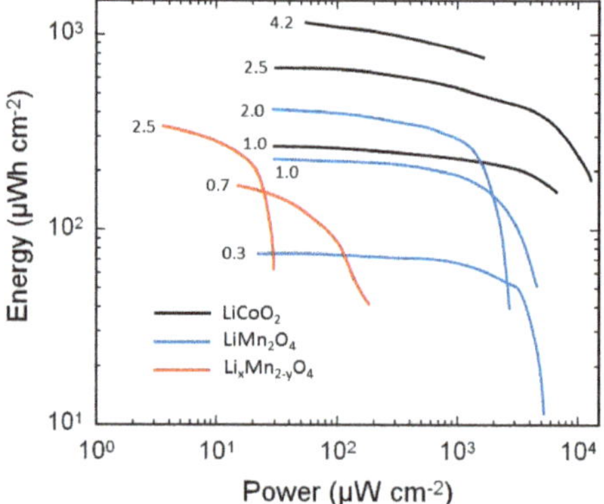

Figure 17. Ragone plot (normalized by the active cell area) for lithium thin-film microbatteries fabricated with crystalline LiCoO$_2$ (black lines), crystalline LiMn$_2$O$_4$ (blue lines), and nanocrystalline Li$_x$Mn$_{2-y}$O$_4$ (red lines) cathode materials with different thicknesses (in μm).

To avoid the poor performance of LIBs derived from hindered lithium-ion diffusion at the interface between the LCO positive electrode and electrolyte, modifications of the cathode surface have been realized by the deposition of a thin layer of a fast-ionic Li$^+$ conductor, such as amorphous Li$_2$WO$_4$ or Li$_3$PO$_4$. This layer reduces the interfacial Li$^+$-ion transfer resistance that results in a rapid charge–discharge rate. The a-Li$_2$WO$_3$/LCO/Pt/Cr/SiO$_2$ electrode cycled at a high rate of 20C with a high capacity retention [84]. Another electrode exhibiting a fast charge–discharge rate as high as 348C has been fabricated by the multilayer PLD technique, but in this case, the LMO thin film exhibited a significant pseudocapacitive behavior (non-diffusion-controlled) instead of a faradaic mechanism. An additional promising improvement is the fabrication of an LCO thin film sandwiched between a PLD-prepared SrRuO$_3$ film as the electronic conductor and the film of Li$_3$PO$_4$ (3.2 nm thick) as the ionic conductor with the result being limited surface structural change in the high voltage range (4.4 V) [71].

The influence of the PLD conditions on the texture of LiMn$_2$O$_4$ thin films has shown that T_s = 500 °C and P_{O_2} = 20 Pa are the optimum values that maintain the Li/Mn ratio close to 1, when an

Li-enriched target is used, and obtains the best mass transfer [122]. It was also noticed that any substrate does not strongly influence the stoichiometry, but affects the out-of-plane preferred texture. The applicability of the PLD-grown V_2O_5 films in lithium microbatteries has been evidenced that, in the range of $200 < T_s < 400\ ^{\circ}C$ under $P_{O_2} = 6$ Pa, the films offer better electrochemical performance than those grown at other temperatures in terms of their structural quality and stability. Only two works were devoted to the fabrication of solid-state thin-film batteries with vanadium oxide as the cathode materials: The $Li/Li_{1.4}B_{2.5}S_{0.1}O_{4.9}/V_2O_5$ cell delivered a capacity of ~400 mC·cm^{-2}·μm^{-1} [221], while the $Li/LiPON/Ag_{0.3}V_2O_5$ maintained a specific capacity of 40 μAh·cm^{-2}·μm^{-1} after 100 cycles [222], but the low current density was due to the poor electronic conductivity of the positive electrode. Recently, a solid-state thin-film battery, $Li/Li_3PO_4/LiMnPO_4$, was successfully fabricated by PLD [48]. Such a cell delivered a modest specific capacity of 10 μAh·cm^{-2}·μm^{-1}, which was limited by the slow chemical diffusion coefficient of the Li^+ ion in the olivine framework (3×10^{-17} cm^2·s^{-1}).

Lithium phosphates, i.e., Li_3PO_4 and LiPON, are the most widely used solid electrolytes in microbatteries; they are easily fabricated by PLD using an ArF excimer laser and show a good ionic conductivity. The LiPON electrolyte is known to exhibit a better chemical stability than Li_3PO_4 [282]. However, the electrochemical stability of PLD-prepared Li_3PO_4 thin films is greater than 4.7 V [265].

Few works have attempted to replace the lithium metal thin-film anode by other lithiated materials (i.e., intercalation compound or alloy) for the fabrication of microbatteries. The most stable insertion compound should be $Li_4Ti_5O_{12}$ spinel with minor volumetric changes but the high voltage plateau of 1.5 V is a great penalty for high energy density. The $In/80Li_2S–20P_2S_5/LiCoO_2$ microbattery developed by Sakuda et al. seems to be promising as the Li-In alloy allows a high specific discharge capacity at moderate current density of 0.13 mA·g^{-1} [42].

7. Concluding Remarks

The results of the intensive research on the growth of thin films by pulsed laser deposition in recent years have been reviewed. Due to careful investigations of the mechanism of the sample preparation, optimized materials with adequate properties for energy storage and conversion have been obtained. The PLD technique is considered to be suitable for improving the density and adhesion properties of films. A huge effort has been mainly concentrated on the deposition of lithiated oxides, which require specific conditions due to the volatile character of lithium vapor species during the PLD process. Due to the outstanding performance of the conventional cathode materials, $LiCoO_2$ and $LiMn_2O_4$, PLD films exhibiting a specific capacity close to the theoretical one are the most popular. The progress concerns mainly the epitaxial films grown with an orientation favorable to a high rate of transport of Li ions at the electrode/electrolyte interface. For instance, the pyramidal-type $LiMn_2O_4$ films cycled at the 3.3C rate demonstrate a specific capacity of 90 mAh·g^{-1} after 1000 cycles.

The PLD technique has proved to also be efficient for the preparation of thin films of anode materials. The best example is the production of LTO, which is a "zero-strain" compound. Other anode thin film materials, such as silicon and conversion-type oxides, are attractive due to their high specific capacity and easy PLD fabrication.

So far, solid-electrolyte thin-films have been fabricated essentially by thermal vacuum evaporation and rf-sputtering. The manufacture of solid-electrolyte thin films by PLD has brought improvements in their intrinsic properties. For example, the electronic conductivity of PLD films is small in comparison with rf-sputtered films. LiPON and Li–V–S–O are the most popular solid-electrolyte films.

In recent years, due to a strong demand for smaller power sources, the interest in rechargeable micro-batteries has gradually increased. The progress on lithium microbatteries is remarkable, mainly due to the PLD growth of high-quality, pinhole-free, solid-state electrolyte thin films, such as $Li_{6.1}V_{0.61}Si_{0.39}O_{5.36}$. The rechargeable thin-film lithium-ion battery designed by the Japanese group at Tohoku University was fabricated using the sequential PLD technique. This microcell delivered a specific capacity of 9.5 Ah cm^{-2} discharged at a current density of 44 μA·cm^{-2} using an Li-Sn alloy film as the anode and showed good reversibility over 100 cycles.

Funding: This research received no external funding.

Conflicts of Interest: The authors declare no conflict of interest.

References

1. Chrisey, D.G.; Hubler, G.K. (Eds.) *Pulsed Laser Deposition of Thin Films*; Wiley: New York, NY, USA, 1994.
2. Bauerle, D. *Laser Processing and Chemistry*; Springer: Berlin, Germany, 1996.
3. Julien, C.; Camacho-Lopez, M.A.; Escobar-Alarcon, L.; Haro-Poniatowski, E. Fabrication of $LiCoO_2$ thin-film cathodes for rechargeable lithium microbatteries. *Mater. Chem. Phys.* **2001**, *68*, 210–216. [CrossRef]
4. Antaya, M.; Dahn, J.R.; Preston, J.S.; Rossen, E.; Reimers, J.N. Preparation and characterization of $LiCoO_2$ thin films by laser ablation deposition. *J. Electrochem. Soc.* **1993**, *140*, 575–578. [CrossRef]
5. Tsuruhama, T.; Hitosugi, T.; Oki, H.; Hirose, Y.; Hasegawa, T. Preparation of layered-rhombohedral $LiCoO_2$ epitaxial thin films using pulsed laser deposition. *Appl. Phys. Express* **2009**, *2*, 085502. [CrossRef]
6. Perkins, J.D.; Bahn, C.S.; McGraw, J.M.; Parilla, P.A.; Ginley, D.S. Pulsed laser deposition and characterization of crystalline lithium cobalt dioxide ($LiCoO_2$) thin films. *J. Electrochem. Soc.* **2001**, *148*, A1301–A1312. [CrossRef]
7. Julien, C.; Haro-Poniatowski, E.; Camacho-Lopez, M.A.; Escobar-Alarcon, L.; Jimenez-Jarquin, J. Growth of $LiMn_2O_4$ thin films by pulsed-laser deposition and their electrochemical properties in lithium microbatteries. *Mater. Sci. Eng. B* **2000**, *72*, 36–46. [CrossRef]
8. Julien, C.; Haro-Poniatowski, E.; Hussain, O.M.; Ramana, C.V. Structure and electrochemistry of thin-film oxides grown by laser-pulsed deposition. *Ionics* **2001**, *7*, 165–171. [CrossRef]
9. Montenegro, M.J.; Lippert, T. Films for electrochemical applications. In *Pulsed Laser Deposition of Thin Films: Applications-Led Growth of Functional Materials*; Eason, R., Ed.; John Wiley & Son Inc.: Hoboken, NJ, USA, 2007; Chapter 22; pp. 563–584.
10. Iriyama, Y.; Inaba, M.; Abe, T.; Ogumi, Z. Preparation of c-axis oriented thin films of $LiCoO_2$ by pulsed laser deposition and their electrochemical properties. *J. Power Sources* **2001**, *94*, 175–182. [CrossRef]
11. Huo, W.; Li, J.; Chen, G.; Wang, Y.; Zou, W.; Rao, Q.; Zhou, A.; Chang, A.; Wang, Q. $LiCoO_2$ thin film cathode fabricated by pulsed laser deposition. *Rare Met.* **2011**, *30* (Suppl. 1), 106–110. [CrossRef]
12. Xia, H.; Tang, S.B.; Lu, L. Thin film microbatteries prepared by pulsed laser deposition. *J. Korean Phys. Soc.* **2007**, *51*, 1055–1062. [CrossRef]
13. Escobar-Alarcon, L.; Haro-Poniatowski, E.; Camacho-Lopez, M.A.; Fernandez-Guasti, M.; Jimenez-Jarquin, J.; Sanchez-Pineda, A. Structural characterization of TiO_2 thin films obtained by pulsed laser deposition. *Appl. Surf. Sci.* **1999**, *137*, 38–44. [CrossRef]
14. Kuwata, N.; Iwagami, N.; Matsuda, Y.; Tanji, Y.; Kawamura, J. Thin film batteries with Li_3PO_4 solid electrolytes fabricated by pulsed laser deposition. *ECS Trans.* **2009**, *16*, 53–60.
15. Julien, C. Solid State Batteries. In *Handbook of Solid State Electrochemistry*; Gellings, P.J., Bouwmeester, H.J.M., Eds.; CRC Press: Boca Raton, FL, USA, 1997; pp. 371–406.
16. Julien, C. Thin film technology and microbatteries. In *Lithium Batteries-New Materials-Development and Perspectives*; Pistoia, G., Ed.; Elsevier: Amsterdam, The Netherlands, 1994; pp. 167–237.
17. Sator, A. Pile réversible dont l'électrolyte est un cristal déposé en lame mince par évaporation. *C. R. Acad. Sci. Paris* **1952**, *234*, 2283–2285.
18. Kennedy, J.H. Thin film electrolyte systems. *Thin Solid Film* **1977**, *43*, 41–92. [CrossRef]
19. Bates, J.B.; Dudney, N.J.; Gruzalski, G.R.; Luck, C.F. Thin Film Battery and Method for Making Same. U.S. Patent 5,338,625, 16 April 1994.
20. Bates, J.B.; Dudney, N.J.; Gruzalski, G.R.; Zuhr, R.A.; Choudhury, A.; Luck, C.F.; Robertson, J.D. Electrical properties of amorphous lithium electrolyte thin films. *Solid State Ion.* **1992**, *53*, 647–654. [CrossRef]
21. Bates, J.B.; Dudney, N.J.; Gruzalski, G.R.; Zuhr, R.A.; Choudhury, A.; Luck, C.F.; Robertson, J.D. Fabrication and characterization of amorphous lithium electrolyte thin films and rechargeable thin-film batteries. *J. Power Sources* **1993**, *43*, 103–110. [CrossRef]
22. Bates, J.B.; Dudney, N.J.; Lubben, D.C.; Gruzalski, G.R.; Kwak, B.S.; Yu, X.H.; Zuhr, R.A. Thin-film rechargeable lithium batteries. *J. Power Sources* **1995**, *54*, 58–62. [CrossRef]
23. Bates, J.B.; Dudney, N.J.; Luck, C.F.; Sales, B.C.; Zuhr, R.A.; Robertson, J.D. Deposition and characterization of $Li_2O-SiO_2-P_2O_5$ thin films. *J. Am. Ceram. Soc.* **1993**, *76*, 929–943. [CrossRef]

24. Bates, J.B.; Gruzalski, G.R.; Dudney, N.J.; Luck, C.F.; Yu, X.H. Rechargeable thin-film lithium batteries. *Solid State Ion.* **1994**, *70*, 619–628. [CrossRef]

25. Julien, C. Lithium microbatteries. In *Materials for Lithium-Ion Batteries*; Julien, C., Stoynov, Z., Eds.; Springer: Dordrecht, The Netherlands, 2000; pp. 381–400.

26. Dudney, N.J. Thin film micro-batteries. *Electrochem. Interface* **2001**, *17*, 44–48.

27. Oudenhoven, J.F.M.; Baggetto, L.; Notten, P.H.L. All-solid-state lithium-ion microbatteries: A review of various three-dimensional concepts. *Adv. Energy Mater.* **2011**, *1*, 10–33. [CrossRef]

28. Wang, Y.; Liu, B.; Li, Q.; Cartmell, S.; Ferrara, S.; Deng, Z.D.; Xiao, J. Lithium and lithium-ion microbatteries for applications in microelectronic devices: A review. *J. Power Sources* **2015**, *286*, 330–345. [CrossRef]

29. Ferrari, S.; Loveridge, M.; Beattie, S.D.; Jahn, M.; Dashwood, R.J.; Bhagat, R. Latest advances in the manufacturing of 3D rechargeable lithium microbatteries. *J. Power Sources* **2015**, *286*, 25–46. [CrossRef]

30. Meng, Y.S.; McGilway, T.; Yang, M.-C.; Gostovic, D.; Wang, F.; Zeng, D.; Zhu, Y.; Graetz, J. In situ analytical electron microscopy for probing nanoscale electrochemistry. *Electrochem. Interface* **2011**, *20*, 49–53. [CrossRef]

31. Jones, S.D.; Akridge, J.R. Athin film solid state microbattery. *Solid State Ion.* **1992**, *53*, 628–634. [CrossRef]

32. Laïk, B.; Ressejac, I.; Venet, C.; Pereira-Ramos, J.P. Comparative study of electrochemical performance of commercial solid-state thin film Li microbatteries. *Thin Solid Films* **2018**, *649*, 69–74. [CrossRef]

33. Enfucell Powers the Internet of Things Revolution. Available online: http://www.enfucell.com/softbattery (accessed on 28 September 2018).

34. Enerchip™ Smart Solid State Batteries. Available online: http://www.cymbet.com/products/enerchip-solid-state-batteries.php (accessed on 1 January 2019).

35. Thin Film Battery Technology. Available online: http://www.excellatron.com./thin-film-battery-technology (accessed on 1 January 2018).

36. Thinergy Micro-Energy Cell Product Family. Available online: https://www.batterypoweronline.com/markets/batteries/thinergy-micro-energy-cell-product-family/ (accessed on 1 January 2019).

37. NanoEnergy™. Available online: http:/www.frontedgetechnology.com./gen.htm (accessed on 1 January 2014).

38. Ulvac Enables Flexible Li-Ion Battery. Available online: https://fr.scribd.com/document/118211050/196729-Ulvac-Enables (accessed on 22 December 2008).

39. Batteries Ultraminces à Durée de vie Etendue de STMicroelectronics Pour L'Alimentation Des Technologies Miniatures de Demain. Available online: https://www.digikey.fr/fr/product-highlight/s/stmicroelectronics/efl700a39-enfilm-rechargeable-battery/ (accessed on 13 January 2016).

40. Thin Film Rechargeable Batteries from Korea. Available online: https://www.idtechex.com/research/articles/thin-film-rechargeable-batteries-from-korea-00003130.ja.asp?donotredirect=true&setlang=ja (accessed on 23 February 2011).

41. Kuwata, N.; Kawamura, J.; Toribami, K.; Hattori, T.; Sata, N. Thin-film lithium-ion battery with amorphous solid electrolyte fabricated by pulsed laser deposition. *Electrochem. Commun.* **2004**, *6*, 417–421. [CrossRef]

42. Sakuda, A.; Hayashi, A.; Ohtomo, T.; Hama, S.; Tatsumisago, M. All-solid-state lithium secondary batteries using $LiCoO_2$ particles with pulsed laser deposition coatings of Li_2S–P_2S_5 solid electrolytes. *J. Power Sources* **2011**, *196*, 6735–6741. [CrossRef]

43. Schichtel, P.; Geiß, M.; Leichtweiß, T.; Sann, J.; Weber, D.A.; Janek, J. On the impedance and phase transition of thin film all-solid-state batteries based on the $Li_4Ti_5O_{12}$ system. *J. Power Sources* **2017**, *360*, 593–604. [CrossRef]

44. Nakagawa, A.; Kuwata, N.; Matsuda, Y.; Kawamura, J. Thin film lithium battery using stable solid electrolyte Li_4SiO_4 fabricated by PLD. *ECS Trans.* **2010**, *25*, 155–161.

45. Fujimoto, D.; Kuwata, N.; Matsuda, Y.; Kawamura, J.; Kang, F. Fabrication of solid-state thin-film batteries using $LiMnPO_4$ thin films deposited by pulsed laser deposition. *Thin Solid Films* **2015**, *579*, 81–88. [CrossRef]

46. Ohzuku, T.; Ueda, A. Solid-state redox reactions of $LiCoO_2$ (R3m) for 4 volt secondary lithium cells. *J. Electrochem. Soc.* **1994**, *141*, 2972–2977. [CrossRef]

47. Striebel, K.A.; Deng, C.Z.; Wen, S.J.; Cairns, E.J. Electrochemical behavior of $LiMn_2O_4$ and $LiCoO_2$ thin films produced with pulsed laser deposition. *J. Electrochem. Soc.* **1996**, *143*, 1821–1827. [CrossRef]

48. Bouwman, P.J.; Boukamp, B.A.; Bouwmeester, J.M.; Notten, P.H.L. Influence of diffusion plane orientation on electrochemical properties of thin film $LiCoO_2$ electrodes. *J. Electrochem. Soc.* **2002**, *149*, A699–A709. [CrossRef]

49. McGraw, J.; Bahn, C.; Parilla, P.; Perkins, J.; Readey, D.; Ginley, D. Li ion diffusion measurements in V_2O_5 and $Li(Co_{1-x}Al_x)O_2$ thin-film battery cathodes. *Electrochim. Acta* **1999**, *45*, 187–196. [CrossRef]

50. Escobar-Alarcon, L.; Haro-Poniatowski, E.; Massot, M.; Julien, C. Physical properties of lithium-cobalt oxides grown by laser ablation. *Mater. Res. Soc. Symp. Proc.* **1999**, *548*, 223–228. [CrossRef]

51. Okada, K.; Ohnishi, T.; Mitsuishi, K.; Takada, K. Epitaxial growth of LiCoO$_2$ thin films with (001) orientation. *AIP Adv.* **2017**, *7*, 115011. [CrossRef]

52. Ohnishi, T.; Hang, B.T.; Xu, X.; Osada, M.; Takada, K. Quality control of epitaxial LiCoO$_2$ thin films grown by pulsed laser deposition. *J. Mater. Res.* **2010**, *25*, 1886–1889. [CrossRef]

53. Ohnishi, T.; Nishio, K.; Takada, K. Composition controlled LiCoO$_2$ epitaxial thin film growth by pulsed laser deposition. *Proc. SPIE* **2015**, *9364*, 93640L.

54. Zhang, Y.; Zhong, Z.Y.; Zhu, M. Effect of deposition condition on structure and morphology of pulsed-laser deposited LiCoO$_2$ films. *J. Funct. Mater.* **2007**, *38*, 1971–1974.

55. Bouwman, P.J.; Boukamp, B.A.; Bouwmeester, H.J.M.; Wondergem, H.J.; Notten, P.H.L. Structural analysis of submicrometer LiCoO$_2$ films. *J. Electrochem. Soc.* **2001**, *148*, A311–A317. [CrossRef]

56. Tang, S.B.; Lu, L.; Lai, M.O. Characterization of a LiCoO$_2$ thin film cathode grown by pulsed laser deposition. *Philos. Mag.* **2005**, *85*, 2831–2842. [CrossRef]

57. Nishio, K.; Ohnishi, T.; Akatsuka, K.; Takada, K. Crystal orientation of epitaxial LiCoO$_2$ films grown on SrTiO$_3$ substrates. *J. Power Sources* **2014**, *247*, 687–691. [CrossRef]

58. Rao, M.C.; Muntaz-Begum, S.K. Substrate effect on microstructure of LiCoO$_2$ thin film cathodes. *Optoelectron. Adv. Mater.* **2012**, *6*, 508–510.

59. Xia, H.; Lu, L.; Ceder, G. Substrate effect on the microstructure and electrochemical properties of LiCoO$_2$ thin films grown by PLD. *J. Alloy Compd.* **2006**, *417*, 304–310. [CrossRef]

60. Nishio, K.; Ohnishi, T.; Mitsuishi, K.; Ohta, N.; Watanabe, K.; Takada, K. Orientation alignment of epitaxial LiCoO$_2$ thin films on vicinal SrTiO$_3$ (100) substrates. *J. Power Sources* **2016**, *325*, 306–310. [CrossRef]

61. Liu, D.; Chao, Y.; Wu, S.; Zhong, S.; Lei, M. Morphology control of LiCoO$_2$ thin film prepared by pulsed laser deposition. In Proceedings of the 2013 International Conference on Materials for Renewable Energy and Environment, Chengdu, China, 19–21 August 2014; Volume 2, pp. 529–531.

62. Ohnishi, T.; Takada, K. High-rate growth of high-crystallinity LiCoO$_2$ epitaxial thin films by pulsed laser deposition. *Appl. Phys. Express* **2012**, *5*, 055502. [CrossRef]

63. Nishio, K.; Ohnishi, T.; Osada, M.; Ohta, N.; Watanabe, K.; Takada, K. Influences of high deposition rate on LiCoO$_2$ epitaxial films prepared by pulsed laser deposition. *Solid State Ion.* **2016**, *285*, 91–95. [CrossRef]

64. Funayama, K.; Nakamura, T.; Kuwata, N.; Kawamura, J.; Kawada, T.; Amezawa, K. Electromotive force measurements of LiCoO$_2$ electrode on a lithium ion-conducting glass ceramics under mechanical stress. *Solid State Ion.* **2016**, *285*, 75–78. [CrossRef]

65. Rao, M.C. Structure and surface morphology of LiCoO$_2$ thin film cathodes prepared by pulsed laser deposition. *J. Intense Pulsed Lasers Appl. Adv. Phys.* **2012**, *2*, 35–38.

66. Rao, M.C. Optical absorption studies of LiCoO$_2$ thin films grown by pulsed laser deposition. *Int. J. Pure Appl. Phys.* **2010**, *6*, 365–370.

67. Rao, M.C.; Hussain, O.M. Optical and electrical properties of laser ablated amorphous LiCoO$_2$ thin film cathodes. In *IOP Conference Series: Materials Science and Engineering*; IOP Publishing: Bristol, UK, 2009; Volume 2, p. 012037.

68. Bouwman, P.J.; Boukamp, B.A.; Bouwmeester, H.J.M.; Notten, P.H.L. Structure-related intercalation behaviour of LiCoO$_2$ films. *Solid State Ion.* **2002**, *152*, 181–188. [CrossRef]

69. Xia, H.; Lunney, J.; Ceder, G. Li diffusion in LiCoO$_2$ thin films prepared by pulsed laser deposition. *J. Power Sources* **2006**, *159*, 1422–1427. [CrossRef]

70. Xia, H.; Lu, L. Texture effect on the electrochemical properties of LiCoO$_2$ thin films prepared by PLD. *Electrochim. Acta* **2007**, *52*, 7014–7020. [CrossRef]

71. Taminato, S.; Hirayama, M.; Suzuki, K.; Tamura, K.; Minato, T.; Arai, H.; Uchimoto, Y.; Ogumi, Z.; Kanno, R. Lithium intercalation and structural changes at the LiCoO$_2$ surface under high voltage battery operation. *J. Power Sources* **2016**, *307*, 599–603. [CrossRef]

72. Vasanthi, R.; Mangani, I.R.; Selladurai, S.; Manoravi, P. Comparison of LiCo$_{0.97}$Mg$_{0.03}$O$_2$ PLD film with bulk for the application of lithium micro-batteries. In Proceedings of the 8th Asian Conference on Solid State Ionics: Trends in the New Millennium, Langkawi, Malaysia, 15–19 December 2002; pp. 163–170.

73. Rao, M.C.; Hussain, O.M. Growth and characterization of tetravalent doped LiCoO$_2$ thin film cathodes. *Indian J. Eng. Mater. Sci.* **2009**, *16*, 335–340.

74. Rao, M.C.; Hussain, O.M. Synthesis and electrochemical properties of Ti doped $LiCoO_2$ thin film cathodes. *J. Alloy. Compd.* **2010**, *491*, 503–506. [CrossRef]

75. Hirayama, M.; Sonoyama, N.; Abe, T.; Minoura, M. Characterization of electrode/electrolyte interface for lithium batteries using in situ synchrotron X-ray reflectometry—A new experimental technique for $LiCoO_2$ model electrode. *J. Power Sources* **2007**, *168*, 493–500. [CrossRef]

76. Yamada, I.; Iriyama, Y.; Abe, T.; Ogumi, Z. Lithium-ion transfer on a Li_xCoO_2 thin film electrode prepared by pulsed laser deposition—Effect of orientation. *J. Power Sources* **2007**, *172*, 933–937. [CrossRef]

77. Tang, S.B.; Lai, M.O.; Lu, L. Li-ion diffusion in highly (003) oriented $LiCoO_2$ thin film cathode prepared by pulsed laser deposition. *J. Alloy. Compd.* **2008**, *449*, 300–303. [CrossRef]

78. Takeuchi, S.; Tan, H.; Bharathi, K.K.; Stafford, G.R.; Shin, J.; Yasui, S.; Takeuchi, I.; Bendersky, L.A. Epitaxial $LiCoO_2$ films as a model system for fundamental electrochemical studies of positive electrodes. *ACS Appl. Mater. Interfaces* **2015**, *7*, 7901–7911. [CrossRef]

79. Tan, H.; Takeuchi, S.; Bharathi, K.K.; Takeuchi, I.; Bendersky, L.A. Microscopy study of structural evolution in epitaxial $LiCoO_2$ positive electrode films during electrochemical cycling. *ACS Appl. Mater. Interfaces* **2016**, *8*, 6727–6735. [CrossRef] [PubMed]

80. Li, Z.; Yasui, S.; Takeuchi, S.; Creuziger, A.; Maruyama, S.; Herzing, A.A.; Takeuchi, I.; Bendersky, L.A. Structural study of epitaxial $LiCoO_2$ films grown by pulsed laser deposition on single crystal $SrTiO_3$ substrates. *Thin Solid Films* **2016**, *612*, 472–482. [CrossRef]

81. Hussain, O.M.; Rao, M.C.; Julien, C.M. Growth of nano-structured TiO_2-doped $LiCoO_2$ PLD films and their electrochemical properties. In *New Trends in Intercalation Compounds for Energy Storage and Conversion*; Zaghib, K., Julien, C.M., Prakash, J., Eds.; The Electrochemical Society: Pennington, NJ, USA, 2003; pp. 627–631.

82. Kato, M.; Hayashi, T.; Hasegawa, G.; Lu, X.; Miyazaki, T.; Matsuda, Y.; Kuwata, N.; Koji Kurihara, K.; Kawamura, J. Electrochemical properties of $LiCoO_2$ thin film surface modified by lithium tantalate and lithium niobate coatings. *Solid State Ion.* **2017**, *308*, 54–60. [CrossRef]

83. Hayashi, T.; Okada, J.; Toda, E.; Kuzuo, R.; Matsuda, Y.; Kuwata, N.; Kawamura, J. Electrochemical effect of lithium tungsten oxide modification on $LiCoO_2$ thin film electrode. *J. Power Sources* **2015**, *285*, 559–567. [CrossRef]

84. Hayashi, T.; Matsuda, Y.; Kuwata, N.; Kawamura, J. High-power durability of $LiCoO_2$ thin film electrode modified with amorphous lithium tungsten oxide. *J. Power Sources* **2017**, *354*, 41–47. [CrossRef]

85. Hayashi, T.; Toda, E.; Kuzuo, R.; Matsuda, Y.; Kuwata, N.; Junichi Kawamura, J. Contribution of randomly oriented Li_2WO_4 with tetragonal symmetry to Li^+ ion transfer resistance reduction in lithium ion batteries. *Int. J. Electrochem. Sci.* **2015**, *10*, 8150–8157.

86. Hayashi, T.; Miyazaki, T.; Matsuda, Y.; Kuwata, N.; Saruwatari, M.; Furuichi, Y.; Kurihara, K.; Kuzuo, R.; Kawamura, J. Effect of lithium-ion diffusibility on interfacial resistance of $LiCoO_2$ thin film electrode modified with lithium tungsten oxides. *J. Power Sources* **2016**, *305*, 46–53. [CrossRef]

87. Kuwata, N.; Ise, K.; Matsuda, Y.; Kawamura, J.; Tsurui, T.; Kamishima, O. Detection of degradation in $LiCoO_2$ thin films by in situ micro Raman microscopy. In *Ionics for Sustainable World, Proceedings of the 13th Asian Conference on Solid State Ionics, Sendai, Japan, 17–20 July 2012*; Chowdari, B.V.R., Kawamura, J., Mizusaki, J., Amezawa, K., Eds.; World Scientific Publishing Co.: Singapore, 2013; pp. 138–143.

88. Matsuda, Y.; Kuwata, N.; Kawamura, J. Thin-film lithium batteries with 0.3–30 μm thick $LiCoO_2$ films fabricated by high-rate pulsed laser deposition. *Solid State Ion.* **2014**, *320*, 38–44. [CrossRef]

89. Shiraki, S.; Oki, H.; Takagi, Y.; Suzuki, T.; Kumatani, A.; Shimizu, R.; Haruta, M.; Ohsawa, T.; Sato, Y.; Ikuhara, Y.; et al. Fabrication of all-solid-state battery using epitaxial $LiCoO_2$ thin films. *J. Power Sources* **2014**, *267*, 881–887. [CrossRef]

90. Wen, S.-J.; von Rottkay, K.; Rubin, M. Electrochromic lithium nickel oxide thin film by pulsed laser deposition. *ECS Symp. Proc.* **1997**, *96*, 54–63.

91. Bouessay, I.; Rougier, A.; Tarascon, J.M. Electrochromic mechanism in nickel oxide thin films grown by pulsed laser deposition. *Electrochem. Soc. Proc.* **2003**, *2003*, 91–102.

92. Jahromi, S.P.; Huang, N.M.; Kamalianfar, A.; Lim, H.N.; Muhamad, M.R.; Yousef, R. Facile synthesis of porous-structured nickel oxide thin film by pulsed laser deposition. *J. Nanomater.* **2012**, *2012*, 173825.

93. Wang, H.; Wang, Y.; Wang, X. Pulsed laser deposition of the porous nickel oxide thin film at room temperature for high-rate pseudocapacitive energy storage. *Electrochem. Commun.* **2012**, *18*, 92–95. [CrossRef]

94. López-Iturbe, J.; Camacho-López, M.A.; Escobar-Alarcón, L.; Camps, E. Synthesis and characterization of LiNiO$_2$ targets for thin film deposition by pulsed laser ablation. *Superficies Y Vacío* **2005**, *18*, 27–30.

95. Rao, M.C.; Ravindranadh, K.; Srikanth, K.S. Microstructural and electrochemical studies on laser ablated LiNiO$_2$ thin films. *J. Chem. Pharm. Res.* **2016**, *8*, 677–684.

96. Yuki, Y.; Uchiyama, T.; Yamamoto, K.; Uchimoto, Y. Preparation of Li$_{1-x}$Ni$_{1+x}$O$_2$ thin films by pulsed laser deposition and the electrochemical performance for oxygen evolution reaction in alkaline media. In *ESC Meeting Abstracts*; The Electrochemical Society: Pennington, NJ, USA, 2018; p. 2585.

97. Razeg, K.H.; Al-Hilli, M.F.; Khalefa, A.A.; Aadim, K.A. Structural and optical properties of (Li$_x$Ni$_{2-x}$O$_2$) thin films deposited by pulsed laser deposited (PLD) technique at different doping ratio. *Int. J. Phys.* **2017**, *5*, 46–52.

98. Julien, C.; El-Farh, L.; Rangan, S.; Massot, M. studies of LiNi$_{0.8}$Co$_{0.2}$O$_2$ cathode material prepared by the citric acid-assisted sol-gel method for lithium batteries. *J. Sol-Gel Sci. Technol.* **1999**, *15*, 63–72. [CrossRef]

99. Ramana, C.V.; Zaghib, K.; Julien, C.M. Synthesis, structural and electrochemical properties of pulsed laser deposited Li(Ni,Co)O$_2$ films. *J. Power Sources* **2006**, *159*, 1310–1315. [CrossRef]

100. Ramana, C.V.; Zaghib, K.; Julien, C.M. Highly oriented growth of pulsed-laser deposited LiNi$_{0.8}$Co$_{0.2}$O$_2$ films for application in microbatteries. *Chem. Mater.* **2006**, *18*, 1397–1400. [CrossRef]

101. Ramana, C.V.; Zaghib, K.; Julien, C.M. Growth and electrochemical properties of pulsed laser deposited Al-doped Li-Ni-Co oxide films. In Proceedings of the Extended Abstracts of the 231st ACS National Meeting, Atlanta, GA, USA, 26–30 March 2006; p. 944631.

102. Hirayama, M.; Sakamoto, K.; Hiraide, T.; Mori, D.; Yamada, A.; Kanno, R.; Sonoyama, N.; Tamura, K.; Mizuki, J. Characterization of electrode/electrolyte interface using in situ X-ray reflectometry and LiNi$_{0.8}$Co$_{0.2}$O$_2$ epitaxial film electrode synthesized by pulsed laser deposition method. *Electrochim. Acta* **2007**, *53*, 871–881. [CrossRef]

103. Wang, G.X.; Lindsay, M.J.; Ionescu, M.; Bradhurst, D.H.; Dou, S.X.; Liu, H.K. Physical and electrochemical characterization of LiNi$_{0.8}$Co$_{0.2}$O$_2$ thin-film electrodes deposited by laser ablation. *J. Power Sources* **2001**, *97*, 298–302. [CrossRef]

104. Imanishi, N.; Shizuka, K.; Matsumura, T.; Hirano, A.; Takeda, Y.; Kanno, R. Impedance analysis of PLD LiNi$_{0.8}$Co$_{0.2}$O$_2$ film electrode. *J. Power Sources* **2007**, *174*, 751–755. [CrossRef]

105. Baskaran, R.; Kuwata, N.; Kamishima, O.; Kawamura, J.; Selvasekarapandian, S. Structural and electrochemical studies on thin film LiNi$_{0.8}$Co$_{0.2}$O$_2$ by PLD for micro battery. *Solid State Ion.* **2009**, *180*, 636–643. [CrossRef]

106. Xia, H.; Lu, L.; Meng, Y.S. Growth of layered LiNi$_{0.5}$Mn$_{0.5}$O$_2$ thin films by pulsed laser deposition for application in microbatteries. *Appl. Phys. Lett.* **2008**, *92*, 011912. [CrossRef]

107. Xia, H.; Lu, L.; Lai, M.O. Li diffusion in LiNi$_{0.5}$Mn$_{0.5}$O$_2$ thin film electrodes prepared by pulsed laser deposition. *Electrochim. Acta* **2009**, *54*, 5986–5991. [CrossRef]

108. Sakamoto, K.; Konishi, H.; Sonoyama, N.; Yamada, A.; Tamura, K.; Mizuki, J.; Kanno, R. Mechanistic study on lithium intercalation using a restricted reaction field in LiNi$_{0.5}$Mn$_{0.5}$O$_2$. *J. Power Sources* **2007**, *174*, 678–682. [CrossRef]

109. Sakamoto, K.; Hirayama, M.; Konishi, H.; Sonoyama, N.; Dupre, N.; Guyomard, D.; Tamura, K.; Mizuki, J.; Kanno, R. Structural changes in surface and bulk LiNi$_{0.5}$Mn$_{0.5}$O$_2$ during electrochemical reaction on epitaxial thin-film electrodes characterized by in situ X-ray scattering. *Phys. Chem. Chem. Phys.* **2010**, *12*, 3815–3823. [CrossRef]

110. Ramana, C.V.; Zaghib, K.; Julien, C.M. Pulsed-laser deposited LiNi$_{0.8}$Co$_{0.15}$Al$_{0.05}$O$_2$ thin films for application in microbatteries. *Appl. Phys. Lett.* **2007**, *90*, 21916. [CrossRef]

111. Deng, J.; Xi, L.; Wang, L.; Wang, Z.; Chung, C.Y.; Han, X.; Zhou, H. Electrochemical performance of LiNi$_{1/3}$Co$_{1/3}$Mn$_{1/3}$O$_2$ thin film electrodes prepared by pulsed laser deposition. *J. Power Sources* **2012**, *217*, 491–497. [CrossRef]

112. Jacob, C.; Lynch, T.; Chen, A.; Jian, J.; Wang, H. Highly textured Li(Ni$_{0.5}$Mn$_{0.3}$Co$_{0.2}$)O$_2$ thin films on stainless steel as cathode for lithium-ion battery. *J. Power Sources* **2013**, *241*, 410–414. [CrossRef]

113. Smyrek, P.; Kim, H.; Zheng, Y.; Seifert, H.J.; Piqué, A.; Pfleging, W. Laser-printing and femtosecond laser-structuring of electrode materials for the manufacturing of 3D lithium-ion micro-batteries. *Proc. SPIE* **2016**, *9738*, 973806.

114. Abe, M.; Iba, H.; Suzuki, K.; Minamishima, H.; Hirayama, M.; Tamura, K.; Mizuki, J.; Saito, T.; Ikuhara, Y.; Kanno, R. Study on the deterioration mechanism of layered rock-salt electrodes using epitaxial thin films: Li(Ni, Co, Mn)O_2 and their Zr-O surface modified electrodes. *J. Power Sources* **2017**, *345*, 108–119. [CrossRef]

115. Bendersky, L.A.; Tan, H.; Karuppanan, K.B.; Li, Z.P.; Johnston-Peck, A.C. Crystallography and growth of epitaxial oxide films for fundamental studies of cathode materials used in advanced Li-ion batteries. *Crystals* **2017**, *7*, 127. [CrossRef]

116. Johnston-Peck, A.C.; Takeuchi, S.; Kamala-Bharathi, K.; Herzing, A.A.; Bendersky, L.A. Domain formation in lithium-rich manganese-nickel-cobalt-oxide epitaxial thin films and implications for interpretation of electrochemical behavior. *Thin Solid Films* **2018**, *647*, 40–49. [CrossRef]

117. Yan, B.; Liu, J.; Song, B.; Xiao, P.; Lu, L. Li-rich thin film cathode prepared by pulsed laser deposition. *Sci. Rep.* **2013**, *3*, 3332. [CrossRef] [PubMed]

118. Taminato, S.; Hirayama, M.; Suzuki, K.; Yamada, N.L.; Yonemura, M.; Son, J.Y.; Kanno, R. Highly reversible capacity at the surface of a lithium-rich manganese oxide: A model study using an epitaxial film system. *Chem. Commun.* **2015**, *51*, 1673–1676. [CrossRef] [PubMed]

119. Zheng, Y.; Taminato, S.; Suzuki, K.; Hirayama, M.; Kanno, R. Fabrication and lithium intercalation properties of epitaxial Li_2RuO_3 thin films. *Thin Solid Films* **2012**, *520*, 4889–4893. [CrossRef]

120. Taminato, S.; Hirayama, M.; Suzuki, K.; Kim, K.; Zheng, Y.; Tamura, K.; Mizuki, J.I.; Kanno, R. Mechanistic studies on lithium intercalation in a lithium-rich layered material using Li_2RuO_3 epitaxial film electrodes and in situ surface X-ray analysis. *J. Mater. Chem. A* **2014**, *2*, 17875–17882. [CrossRef]

121. Kuwata, N.; Kumar, R.; Toribami, K.; Suzuki, T.; Hattori, T.; Kawamura, J. Thin film lithium ion batteries prepared only pulsed laser deposition. *Solid State Ion.* **2006**, *177*, 2827–2832. [CrossRef]

122. Morcrette, M.; Barboux, P.; Perrière, J.; Brousse, T. $LiMn_2O_4$ thin films for lithium ion sensors. *Solid State Ion.* **1998**, *112*, 249–254. [CrossRef]

123. Camacho-Lopez, M.A.; Escobar-Alarcon, L.; Haro-Poniatowski, E.; Julien, C. $LiMn_2O_4$ films grown by pulsed-laser deposition. *Ionics* **1999**, *5*, 244–250. [CrossRef]

124. Singh, D.; Houriet, R.; Vacassy, R.; Hofinann, H.; Craciun, V.; Singh, R.K. Pulsed laser deposition and characterization of $LiMn_2O_4$ thin films for applications in Li-ion rechargeable battery systems. *Mater. Res. Soc. Symp. Proc.* **1999**, *575*, 83–89. [CrossRef]

125. Camacho-Lopez, M.A.; Escobar-Alarcon, L.; Haro-Poniatowski, E.; Julien, C. Physico-chemical properties of $LiMn_2O_4$ films grown by laser ablation. In *Materials for Lithium-Ion Batteries*; Julien, C., Stoynov, Z., Eds.; NATO Science Series; Kluwer Academic Publishers: Dordrecht, The Netherlands, 2000; Volume 3–85, pp. 535–541.

126. Camacho-Lopez, M.A.; Escobar-Alarcon, L.; Haro-Poniatowski, E.; Julien, C. Raman studies of $LiMn_2O_4$ films grown by laser ablation. *Mater. Res. Soc. Symp. Proc.* **2000**, *617*. [CrossRef]

127. Simmen, F.; Lippert, T.; Novák, P.; Neuenschwander, B.; Döbeli, M.; Mallepell, M.; Wokaun, A. The influence of lithium excess in the target on the properties and compositions of $Li_{1+x}Mn_2O_{4-\delta}$ thin films prepared by PLD. *Appl. Phys. A* **2008**, *93*, 711–716. [CrossRef]

128. Gao, X.; Ikuhara, Y.H.; Fisher, C.A.; Moriwake, H.; Kuwabara, A.; Oki, H.; Kohama, K.; Yoshida, R.; Huang, R.; Ikuhara, Y. Structural distortion and compositional gradients adjacent to epitaxial $LiMn_2O_4$ thin film interfaces. *Adv. Mater. Interfaces* **2014**, *1*, 1400143. [CrossRef]

129. Hendriks, R.; Monteiro Cunha, D.; Singh, D.P.; Huijben, M. Enhanced lithium transport by control of crystal orientation in spinel $LiMn_2O_4$ thin film cathodes. *ACS Appl. Energy Mater.* **2018**, *1*, 7046–7051. [CrossRef] [PubMed]

130. Sonoyama, N.; Iwase, K.; Takatsuka, H.; Matsumura, T.; Imanishi, N.; Takeda, Y.; Kanno, R. Electrochemistry of $LiMn_2O_4$ epitaxial films deposited on various single crystal substrates. *J. Power Sources* **2009**, *189*, 561–565. [CrossRef]

131. Canulescu, S.; Papadopoulou, E.L.; Anglos, D.; Lippert, T.; Schneider, C.W.; Wokaun, A. Mechanisms of the laser plume expansion during the ablation of $LiMn_2O_4$. *J. Appl. Phys.* **2009**, *105*, 063107. [CrossRef]

132. Hussain, O.M.; Hari Krishna, K.; Kalai-Vani, V.; Julien, C.M. Pulsed laser deposition of highly oriented $LiCoO_2$ and $LiMn_2O_4$ thin films for microbattery applications. In Proceedings of the 10th Asian Conference on Solid State Ionics, Kandy, Sri Lanka, 12–16 June 2006; pp. 136–145.

133. Hussain, O.M.; Hari-Krishna, K.; Kalai-Vani, V.; Julien, C.M. Structural and electrical properties of lithium manganese oxide thin films grown by pulsed laser deposition. *Ionics* **2007**, *13*, 455–459. [CrossRef]

134. Tang, S.B.; Lai, M.O.; Lu, L.; Tripathy, S. Comparative study of $LiMn_2O_4$ thin film cathode grown at high medium and low temperatures by pulsed laser deposition. *J. Solid State Chem.* **2006**, *179*, 3831–3838. [CrossRef]

135. Tang, S.B.; Xia, H.; Lai, M.O.; Lu, L. Characterization of $LiMn_2O_4$ thin films grown on Si substrates by pulsed laser deposition. *J. Alloy. Compd.* **2008**, *449*, 322–325. [CrossRef]

136. Morcrette, M.; Barboux, P.; Perrière, J.; Brousse, T.; Traverse, A.; Boilot, J.P. Non-stoichiometry in $LiMn_2O_4$ thin films by laser ablation. *Solid State Ion.* **2001**, *138*, 213–219. [CrossRef]

137. Yamada, I.; Abe, T.; Iriyama, Y.; Ogumi, Z. Lithium-ion transfer at $LiMn_2O_4$ thin film electrode prepared by pulsed laser deposition. *Electrochem. Commun.* **2003**, *5*, 502–505. [CrossRef]

138. Albrecht, D.; Wulfmeier, H.; Fritze, H. Preparation and characterization of c-$LiMn_2O_4$ thin films prepared by pulsed laser deposition for lithium-ion batteries. *Energy Technol.* **2016**, *4*, 1558–1564. [CrossRef]

139. Yu, X.; Chen, X.; Buchholz, D.B.; Li, Q.; Wu, J.; Fenter, P.A.; Bedzyk, M.J.; Dravid, V.P.; Barnett, S.A. Pulsed laser deposition and characterization of heteroepitaxial $LiMn_2O_4/La_{0.5}Sr_{0.5}CoO_3$ bilayer thin films as model lithium Ion battery cathodes. *ACS Appl. Nano Mater.* **2018**, *1*, 642–653. [CrossRef]

140. Singh, D.; Houriet, R.; Giovannini, R.; Hofmann, H.; Cruciun, V.; Singh, R.K. Challenges in making of thin films for $Li_xMn_yO_4$ rechargeable lithium batteries for MEMS. *J. Power Sources* **2001**, *97*, 826–831. [CrossRef]

141. Rao, M.C. Thermal properties of laser ablated $LiMn_2O_4$ thin films. *J. Chem. Biol. Phys. Sci. C* **2014**, *4*, 2555–2559.

142. Rougier, A.; Striebel, K.A.; Wen, S.J.; Cairns, E.J. Cyclic voltammetry of pulsed laser deposited $Li_xMn_2O_4$ thin films. *J. Electrochem. Soc.* **1998**, *145*, 2975–2980. [CrossRef]

143. Singh, D.; Singh, R.K. Lithium-Based Rechargeable Batteries. U.S. Patent 20030186128A1, 2 October 2003.

144. Rao, M.C.; Ravindranadh, K.; Ramesh, K.K.D.; Srikanth, K.S. Morphological and electrochemical properties of $LiMn_2O_4$ thin films. *Der Pharma Chemica* **2016**, *8*, 74–79.

145. Suzuki, K.; Kim, K.S.; Taminato, S.; Komo, M.; Hagiwara, A.; Son, J.Y.; Inami, T.; Konishi, H.; Tamura, K.; Mizuki, J.; et al. Modification effects on structural changes of $LiMn_2O_4$ electrode during the electrochemical process. In Proceedings of the Extended Abstracts of the 222nd Electrochem. Soc. Meeting, Honolulu, HI, USA, 7–12 October 2012; p. 216.

146. Striebel, K.; Sakai, E.; Cairns, E. Impedance behavior of the $LiMn_2O_4/LiPF_6$-DMC-EC interface during cycling. In *Workshop on Interfaces, Phenomena, and Nanostructures in Lithium Batteries*; Kingler, R., Landgrebe, A., Maupin, P., Eds.; The Electrochemical Society: Pennington, NJ, USA, 2000; Volume 2000-36, pp. 61–67.

147. Hirayama, M. Interfacial reactions of intercalation electrodes in lithium ion batteries. *Electrochemistry* **2015**, *83*, 701–706. [CrossRef]

148. Hirayama, M.; Ido, H.; Kim, K.S.; Cho, W.; Tamura, K.; Mizuki, J.; Kanno, R. Dynamic structural changes at $LiMn_2O_4$/electrolyte interface during lithium battery reaction. *J. Am. Chem. Soc.* **2010**, *132*, 15268–15276. [CrossRef] [PubMed]

149. Inaba, M.; Doi, T.; Iriyama, Y.; Abe, T.; Ogumi, Z. Electrochemical STM observation of $LiMn_2O_4$ thin films prepared by pulsed laser deposition. *J. Power Sources* **1999**, *81*, 554–557. [CrossRef]

150. Tang, S.B.; Lai, M.O.; Lu, L. Electrochemical studies of low-temperature processed nano-crystalline $LiMn_2O_4$ thin film cathode at 55 °C. *J. Power Sources* **2007**, *164*, 372–378. [CrossRef]

151. Xie, J.; Kohno, K.; Matsumura, T.; Imanishi, N.; Hirano, A.; Takeda, Y.; Yamamoto, O. Li-ion diffusion kinetics in $LiMn_2O_4$ thin films prepared by pulsed laser deposition. *Electrochim. Acta* **2008**, *54*, 376–381. [CrossRef]

152. Tang, S.B.; Lai, M.O.; Lu, L. Study on Li+-ion diffusion in nano-crystalline $LiMn_2O_4$ thin film cathode grown by pulsed laser deposition using CV, EIS, PITT techniques. *Mater. Chem. Phys.* **2008**, *111*, 149–153. [CrossRef]

153. Ishiyama, H.; Jeong, S.C.; Watanabe, Y.X.; Hirayama, Y.; Imai, N.; Jung, H.S.; Miyatake, H.; Oyaizu, M.; Osa, A.; Otokawa, Y.; et al. Direct measurement of nanoscale lithium diffusion in solid battery materials using radioactive tracer of ^8Li. *Nucl. Instrum. Methods Phys. Res. B* **2016**, *376*, 379–381. [CrossRef]

154. Kuwata, N.; Nakane, M.; Miyazaki, T.; Mitsuishi, K.; Kawamura, J. Lithium diffusion coefficient in $LiMn_2O_4$ thin films measured by secondary ion mass spectrometry with ion-exchange method. *Solid State Ion.* **2018**, *320*, 266–271. [CrossRef]

155. Marchini, F.; Rubi, D.; del Pozo, M.; Federcio, W.J.; Calvo, E.J. Surface chemistry and lithium-ion exchange in $LiMn_2O_4$ for the electrochemical selective extraction of LiCl from natural salt lake brines. *J. Phys. Chem. C* **2016**, *120*, 15875–15883. [CrossRef]

156. Matsumura, T.; Imanishi, N.; Hirano, A.; Sonoyama, N.; Takeda, Y. Electrochemical performances for preferred oriented PLD thin-film electrodes of $LiNi_{0.8}Co_{0.2}O_2$, $LiFePO_4$ and $LiMn_2O_4$. *Solid State Ion.* **2008**, *179*, 2011–2015. [CrossRef]

157. Tang, S.B.; Lai, M.O.; Lu, L. Properties of nano-crystalline $LiMn_2O_4$ thin films deposited by pulsed laser deposition. *Electrochim. Acta* **2006**, *52*, 1161–1168. [CrossRef]

158. Hirayama, M.; Sonoyama, N.; Ito, M.; Minoura, M.; Mori, D.; Yamada, A.; Tamura, K.; Mizuki, J.; Kanno, R. Characterization of electrode electrolyte interface with X-ray reflectometry and epitaxial-film $LiMn_2O_4$ electrode. *J. Electrochem. Soc.* **2007**, *154*, A1065–A1072. [CrossRef]

159. Shin, D.W.; Choi, J.W.; Ahn, J.P.; Choi, W.K.; Cho, Y.S.; Yoon, S.J. ZrO_2-modified $LiMn_2O_4$ thin-film cathodes prepared by pulsed laser deposition. *J. Electrochem. Soc.* **2010**, *157*, A567–A570. [CrossRef]

160. Fehse, M.; Trocoli, R.; Hernandez, E.; Ventosa, E.; Sepulveda, A.; Morata, A.; Tarancon, A. An innovative multi-layer pulsed laser deposition approach for $LiMn_2O_4$ thin film cathodes. *Thin Solid Films* **2018**, *648*, 108–112. [CrossRef]

161. Kim, S.; Hirayama, M.; Suzuki, K.; Kanno, R. Hetero-epitaxial growth of $Li_{0.17}La_{0.61}TiO_3$ solid electrolyte on $LiMn_2O_4$ electrode for all solid-state batteries. *Solid State Ion.* **2014**, *262*, 578–581. [CrossRef]

162. Suzuki, K.; Kim, K.; Taminato, S.; Hirayama, M.; Kanno, R. Fabrication and electrochemical properties of $LiMn_2O_4$/$SrRuO_3$ multi-layer epitaxial thin film electrodes. *J. Power Sources* **2013**, *226*, 340–345. [CrossRef]

163. Yim, H.; Shin, D.W.; Choi, J.W. $LiMn_2O_4$-based cathode thin films for Li thin-film batteries. *J. Korean Phys. Soc.* **2016**, *68*, 41–53. [CrossRef]

164. Fehse, M.; Trocoli, R.; Hernandez, E.; Ventosa, E.; Hernandez, E.; Sepulveda, A.; Morata, A.; Tarancon, A. Ultrafast dischargeable $LiMn_2O_4$ thin-film electrodes with pseudocapacitive properties for microbatteries. *ACS Appl. Mater. Interfaces* **2017**, *9*, 5295–5301. [CrossRef] [PubMed]

165. Striebel, K.A.; Rougier, A.; Horne, C.R.; Reade, R.P.; Cairns, E.J. Electrochemical studies of substituted spinel thin films. *J. Electrochem. Soc.* **1999**, *146*, 4339–4347. [CrossRef]

166. Xia, H.; Tang, S.B.; Lu, L.; Meng, Y.S.; Ceder, G. The influence of preparation conditions on electrochemical properties of $LiNi_{0.5}Mn_{1.5}O_4$ thin film electrodes by PLD. *Electrochim. Acta* **2007**, *52*, 1120–1126. [CrossRef]

167. Xia, H.; Meng, Y.S.; Lu, L.; Ceder, G. Electrochemical properties of nonstoichiometric $LiNi_{0.5}Mn_{1.5}O_{4-\delta}$ thin-film electrodes prepared by pulsed laser deposition. *J. Electrochem. Soc.* **2007**, *154*, A737–A743. [CrossRef]

168. Xia, H.; Lu, L. Li diffusion in spinel $LiNi_{0.5}Mn_{1.5}O_4$ thin films prepared by pulsed laser deposition. *Phys. Scr.* **2007**, *2007*, 43–48. [CrossRef]

169. Konishi, H.; Suzuki, K.; Taminato, S.; Kim, K.; Kim, S.; Lim, J.; Hirayama, M.; Kanno, R. Structure and electrochemical properties of $LiNi_{0.5}Mn_{1.5}O_4$ epitaxial thin film electrodes. *J. Power Sources* **2014**, *246*, 365–370. [CrossRef]

170. Liu, D.; Zhu, W.; Trottier, J.; Gagnon, C.; Barray, F.; Gariépy, V.; Guerfi, A.; Mauger, A.; Groult, H.; Julien, C.M.; et al. Spinel materials for high-voltage cathodes in Li-ion batteries. *RSC Adv.* **2014**, *4*, 154–167. [CrossRef]

171. Kuwata, N.; Kudo, S.; Matsuda, Y.; Kawamura, J. Fabrication of thin-film lithium batteries with 5-V-class $LiCoMnO_4$ cathodes. *Solid State Ion.* **2014**, *262*, 165–169. [CrossRef]

172. Lim, J.; Lee, S.; Suzuki, K.; Kim, K.; Kim, S.; Taminato, S.; Hirayama, M.; Oshima, Y.; Takayanagi, K.; Kanno, R. Synthesis, structure and electrochemical properties of novel Li–Co–Mn–O epitaxial thin-film electrode using layer-by-layer deposition process. *J. Power Sources* **2015**, *279*, 502–509. [CrossRef]

173. Yang, D. Pulsed laser deposition of manganese oxide thin films for supercapacitor applications. *J. Power Sources* **2011**, *196*, 8843–8849. [CrossRef]

174. Xia, H.; Lai, M.O.; Lu, L. Nanoporous MnO_x thin-film electrodes synthesized by electrochemical lithiation/delithiation for supercapacitors. *J. Power Sources* **2011**, *196*, 2398–2402. [CrossRef]

175. Yang, D. Pulsed laser deposition of cobalt-doped manganese oxide thin films for supercapacitor applications. *J. Power Sources* **2012**, *198*, 416–422. [CrossRef]

176. Yang, D. Pulsed laser deposition of vanadium-doped manganese oxide thin films for supercapacitor applications. *J. Power Sources* **2013**, *228*, 89–96. [CrossRef]

177. Lu, Z.G.; Cheng, H.; Lo, M.F.; Chung, C.Y. Pulsed laser deposition and electrochemical characterization of $LiFePO_4$–Ag composite thin films. *Adv. Funct. Mater.* **2007**, *17*, 3885–3896. [CrossRef]

178. Lu, Z.G.; Lo, M.F.; Chung, C.Y. Pulse laser deposition and electrochemical characterization of $LiFePO_4$–C composite thin films. *J. Phys. Chem. C* **2008**, *112*, 7069–7078. [CrossRef]

179. Tang, K.; Yu, X.; Sun, J.; Li, H.; Huang, X. Kinetic analysis on LiFePO$_4$ thin films by CV, GITT, and EIS. *Electrochim. Acta* **2011**, *56*, 4869–4875. [CrossRef]

180. Iriyama, Y.; Yokoyama, M.; Yada, C.; Jeong, S.K.; Yamada, I.; Abe, T.; Inaba, M.; Ogumi, Z. Preparation of LiFePO$_4$ thin films by pulsed laser deposition and their electrochemical properties. *Electrochem. Solid State Lett.* **2004**, *7*, A340–A342. [CrossRef]

181. Yada, C.; Iriyama, Y.; Jeong, S.K.; Abe, T.; Inaba, M.; Ogumi, Z. Electrochemical properties of LiFePO$_4$ thin films prepared by pulsed laser deposition. *J. Power Sources* **2005**, *146*, 559–564. [CrossRef]

182. Song, S.W.; Reade, R.P.; Kostecki, R.; Striebel, K.A. Electrochemical studies of the LiFePO$_4$ thin films prepared with pulsed laser deposition. *J. Electrochem. Soc.* **2006**, *153*, A12–A19. [CrossRef]

183. Sun, J.; Tang, K.; Yu, X.; Li, H.; Huang, X. Needle-like LiFePO$_4$ thin films prepared by an off-axis pulsed laser deposition technique. *Thin Solid Films* **2009**, *517*, 2618–2622. [CrossRef]

184. Palomares, V.; Ruiz de Larramendi, I.; Alonso, J.; Bengoechea, M.; Goñi, A.; Miguel, O.; Rojo, T. LiFePO$_4$ thin films grown by pulsed laser deposition: Effect of the substrate on the film structure and morphology. *Appl. Surf. Sci.* **2010**, *256*, 2563–2568. [CrossRef]

185. Tang, K.; Sun, J.; Yu, X.; Li, H.; Huang, X. Electrochemical performance of LiFePO$_4$ thin films with different morphology and crystallinity. *Electrochim. Acta* **2009**, *54*, 6565–6569. [CrossRef]

186. Xue, M.Z.; Fu, Z.W. Electrochemical properties of LiFePO$_4$ cathode thin film fabricated by pulsed laser deposition. *Acta Phys. Chem. Sin.* **2005**, *21*, 707–710.

187. Sauvage, F.; Baudrin, E.; Morcrette, M.; Tarascon, J.M. Pulsed laser deposition and electrochemical properties of LiFePO$_4$ thin films. *Electrochem. Solid-State Lett.* **2004**, *7*, A15–A18. [CrossRef]

188. Sauvage, F.; Baudrin, E.; Gengembre, L.; Tarascon, J.M. Effect of texture on the electrochemical properties of LiFePO$_4$ thin films. *Solid State Ion.* **2005**, *176*, 1869–1876. [CrossRef]

189. Sauvage, F.; Baudrin, E.; Laffont, L.; Tarascon, J.M. Origin of electrochemical reactivity enhancement of post-annealed LiFePO$_4$ thin films: Preparation of heterosite-type FePO$_4$. *Solid State Ion.* **2007**, *178*, 145–152. [CrossRef]

190. Sauvage, F.; Laffont, L.; Tarascon, J.M.; Baudrin, E. Factors affecting the electrochemical reactivity vs. lithium of carbon-free LiFePO$_4$ thin films. *J. Power Sources* **2008**, *175*, 495–501. [CrossRef]

191. Raveendran, N.R.; Kumary, T.G.; Rajkumari, S.; Pandian, R.; Janaki, J.; Mani, A. Synthesis and study of FeSe thin films and LiFeO$_2$/FeSe bi-layers. *AIP Conf. Proc.* **2017**, *1832*, 130034.

192. Mauger, A.; Julien, C.M. V$_2$O$_5$ thin films for energy storage and conversion. *AIMS Mater. Sci.* **2018**, *5*, 349–401. [CrossRef]

193. McGraw, J.; Perkins, J.; Zhang, J.G.; Liu, P.; Parilla, P.; Turner, J.A.; Schulz, D.L.; Curtis, C.J.; Ginley, D.S. Next generation V$_2$O$_5$ cathode materials for Li rechargeable batteries. *Solid State Ion.* **1998**, *113*, 407–413. [CrossRef]

194. Julien, C.; Haro-Poniatowski, E.; Camacho-Lopez, M.A.; Escobar-Alarcon, L.; Jimenez-Jarquin, J. Applications of V$_2$O$_5$ thin films obtained by pulsed laser deposition in solid state batteries. *Bull. Am. Phys. Soc.* **1999**, *44*, 1274.

195. Julien, C.; Haro-Poniatowski, E.; Camacho-Lopez, M.A.; Escobar-Alarcon, L.; Jimenez-Jarquin, J. Growth of V$_2$O$_5$ thin films by pulsed laser deposition and their applications in lithium microbatteries. *Mater. Sci. Eng. B* **1999**, *65*, 170–176. [CrossRef]

196. Ramana, C.V.; Smith, R.J.; Hussain, O.M.; Julien, C. On the growth mechanism of pulsed-laser deposited vanadium oxide thin films. *Mater. Sci. Eng. B* **2004**, *111*, 218–225. [CrossRef]

197. McGraw, J.; Perkins, J.; Zhang, J.G.; Parilla, P.; Ciszek, T.F.; Fu, M.L.; Trickett, D.M.; Turner, J.A.; Ginley, D. Structural phase transformations in V$_2$O$_5$ thin-film cathode material for rechargeable batteries. *Mater. Res. Soc. Symp. Proc.* **1998**, *496*, 373–378. [CrossRef]

198. McGraw, J.; Bahn, C.; Parilla, P.; Perkins, J.; Readey, D.; Ginley, D. Li ion diffusion measurements in crystalline and amorphous V$_2$O$_5$ thin-film battery cathodes. *Mater. Res. Soc. Symp. Proc.* **1999**, *575*, 103–108. [CrossRef]

199. McGraw, J.M.; Perkins, J.D.; Hasoon, F.; Parilla, P.A.; Warmsingh, C.; Ginley, D.S. Pulsed laser deposition of oriented V$_2$O$_5$ thin films. *J. Mater. Res.* **2000**, *15*, 2249–2265. [CrossRef]

200. Zhang, J.G.; McGraw, J.M.; Turner, J.; Ginley, D. Charging capacity and cycling stability of VO$_x$ films prepared by pulsed laser deposition. *J. Electrochem. Soc.* **1997**, *144*, 1630–1634. [CrossRef]

201. Madhuri, K.V.; Rao, K.S.; Naidu, B.S.; Hussain, O.M.; Pinto, R. Characterization of laser-ablated V$_2$O$_5$ thin films. *J. Mater. Sci. Mater. Electron.* **2002**, *13*, 425–432. [CrossRef]

202. Iida, Y.; Kaneko, Y.; Kanno, Y. Fabrication of pulsed-laser deposited V_2O_5 thin films for electrochromic devices. *J. Mater. Process. Technol.* **2008**, *197*, 261–267. [CrossRef]
203. Ramana, C.V.; Hussain, O.M.; Pinto, R.; Julien, C.M. Microstructure features of pulsed-laser deposited V_2O_5 thin films. *Appl. Surf. Sci.* **2003**, *207*, 135–138. [CrossRef]
204. Ramana, C.V.; Smith, R.J.; Hussain, O.M.; Julien, C.M. Growth and surface characterization of V_2O_5 thin films made by pulsed laser deposition. *J. Vac. Sci. Technol. A* **2004**, *22*, 2453–2458. [CrossRef]
205. Ramana, C.V.; Smith, R.J.; Hussain, O.M.; Chusuei, C.C.; Julien, C.M. Correlation between growth conditions, microstructure and optical properties in pulsed-laser-deposited V_2O_5 thin films. *Chem. Mater.* **2005**, *17*, 1213–1219. [CrossRef]
206. Iida, Y.; Kanno, Y. Doping effect of *M* (*M* = Nb, Ce, Nd, Dy, Sm, Ag, and/or Na) on the growth of pulsed-laser deposited V_2O_5 thin films. *J. Mater. Process. Technol.* **2009**, *209*, 2421–2427. [CrossRef]
207. Beke, S.; Giorgio, S.; Korosi, L.; Nanai, L.; Marine, W. Structural and optical properties of pulsed laser deposited V_2O_5 thin films. *Thin Solid Films* **2008**, *516*, 4659–4664. [CrossRef]
208. Bowman, R.M.; Gregg, J.M. VO_2 thin films: Growth and the effect of applied strain on their resistance. *J. Mater. Sci.-Mater. Electron.* **1998**, *9*, 187–191. [CrossRef]
209. Deng, Y.; Pelton, A.; Mayanovic, R. Comparison of vanadium oxide thin films prepared using femtosecond and nanosecond pulsed laser deposition. *MRS Adv.* **2016**, *1*, 2737–2742. [CrossRef]
210. Ramana, C.V.; Hussain, O.M.; Pinto, R.; Julien, C.M. Surface and material properties of nanostructured vanadium oxide films made by pulsed-laser ablation. In Proceedings of the 203rd ECS Meeting, Paris, France, 27 April–2 May 2003.
211. Ramana, C.V.; Smith, R.J.; Hussain, O.M.; Massot, M.; Julien, C.M. Surface analysis of pulsed-laser-deposited V_2O_5 films and their lithium intercalated products studied by Raman spectroscopy. *Surf. Interface Anal.* **2005**, *37*, 406–411. [CrossRef]
212. Iida, Y.; Takashima, H.; Kanno, Y. Optical and electrical properties of V_2O_5 thin films fabricated by pulsed laser deposition. *Jpn. J. Appl. Phys.* **2008**, *47*, 633–636. [CrossRef]
213. Shibuya, K.; Sawa, A. Optimization of conditions for growth of vanadium dioxide thin films on silicon by pulsed-laser deposition. *AIP Adv.* **2015**, *5*, 107118. [CrossRef]
214. Fang, G.J.; Liu, Z.L.; Wang, Y.Q.; Liu, H.H.; Yao, K.L. Orientated growth of V_2O_5 electrochromic thin films on transparent conductive glass by pulsed excimer laser ablation technique. *J. Phys. D Appl. Phys.* **2000**, *33*, 3018–3021. [CrossRef]
215. Fang, G.J.; Yao, K.L.; Liu, Z.L. Fabrication and electrochromic properties of double layer $WO_3(V)/V_2O_5(Ti)$ thin films prepared by pulsed laser ablation technique. *Thin Solid Films* **2001**, *394*, 64–71. [CrossRef]
216. Fang, G.J.; Liu, Z.L.; Wang, Y.; Liu, Y.H.; Yao, K.L. Synthesis and structural, electrochromic characterization of pulsed laser deposited vanadium oxide thin films. *J. Vac. Sci. Technol. A* **2001**, *19*, 887–892. [CrossRef]
217. Rougier, A.; Blyr, A. Electrochromic properties of vanadium tungsten oxide thin films grown by pulsed laser deposition. *Electrochim. Acta* **2001**, *46*, 1945–1950. [CrossRef]
218. Huotari, J.; Lappalainen, J.; Eriksson, J.; Bjorklund, R.; Heinonen, E.; Miinalainen, I.; Puustinen, J.; Spetz, A.L. Synthesis of nanostructured solid-state phases of V_7O_{16} and V_2O_5 compounds for ppb-level detection of ammonia. *J. Alloy. Compd.* **2016**, *675*, 433–440. [CrossRef]
219. Huotari, J.; Bjorklund, R.; Lappalainen, J.; Lloyd Spetz, A. Nanostructured mixed phase vanadium oxide thin films as highly sensitive ammonia sensor material. *Procedia Eng.* **2014**, *87*, 1035–1038. [CrossRef]
220. Ramana, C.V.; Julien, C.M. Electrochemical performance of pulsed-laser deposited V_2O_5 films. In *ESC Meeting Abstracts*; The Electrochemical Society: Pennington, NJ, USA, 2006; Volume 210, p. 221.
221. Julien, C. Technology of microbatteries and its applications. In *Trends in Materials Science*; Radhakrishna, S., Ed.; Narosa: New Delhi, India, 1997; pp. 24–43.
222. Zhao, S.L.; Wen, J.B.; Zhu, Y.M.; Qin, Q.Z. Fabrication and electrochemical properties of all solid state $0.3Ag_0V_2O_5/Lipon/Li$ thin film battery. *J. Funct. Mater.* **2008**, *39*, 91–94.
223. Petnikota, S.; Chua, R.; Zhou, Y.; Edison, E.; Srinivasan, M. Amorphous vanadium oxide thin films as stable performing cathodes of lithium and sodium-ion batteries. *Nanoscale Res. Lett.* **2018**, *13*, 363. [CrossRef] [PubMed]
224. West, K.; Zachau-Christiansen, B.; Jacobsen, T.; Atlung, S. V_6O_{13} as cathode material for lithium cells. *J. Power Sources* **1985**, *14*, 235–245. [CrossRef]

225. Makimura, Y.; Rougier, A.; Tarascon, J.M. Pulsed laser deposited iron fluoride thin films for lithium-ion batteries. *Appl. Surf. Sci.* **2006**, *252*, 4587–4592. [CrossRef]

226. Makimura, Y.; Rougier, A.; Laffont, L.; Womes, M.; Jumas, J.C.; Leriche, J.B.; Tarascon, J.M. Electrochemical behaviour of low temperature grown iron fluoride thin films. *Electrochem. Commun.* **2006**, *8*, 1769–1774. [CrossRef]

227. Santos-Ortiz, R.; Volkov, V.; Schmid, S.; Kuo, F.L.; Kisslinger, K.; Nag, S.; Banerjee, R.; Zhu, Y.; Shepherd, N.D. Microstructure and electronic band structure of pulsed laser deposited iron fluoride thin film for battery electrodes. *ACS Appl. Mater. Interfaces* **2013**, *5*, 2387–2391. [CrossRef] [PubMed]

228. Al-Khateeb, S.; Lind, A.G.; Santos-Ortiz, R.; Shepherd, N.D.; Jones, K.S. Cycling performance and morphological evolution of pulsed laser-deposited FeF_2 thin film cathodes for Li-ion batteries. *J. Mater. Sci.* **2015**, *50*, 5174–5182. [CrossRef]

229. Julien, C.M.; Mauger, A.; Vijh, A.; Zaghib, K. *Lithium Batteries: Science and Technology*; Springer: Cham, Switzerland, 2016.

230. Haro-Poniatowski, E.; Jouanne, M.; Morhange, J.F.; Julien, C.; Diamant, R.; Fernandez-Guasti, M.; Fuentes, G.A.; Alonso, J.C. Micro-Raman characterization of WO_3 and MoO_3 thin films obtained by pulsed laser irradiation. *Appl. Surf. Sci.* **1998**, *127*, 674–678. [CrossRef]

231. Ramana, C.V.; Hussain, O.M.; Julien, C.M. Effect of growth temperature on the optical properties of pulsed-laser deposited MoO_3 thin films. *Mater. Sci. Indian J.* **2005**, *1*, 10–15.

232. Ramana, C.V.; Hussain, O.M.; Julien, C.M. Electronic properties of MoO_3 thin films grown by PLD and ARE techniques and their performance upon lithium intercalation. In Proceedings of the Extended Abstract of the 208th ECS Meeting, Los Angeles, CA, USA, 16–21 October 2005; p. 885.

233. Al-Kuhaili, M.; Durrani, S.; Bakhtiari, I. Pulsed laser deposition of molybdenum oxide thin films. *Appl. Phys. A Mater. Sci. Process.* **2010**, *98*, 609–615. [CrossRef]

234. Ramana, C.V.; Julien, C.M. Chemical and electrochemical properties of molybdenum oxide thin films prepared by reactive pulsed-laser assisted deposition. *Chem. Phys. Lett.* **2006**, *428*, 114–118. [CrossRef]

235. Ramana, C.V.; Atuchin, V.V.; Pokrovsky, L.D.; Beker, U.; Julien, C.M. Structure and chemical properties of molybdenum oxide thin films. *J. Vac. Sci. Technol. A* **2007**, *25*, 1166–1171. [CrossRef]

236. Ramana, C.V.; Hussain, O.M.; Julien, C.M. Electronic properties and performance upon lithium intercalation of MoO_3 thin films grown by PLD. *ECS Trans.* **2006**, *1*, 1–7.

237. Puppala, H.K.; Pelton, A.T.; Mayanovic, R.A. A comparative characterization study of molybdenum oxide thin films grown using femtosecond and nanosecond pulsed laser deposition. *Mater. Res. Soc. Adv.* **2016**, *1*, 2585–2590. [CrossRef]

238. Sunu, S.S.; Prabhu, E.; Jayaraman, V.; Gnanasekar, K.I.; Gnanasekaran, T. Gas sensing properties of PLD made MoO_3 films. *Sens. Actuators B Chem.* **2003**, *94*, 189–196. [CrossRef]

239. Ashrafi, M.A.; Ranjbar, M.; Kalhori, H.; Salamati, H. Pulsed laser deposition of Mo-V-O thin films for chromogenic applications. *Thin Solid Films* **2017**, *621*, 220–228. [CrossRef]

240. Chang, C.C.; Luo, J.Y.; Chen, T.K.; Yeh, K.W.; Huang, T.W.; Hsu, C.H.; Chao, W.H.; Ke, C.T.; Hsu, P.C.; Wang, M.J.; et al. Pulsed laser deposition of $(MoO_3)_{1-x}(V_2O_5)_x$ thin films: Preparation, characterization and gasochromic studies. *Thin Solid Films* **2010**, *519*, 1552–1557. [CrossRef]

241. Rougier, A.; Portemer, F.; Quédé, A.; El Marssi, M. Characterization of pulsed laser deposited WO_3 thin films for electrochromic devices. *Appl. Surf. Sci.* **1999**, *153*, 1–9. [CrossRef]

242. Qiu, H.; Lu, Y.F. Studies of pulsed laser deposition mechanism of WO_3 thin films. *Jpn. J. Appl. Phys.* **2001**, *40*, 183–187. [CrossRef]

243. Ramana, C.V.; Utsunomiya, S.; Ewing, R.C.; Julien, C.M.; Becker, U. Electron microscopy investigation of structural transformations in tungsten oxide (WO_3) thin films. *Phys. Status Solidi A* **2005**, *202*, R108–R110. [CrossRef]

244. Ramana, C.V.; Utsunomiya, S.; Ewing, R.C.; Julien, C.M.; Becker, U. Structural stability and phase transitions in WO_3 thin films. *J. Phys. Chem. B* **2006**, *110*, 10430–10435. [CrossRef] [PubMed]

245. Mitsugi, F.; Hiraiwa, E.; Ikegami, T.; Ebihara, K.; Thareja, R.K. WO_3 thin films prepared by pulsed laser deposition. *Jpn. J. Appl. Phys.* **2002**, *41*, 5372–5375. [CrossRef]

246. Hussain, O.M.; Swapnasmitha, A.S.; John, J.; Pinto, R. Structure and morphology of laser-ablated WO_3 thin films. *Appl. Phys. A* **2005**, *81*, 1291–1297. [CrossRef]

247. Suda, Y.; Kawasaki, H.; Ohshima, T.; Yagyu, Y. Characteristics of tungsten oxide thin films prepared on the flexible substrates using pulsed laser deposition. *Thin Solid Films* **2008**, *516*, 4397–4401. [CrossRef]

248. Fang, G.J.; Liu, Z.L.; Sun, G.C.; Yao, K.L. Preparation and electrochromic properties of nanocrystalline WO_3 thin films prepared by pulsed excimer laser ablation technique. *Phys. Status Solidi A* **2001**, *184*, 129–137. [CrossRef]

249. György, E.; Socol, G.; Mihailescu, I.N.; Ducu, C.; Ciuca, S. Structural and optical characterization of WO_3 thin films for gas sensor applications. *J. Appl. Phys.* **2005**, *97*, 093527. [CrossRef]

250. Caruana, A.J.; Cropper, M.D. Orientationally textured thin films of WO_x deposited by pulsed laser deposition. *Mater. Res. Express* **2014**, *1*, 046408. [CrossRef]

251. Impey, S.A.; Garcia-Miguel, J.L.; Allen, S.; Blyr, A.; Boessay, I.; Rougier, A. Colour neutrality for thin oxide films from pulsed laser deposition and sol-gel. *Electrochem. Soc. Proc.* **2003**, *2003–17*, 103–118.

252. Boyadjiev, S.I.; Stefan, N.; Szilágyi, I.M.; Mihailescu, N.; Visan, A.; Mihailescu, I.N.; Stan, G.E.; Besleaga, C.; Iliev, M.T.; Gesheva, K.A. Characterization of MAPLE deposited WO_3 thin films for electrochromic applications. *J. Phys. Conf. Ser.* **2017**, *780*, 012013. [CrossRef]

253. Kawasaki, H.; Ueda, T.; Suda, T.; Ohshima, T. Properties of metal doped tungsten oxide thin films for NO_x gas sensors grown by PLD method combined with sputtering process. *Sens. Actuators B* **2004**, *100*, 266–269. [CrossRef]

254. Leidinger, M.; Huotari, J.; Sauerwald, T.; Lappalainen, J.; Schütze, A. Selective detection of naphthalene with nanostructured WO_3 gas sensors prepared by pulsed laser deposition. *J. Sens. Sens. Syst.* **2016**, *5*, 147–156. [CrossRef]

255. Boyadjiev, S.I.; Georgieva, V.; Stefan, N.; Stan, G.E.; Mihailescu, N.; Visan, A.; Mihailescu, I.N.; Besleag, C.; Szilágyi, I.M. Characterization of PLD grown WO_3 thin films for gas sensing. *Appl. Surf. Sci.* **2017**, *417*, 218–223. [CrossRef]

256. Xia, H.; Wang, H.L.; Xiao, W.; Lai, M.O.; Li, L. Thin film Li electrolytes for all-solid-state micro-batteries. *Int. J. Surf. Sci. Eng.* **2009**, *3*, 23–43. [CrossRef]

257. Ren, Y.; Kai Chen, K.; Chen, R.; Liu, T.; Zhang, Y.; Nan, C.W. Oxide electrolytes for lithium batteries. *J. Am. Ceram. Soc.* **2015**, *98*, 3603–3623. [CrossRef]

258. Cao, C.; Li, Z.B.; Wang, X.L.; Zhao, X.B.; Han, W.Q. Recent advances in inorganic solid electrolytes for lithium batteries. *Front. Energy Res.* **2014**, *2*, 25. [CrossRef]

259. Zhao, S.L.; Fu, Z.W.; Qin, Q.Z. A solid-state electrolyte lithium phosphorous oxynitride film prepared by pulsed laser deposition. *Thin Solid Films* **2002**, *415*, 108–113. [CrossRef]

260. Ohta, N.; Takada, K.; Osada, M.; Zhang, L.Q.; Sasaki, T.; Watanabe, M. Solid electrolyte, thio-LiSICON, thin film prepared by pulsed laser deposition. *J. Power Sources* **2005**, *146*, 707–710. [CrossRef]

261. Ahn, J.K.; Yoon, S.G. Characteristics of amorphous lithium lanthanum titanate electrolyte thin films grown by PLD for use in rechargeable lithium microbatteries. *Electrochem. Solid-State Lett.* **2005**, *8*, A75–A78. [CrossRef]

262. Furusawa, S.I.; Kamiyama, A.; Tsurui, T. Fabrication and ionic conductivity of amorphous lithium meta-silicate thin film. *Solid State Ion.* **2008**, *179*, 536–542. [CrossRef]

263. Furusawa, S.I.; Shimizu, S.; Sekine, K.; Tabuchi, H. Preparation and ionic conductivity of β-$LiAlSiO_4$ thin film. *Solid State Ion.* **2004**, *167*, 325–329.

264. Reinacher, J.; Berendts, S.; Janek, J. Preparation and electrical properties of garnet-type $Li_6BaLa_2Ta_2O_{12}$ lithium solid electrolyte thin films prepared by pulsed laser deposition. *Solid State Ion.* **2014**, *258*, 1–7. [CrossRef]

265. Yu, X.-H.; Bates, J.B.; Jellison, G.E.; Hart, F.X. A stable thin-film lithium electrolyte: Lithium phosphorus oxynitride. *J. Electrochem. Soc.* **1997**, *144*, 524–532. [CrossRef]

266. Zhao, S.L.; Wen, J.B.; Fan, L.M.; Qin, Q.Z. preparation of thin film battery by pulse laser deposition. *J. Funct. Mater.* **2006**, *37*, 172–177.

267. West, W.C.; Hood, Z.D.; Adhikari, S.P.; Liang, C.; Lachgar, A.; Motoyama, M.; Iriyama, Y. Reduction of charge-transfer resistance at the solid electrolyte-electrode interface by pulsed laser deposition of films from a crystalline Li_2PO_2N source. *J. Power Sources* **2016**, *312*, 116–122. [CrossRef]

268. Morcrette, M.; Llorente, A.G.; Laurent, A.; Perrière, J.; Barboux, P.; Boilot, J.P.; Raymond, O.; Brousse, T. Growth by laser ablation of Ti-based oxide films with different valency states. *Appl. Phys. A* **1998**, *67*, 425–428. [CrossRef]

269. Ahn, J.K.; Yoon, S.G. Characteristics of perovskite ($Li_{0.5}La_{0.5}$)TiO_3 solid electrolyte thin films grown by pulsed laser deposition for rechargeable lithium microbattery. *Electrochim. Acta* **2004**, *50*, 371–374. [CrossRef]

270. Furusawa, S.; Tabuchi, H.; Sugiyama, T.; Tao, S.; Irvine, J.T.S. Ionic conductivity of amorphous lithium lanthanum titanate thin film. *Solid State Ion.* **2005**, *176*, 553–558. [CrossRef]

271. Maqueda, O.; Sauvage, F.; Laffont, L.; Martinez-Sarrion, M.L.; Mestres, L.; Baudrin, E. Structural, microstructural and transport properties study of lanthanum lithium titanium perovskite thin films grown by pulsed laser deposition. *Thin Solid Films* **2008**, *516*, 1651–1655. [CrossRef]

272. Ohnishi, T.; Takada, K. Synthesis and orientation control of Li-ion conducting epitaxial $Li_{0.33}La_{0.56}TiO_3$ solid electrolyte thin films by pulsed laser deposition. *Solid State Ion.* **2012**, *228*, 80–82. [CrossRef]

273. Aguesse, F.; Roddatis, V.; Roqueta, J.; García, P.; Pergolesi, D.; Santiso, J.; Kilner, J.A. Microstructure and ionic conductivity of LLTO thin films: Influence of different substrates and excess lithium in the target. *Solid State Ion.* **2015**, *272*, 1–8. [CrossRef]

274. Lee, J.L.; Wang, Z.; Xin, H.L.; Wynn, T.A.; Meng, Y.S. Amorphous lithium lanthanum titanate for solid-state microbatteries. *J. Electrochem. Soc.* **2017**, *164*, A6268–A6273. [CrossRef]

275. Thangadurai, V.; Kaack, H.; Weppner, W. Novel fast lithium ion conduction in garnet-type $Li_5La_3M_2O_{12}$ (M = Nb, Ta). *J. Am. Ceram. Soc.* **2003**, *86*, 437–440. [CrossRef]

276. Saccoccio, M.; Yu, J.; Lu, Z.; Kwok, S.C.T.; Wang, J.; Yeung, K.K.; Yuen, M.M.F.; Ciucci, F. Low temperature pulsed laser deposition of garnet $Li_{6.4}La_3Zr_{1.4}Ta_{0.6}O_{12}$ films as all solid-state lithium battery electrolytes. *J. Power Sources* **2017**, *365*, 43–52. [CrossRef]

277. Garbayo, I.; Struzik, M.; Bowman, W.J.; Pfenninger, R.; Stilp, E.; Rupp, J.L.M. Glass-type polyamorphism in Li-garnet thin film solid state battery conductors. *Adv. Energy Mater.* **2018**, *8*, 1702265. [CrossRef]

278. Tan, J.; Tiwari, A. Fabrication and characterization of $Li_7La_3Zr_2O_{12}$ thin films for lithium ion battery. *ECS Solid State Lett.* **2012**, *1*, Q57–Q60. [CrossRef]

279. Park, J.S.; Cheng, L.; Zorba, V.; Mehta, A.; Cabana, J.; Chen, G.; Doeff, M.M.; Richardson, T.J.; Park, J.H.; Son, J.-W.; et al. Effects of crystallinity and impurities on the electrical conductivity of Li–La–Zr–O thin films. *Thin Solid Films* **2015**, *576*, 55–60. [CrossRef]

280. Kim, S.; Hirayama, M.; Taminato, S.; Kanno, R. Epitaxial growth and lithium ion conductivity of lithium-oxide garnet for an all solid-state battery electrolyte. *Dalton Trans.* **2013**, *42*, 13112. [CrossRef]

281. Rawlence, M.; Garbayo, I.; Buecheler, S.; Rupp, J.L.M. On the chemical stability of post-lithiated garnet Al-stabilized $Li_7La_3Zr_2O_{12}$ solid state electrolyte thin films. *Nanoscale* **2016**, *8*, 14746–14753. [CrossRef]

282. Kuwata, N.; Iwagami, N.; Tanji, Y.; Matsuda, Y.; Kawamura, J. Characterization of thin-film lithium batteries with stable thin-film Li_3PO_4 solid electrolytes fabricated by ArF excimer laser deposition. *J. Electrochem. Soc.* **2011**, *157*, A521–A527. [CrossRef]

283. Sakurai, Y.; Sakuda, A.; Hayashi, A.; Tatsumisago, M. Preparation of amorphous Li_4SiO_4-Li_3PO_4 thin films by pulsed laser deposition for all-solid-state lithium secondary batteries. *Solid State Ion.* **2011**, *182*, 59–63. [CrossRef]

284. Nakagawa, A.; Kuwata, N.; Matsuda, Y.; Kawamura, J. Characterization of stable solid electrolyte lithium silicate for thin film lithium battery. *J. Phys. Soc. Jpn. A* **2010**, *79*, 98–101. [CrossRef]

285. Furusawa, S.I.; Kasahara, T.; Kamiyama, A. Fabrication and ionic conductivity of Li_2SiO_3 thin film. *Solid State Ion.* **2009**, *180*, 649–653. [CrossRef]

286. Sakuda, A.; Hayashi, A.; Hama, S.; Tatsumisago, M. Preparation of highly lithium-ion conductive $80Li_2S$-$20P_2O_5$ thin-film electrolytes using pulsed laser deposition. *J. Am. Ceram. Soc.* **2010**, *93*, 765–768. [CrossRef]

287. Quan, Z.; Hirayama, M.; Sato, D.; Zheng, Y.; Yano, T.; Hara, K.; Suzuki, K.; Hara, M.; Kanno, R. Effect of excess Li_2S on electrochemical properties of amorphous Li_3PS_4 films synthesized by pulsed laser deposition. *J. Am. Ceram. Soc.* **2017**, *100*, 746–753. [CrossRef]

288. Konishi, H.; Suzuki, K.; Taminato, S.; Kim, K.; Zheng, Y.; Kim, S.; Lim, J.; Hirayama, M.; Son, J.Y.; Cui, Y.; et al. Effect of surface Li_3PO_4 coating on $LiNi_{0.5}Mn_{1.5}O_4$ epitaxial thin film electrodes synthesized by pulsed laser deposition. *J. Power Sources* **2014**, *269*, 293–298. [CrossRef]

289. Yubuchi, S.; Ito, Y.; Matsuyama, T.; Hayashi, A.; Tatsumisago, M. 5 V class $LiNi_{0.5}Mn_{1.5}O_4$ positive electrode coated with Li_3PO_4 thin film for all-solid-state batteries using sulfide solid electrolyte. *Solid State Ion.* **2016**, *285*, 79–82. [CrossRef]

290. Aso, K.; Sakuda, A.; Hayashi, A.; Tatsumisago, M. All-solid-state lithium secondary batteries using NiS-carbon fiber composite electrodes coated with Li_2S–P_2S_5 solid electrolytes by pulsed laser deposition. *ACS Appl. Mater. Interfaces* **2013**, *5*, 686–690. [CrossRef]

291. Ito, Y.; Sakuda, A.; Ohtomo, T.; Hayashi, A.; Tatsumisago, M. Preparation of $Li_2S–GeS_2$ solid electrolyte thin films using pulsed laser deposition. *Solid State Ion.* **2013**, *236*, 1–4. [CrossRef]

292. Zhang, Z.; Chou, S.; Gu, Q.; Liu, H.; Li, H.; Ozawa, K.; Wang, J. Enhancing the high rate capability and cycling stability of $LiMn_2O_4$ by coating of solid-state electrolyte $LiNbO_3$. *ACS Appl. Mater. Interfaces* **2014**, *6*, 22155–22165. [CrossRef]

293. Uemura, T.; Goto, K.; Ogawa, M.; Harada, K. High-power all-solid secondary battery with high heat resistance. *SEI Technol. Rev.* **2013**, *77*, 132–136.

294. Zhao, S.; Qin, Q. Li–V–Si–O thin film electrolyte for all-solid-state Li-ion battery. *J. Power Sources* **2003**, *122*, 174–180. [CrossRef]

295. Kawamura, J.; Kuwata, N.; Toribami, K.; Sata, N.; Kamishima, O.; Hattori, T. Preparation of amorphous thin films by pulsed laser deposition. *Solid State Ion.* **2004**, *175*, 273–276. [CrossRef]

296. Glass, A.; Nassau, K.; Negran, T. Ionic Conductivity of quenched alkali niobate and tantalate glasses. *J. Appl. Phys.* **1978**, *49*, 4808–4811. [CrossRef]

297. Ohta, N.; Takada, K.; Sakaguchi, I.; Zhang, L.; Ma, R.; Fukuda, K.; Osada, M.; Sasaki, T. $LiNbO_3$-coated $LiCoO_2$ as cathode material for all solid-state lithium secondary batteries. *Electrochem. Commun.* **2007**, *9*, 1486–1490. [CrossRef]

298. Marsh, A.M.; Harkness, S.D.; Qian, F.; Singh, R.K. Pulsed laser deposition of high quality $LiNbO_3$ films on sapphire substrates. *Appl. Phys. Lett.* **1993**, *62*, 952–954. [CrossRef]

299. Perea, A.; Gonzalo, J.; Afonso, C.N.; Vivien, C.; Leborgne, C. Fast imaging of the plume expansion produced by laser ablation of $LiNbO_3$. *Appl. Phys. A* **1999**, *69*, S141–S144. [CrossRef]

300. Madian, M.; Eychmüller, A.; Giebeler, L. Current advances in TiO_2-based nanostructure electrodes for high performance lithium ion batteries. *Batteries* **2018**, *4*, 7. [CrossRef]

301. Murugesan, S.; Kuppusami, P.; Parvathavarthini, N.; Mohandas, E. Pulsed laser deposition of anatase and rutile TiO_2 thin films. *Surf. Coat. Technol.* **2007**, *201*, 7713–7719. [CrossRef]

302. Suda, Y.; Kawasaki, H.; Ueda, T.; Ohshima, T. Preparation of high-quality nitrogen doped TiO_2 thin film as a photocatalyst using a pulsed laser deposition method. *Thin Solid Films* **2004**, *453*, 162–166. [CrossRef]

303. Kim, J.H.; Seonghoon Lee, S.; Im, H.S. The effect of target density and its morphology on TiO_2 thin films grown on Si(100) by PLD. *Appl. Surf. Sci.* **1999**, *151*, 6–16. [CrossRef]

304. Yamamoto, S.; Sumita, T.; Sugiharuto, M.; Miyashita, A.; Naramoto, H. Preparation of epitaxial TiO_2 films by pulsed laser deposition technique. *Thin Solid Films* **2001**, *401*, 88–93. [CrossRef]

305. Inoue, N.; Yuasa, H.; Okoshi, M. TiO_2 thin films prepared by PLD for photocatalytic applications. *Appl. Surf. Sci.* **2002**, *197*, 393–397. [CrossRef]

306. Choi, Y.; Yamamoto, S.; Umebayashi, T.; Yoshikawa, M. Fabrication and characterization of anatase TiO_2 thin film on glass substrate grown by pulsed laser deposition. *Solid State Ion.* **2004**, *172*, 105–108. [CrossRef]

307. Ohzuku, T.; Ueda, A.; Yamamoto, N. Zero-strain insertion material of $Li[Li_{1/3}Ti_{5/3}]O_4$ for rechargeable lithium cells. *J. Electrochem. Soc.* **1995**, *142*, 1431–1435. [CrossRef]

308. Deng, J.Q.; Lu, Z.G.; Belharouak, I.; Amine, K.; Chung, C.Y. Preparation and electrochemical properties of $Li_4Ti_5O_{12}$ thin film electrodes by pulsed laser deposition. *J. Power Sources* **2009**, *193*, 816–821. [CrossRef]

309. Yu, X.; Wang, R.; He, Y.; Hu, Y.; Li, H.; Huang, X. Electrochromic behavior of transparent $Li_4Ti_5O_{12}$/FTO electrode. *Electrochem. Solid-State Lett.* **2010**, *13*, J99–J101. [CrossRef]

310. Hirayama, M.; Kim, K.; Toujigamori, T.; Cho, W.; Kanno, R. Epitaxial growth and electrochemical properties of $Li_4Ti_5O_{12}$ thin-film lithium battery anodes. *Dalton Trans.* **2011**, *40*, 2882–2887. [CrossRef] [PubMed]

311. Kim, K.; Toujigamori, T.; Suzuki, K.; Taminato, S.; Tamura, K.; Mizuki, J.; Hirayama, M.; Kanno, R. Characterization of nano-sized epitaxial $Li_4Ti_5O_{12}$(110) film electrode for lithium batteries. *Electrochemistry* **2012**, *80*, 800–803. [CrossRef]

312. Deng, J.; Lu, Z.; Chung, C.Y.; Hanc, X.; Wang, Z.; Zhou, H. Electrochemical performance and kinetic behavior of lithium ion in $Li_4Ti_5O_{12}$ thin film electrodes. *Appl. Surf. Sci.* **2014**, *314*, 936–941. [CrossRef]

313. Sugiyama, J.; Umegaki, I.; Uyama, T.; McFadden, R.M.L.; Shiraki, S.; Hitosugi, T.; Salman, Z.; Saadaoui, H.; Morris, G.D.; MacFarlane, W.A.; et al. Lithium diffusion in spinel $Li_4Ti_5O_{12}$ and $LiTi_2O_4$ films detected with ^8Li β-NMR. *Phys. Rev. B* **2017**, *96*, 094402. [CrossRef]

314. Zhao, M.; Lian, J.; Jia, Y.; Jin, K.; Xu, L.; Hu, Z.; Yang, X.; Kang, S. Investigation of the optical properties of $LiTi_2O_4$ and $Li_4Ti_5O_{12}$ spinel films by spectroscopic ellipsometry. *Opt. Mater. Express* **2016**, *6*, 3366–3374. [CrossRef]

315. Pfenninger, R.; Afyon, S.; Garbayo, I.; Struzik, M.; Rupp, J.L.M. Lithium titanate anode thin films for Li-ion solid state battery based on garnets. *Adv. Funct. Mater.* **2018**, *28*, 1800879. [CrossRef]

316. Mesoraca, S.; Kleibeuker, J.E.; Prasad, B.; MacManus-Driscoll, J.L.; Blamire, M.G. Lithium outdiffusion in $LiTi_2O_4$ thin films grown by pulsed laser deposition. *J. Cryst. Growth* **2016**, *454*, 134–138. [CrossRef]

317. Chopdekar, R.V.; Wong, F.J.; Takamura, Y.; Arenholz, E.; Suzuki, Y. Growth and characterization of superconducting spinel oxide $LiTi_2O_4$ thin films. *Phys. C Supercond.* **2009**, *469*, 1885–1891. [CrossRef]

318. Kumatani, A.; Ohsawa, T.; Shimizu, R.; Takagi, Y.; Shiraki, S.; Hitosugi, T. Growth processes of lithium titanate thin films deposited by using pulsed laser deposition. *Appl. Phys. Lett.* **2012**, *101*, 123103. [CrossRef]

319. Cunha, D.M.; Hendriks, T.A.; Vasileiadis, A.; Vos, C.M.; Verhallen, T.; Singh, D.P.; Wagemaker, M.; Huijben, M. Doubling reversible capacities in epitaxial Li4Ti5O12 thin film anodes for microbatteries. *ACS Appl. Energy Mater.* **2019**, *2*, 3410–3418. [CrossRef]

320. Tang, S.B.; Xia, H.; Lai, M.O.; Lu, L. Amorphous $LiNiVO_4$ thin-film anode for microbatteries grown by pulsed laser deposition. *J. Power Sources* **2006**, *159*, 685–689. [CrossRef]

321. Han, T.; Goodenough, J.B. 3-V full cell performance of anode framework $TiNb_2O_7$/spinel $LiNi_{0.5}Mn_{1.5}O_4$. *Chem. Mater.* **2011**, *23*, 3404–3407. [CrossRef]

322. Daramalla, V.; Rao Penki, T.; Munichandraiah, N.; Krupanidhi, S.B. Fabrication of $TiNb_2O_7$ thin film electrodes for Li-ion micro-batteries by pulsed laser deposition. *Mater. Sci. Eng. B* **2016**, *213*, 90–97. [CrossRef]

323. Daramalla, V.; Venkatesh, G.; Kishore, B.; Munichandraiah, N.; Krupanidhi, S.B. Superior electrochemical performance of amorphous titanium niobium oxide thin films for Li-ion thin film batteries. *J. Electrochem. Soc.* **2018**, *165*, A764–A772. [CrossRef]

324. Mauger, A.; Xie, H.; Julien, C.M. Composite anodes for lithium-ion batteries: Status and trends. *AIMS Mater. Sci.* **2016**, *3*, 1054–1106. [CrossRef]

325. Chen, L.B.; Xie, J.Y.; Yu, H.C.; Wang, T.H. An amorphous thin film Si anode with high capacity and long cycling life for lithium ion batteries. *J. Appl. Electrochem.* **2009**, *39*, 1157–1162. [CrossRef]

326. Ohara, S.; Suzuki, J.; Sekine, K.; Takamura, T. A thin film silicon anode for Li-ion batteries having a very large specific capacity and long cycle life. *J. Power Sources* **2004**, *136*, 303–306. [CrossRef]

327. Song, S.W.; Sriebel, K.A.; Reade, R.P.; Roberts, G.A.; Cairns, E.J. Electrochemical studies of nanoncrystalline Mg_2Si thin film electrodes prepared by pulsed laser deposition. *J. Electrochem. Soc.* **2003**, *150*, A121–A127. [CrossRef]

328. Park, M.S.; Wang, G.X.; Liu, H.K.; Dou, S.X. Electrochemical properties of Si thin film prepared by pulsed laser deposition for lithium ion micro-batteries. *Electrochim. Acta* **2006**, *51*, 5246–5249. [CrossRef]

329. Xia, H.; Tang, S.; Lu, L. Properties of amorphous Si thin film anodes prepared by pulsed laser deposition. *Mater. Res. Bull.* **2007**, *42*, 1301–1309. [CrossRef]

330. Chou, S.L.; Zhao, Y.; Wang, J.Z.; Chen, Z.X.; Liu, H.K.; Dou, S.X. Silicon/single-walled carbon nanotube composite paper as a flexible anode material for lithium ion batteries. *J. Phys. Chem. C* **2010**, *114*, 15862–15867. [CrossRef]

331. Radhakrishnan, G.; Adams, P.M.; Foran, B.; Quinzio, M.V.; Brodie, M.J. Pulsed laser deposited Si on multilayer graphene as anode material for lithium ion batteries. *APL Mater.* **2013**, *1*, 062103. [CrossRef]

332. Biserni, E.; Xie, M.; Brescia, R.; Scarpellini, A.; Hashempour, M.; Movahed, P.; George, S.M.; Bestetti, M.; Li Bassi, A.; Bruno, P. Silicon algae with carbon topping as thin-film anodes for lithium-ion microbatteries by a two-step facile method. *J. Power Sources* **2015**, *274*, 252–259. [CrossRef]

333. Garino, N.; Biserni, E.; Bassi, A.L.; Bruno, P.; Gerbaldi, C. Mesoporous Si and multi-layered Si/C films by pulsed laser deposition as Li-ion microbattery anodes. *J. Electrochem. Soc.* **2015**, *162*, A1816–A1822. [CrossRef]

334. Suzuki, N.; Butch-Cervera, R.; Ohnishi, T.; Takada, K. Silicon nitride thin film electrode for lithium-ion batteries. *J. Power Sources* **2013**, *231*, 186–189. [CrossRef]

335. Bleu, Y.; Bourquard, F.; Tite, T.; Loir, A.-S.; Maddi, C.; Donnet, C.; Garrelie, F. Review of graphene growth from a solid carbon source by pulsed laser deposition (PLD). *Front. Chem.* **2018**, *6*, 574. [CrossRef] [PubMed]

336. Kumar, S.R.; Nayak, P.K.; Hedhili, M.N.; Khan, M.A.; Alshareef, H.N. In situ growth of p and n-type graphene thin films and diodes by pulsed laser deposition. *Appl. Phys. Lett.* **2013**, *103*, 192109. [CrossRef]

337. Ren, P.; Pu, E.; Liu, D.; Wang, Y.; Xiang, B.; Ren, X. Fabrication of nitrogen-doped graphenes by pulsed laser deposition and improved chemical enhancement for Raman spectroscopy. *Mater. Lett.* **2017**, *204*, 65–68. [CrossRef]

338. Fortgang, P.; Tite, T.; Barnier, V.; Zehani, N.; Maddi, C.; Lagarde, F.; Loir, A.-S.; Jaffrezic-Renault, N.; Donnet, C.; Garrelie, F.; et al. Robust electrografting on self-organized 3D graphene electrodes. *ACS Appl. Mater. Interfaces* **2015**, *8*, 1424–1433. [CrossRef] [PubMed]

339. Bourquard, F.; Bleu, Y.; Loi, A.-S.; Caja-Munoz, B.; Avila, J.; Asensio, M.-C.; Riamondi, G.; Shokouhi, M.; Rassas, I.; Farre, C.; et al. Electroanalytical performance of nitrogen-doped graphene films processed in one step by pulsed laser deposition directly coupled with thermal annealing. *Materials* **2019**, *12*, 666. [CrossRef]

340. Maddi, C.; Bourquard, F.; Barnier, V.; Avila, J.; Asensio, M.C.; Tite, T.; Donnet, C.; Garrelie, F. Nano-architecture of nitrogen-doped graphene films synthesized from a solid CN source. *Sci. Rep.* **2018**, *8*, 3247. [CrossRef]

341. Yu, S.H.; Feng, X.; Zhang, N.; Seok, J.; Abruna, H.D. Understanding conversion-type electrodes for lithium rechargeable batteries. *Acc. Chem. Res.* **2018**, *51*, 273–281. [CrossRef]

342. Wang, K.X.; Li, Y.; Wu, X.Y.; Chen, J.S. Carbon nanocolumn arrays prepared by pulsed laser deposition for lithium ion batteries. *J. Power Sources* **2012**, *203*, 140–144. [CrossRef]

343. Wei, K.; Zhao, Y.; Cui, Y.; Wang, J.; Cui, Y.; Zhu, R.; Zhuang, Q.; Xue, M. Lithium phosphorous oxynitride (LiPON) coated $NiFe_2O_4$ anode material with enhanced electrochemical performance for lithium ion batteries. *J. Alloy Compd.* **2018**, *769*, 110–119. [CrossRef]

344. Pralong, V.; Leriche, J.B.; Beaudoin, B.; Naudin, E.; Morcrette, M.; Tarascon, J.M. Electrochemical study of nanometer Co_3O_4, Co, $CoSb_3$ and Sb thin films toward lithium. *Solid State Ion.* **2004**, *166*, 295–305. [CrossRef]

345. Song, S.W.; Reade, R.P.; Cairns, E.; Vaughey, J.T.; Thackeray, M.M.; Striebel, K.A. Cu_2Sb thin-film electrodes prepared by pulsed laser deposition for lithium batteries. *J. Electrochem. Soc.* **2004**, *151*, A1012–A1019. [CrossRef]

346. Teng, X.; Qin, Y.; Wang, X.; Li, H.; Shang, X.; Fan, S.; Li, Q.; Xu, J.; Cao, D.; Li, S. A nanocrystalline Fe_2O_3 film anode prepared by pulsed laser deposition for lithium-ion batteries. *Nanoscale Res. Lett.* **2018**, *13*, 60. [CrossRef] [PubMed]

347. Fu, Z.W.; Wang, Y.; Zhang, Y.; Qin, Q.Z. Electrochemical reaction of nanocrystalline Co_3O_4 thin film with lithium. *Solid State Ion.* **2004**, *170*, 105–109. [CrossRef]

348. Wang, Y.; Wang, H.; Wang, X. The cobalt oxide/hydroxide nanowall array film prepared by pulsed laser deposition for supercapacitors with superb-rate capability. *Electrochim. Acta* **2013**, *92*, 298–303. [CrossRef]

349. Zhou, Y.; Zhang, H.; Xue, M.; Wu, C.; Wu, X.; Fu, Z. The electrochemistry of nanostructured In_2O_3 with lithium. *J. Power Sources* **2006**, *162*, 1373–1378. [CrossRef]

350. Xue, M.Z.; Fu, Z.W. Pulsed laser deposited Sb_2Se_3 anode for lithium-ion batteries. *J. Alloy. Compd.* **2008**, *458*, 351–356. [CrossRef]

351. Yu, L.; Wang, H.X.; Liu, Z.Y.; Fu, Z.W. Pulsed laser deposited FeOF as negative electrodes for rechargeable Li batteries. *Electrochim. Acta* **2010**, *56*, 767–775. [CrossRef]

352. Rambabu, A.; Senthilkumar, B.; Dayamani, A.; Krupanidhi, S.B.; Barpanda, P. Preferentially oriented $SrLi_2Ti_6O_{14}$ thin film anode for Li-ion microbatteries fabricated by pulsed laser deposition. *Electrochim. Acta* **2018**, *269*, 212–216. [CrossRef]

353. Yu, X.Q.; He, Y.; Sun, J.P.; Tang, K.; Li, H.; Chen, L.Q.; Huang, X.J. Nanocrystalline MnO thin film anode for lithium ion batteries with low overpotential. *Electrochem. Commun.* **2009**, *11*, 791–794. [CrossRef]

354. Wu, H.; Wei, K.; Tang, B.; Cui, Y.; Zhao, Y.; Xue, M.; Li, C.; Cui, Y. A novel Li_3P-VP nanocomposite fabricated by pulsed laser deposition as anode material for high-capacity lithium ion batteries. *J. Electroanal. Chem.* **2019**, *841*, 21–25. [CrossRef]

© 2019 by the authors. Licensee MDPI, Basel, Switzerland. This article is an open access article distributed under the terms and conditions of the Creative Commons Attribution (CC BY) license (http://creativecommons.org/licenses/by/4.0/).

Article

Macrophage In Vitro Response on Hybrid Coatings Obtained by Matrix Assisted Pulsed Laser Evaporation

Madalina Icriverzi [1,2], Laurentiu Rusen [3], Simona Brajnicov [3,4], Anca Bonciu [3,5], Maria Dinescu [3], Anisoara Cimpean [2], Robert W. Evans [6], Valentina Dinca [3,*] and Anca Roseanu [1,*]

[1] Institute of Biochemistry, Romanian Academy, 296 Splaiul Independentei, 060031 Bucharest, Romania; radu_mada@yahoo.co.uk
[2] Faculty of Biology, University of Bucharest, Splaiul Independentei 91-95, 050095 Bucharest, Romania; anisoara.cimpean@bio.unibuc.ro
[3] National Institute for Lasers, Plasma and Radiation Physics, Atomistilor 409, 077125 Magurele, Bucharest, Romania; laurentiu.rusen@inflpr.ro (L.R.); brajnicov.simona@inflpr.ro (S.B.); anca.bonciu@inflpr.ro (A.B.); maria.dinescu@inflpr.ro (M.D.)
[4] Faculty of Mathematics and Natural Sciences, University of Craiova, 200585 Craiova, Romania
[5] Faculty of Physics, University of Bucharest, RO-077125 Magurele, Bucharest, Romania
[6] School of Engineering and Design, Brunel University, London UB8 3PH, UK; robertwevans49@gmail.com
* Corresponding: valentina.dinca@inflpr.ro (V.D.); roseanua@gmail.com (A.R.)

Received: 28 February 2019; Accepted: 2 April 2019; Published: 4 April 2019

Abstract: The improvement in the research area of the implant by surface functionalization when correlated with the biological response is of major interest in the biomedical field. Based on the fact that the inflammatory response is directly involved in the ultimate response of the implant within the body, it is essential to study the macrophage-material interactions. Within this context, we have investigated the composite material-macrophage cell interactions and the inflammatory response to these composites with amorphous hydroxyapatite (HA), Lactoferrin (Lf), and polyethylene glycol-polycaprolactone (PEG-PCL) copolymer. All materials are obtained by Matrix Assisted Pulsed Laser Evaporation (MAPLE) technique and characterized by Atomic Force Microscopy and Scanning Electron Microscopy. Macrophage-differentiated THP-1 cells proliferation and metabolic activity were assessed by qualitative and quantitative methods. The secretion of tumor necrosis factor alpha (TNF-α) and interleukin 10 (IL-10) cytokine, in the presence and absence of the inflammatory stimuli (bacterial endotoxin; lipopolysaccharide (LPS)), was measured using an ELISA assay. Our results revealed that the cellular response depended on the physical-chemical characteristics of the coatings. Copolymer-HA-Lf coatings led to low level of pro-inflammatory TNF-α, the increased level of anti-inflammatory IL-10, and the polarization of THP-1 cells towards an M2 pro-reparative phenotype in the presence of LPS. These findings could have important potential for the development of composite coatings in implant applications.

Keywords: composite coatings; MAPLE; Lactoferrin; macrophage interactions

1. Introduction

The success of a bone implant material is dependent on the host immune response. Following injury or implantation of a biomaterial, there is an infiltration of inflammatory cells at the site of the wound. Monocyte-derived macrophages [1] are among the first cells that interact and react to implanted biomaterials, playing a role in the inflammatory response and orchestration of tissue repair [2–6]. They are key cells involved in the control and modulation of the inflammatory response associated with the host tissue response to foreign bodies [7–9]. The cellular response may range from the immune stimulator to the immune suppressor depending on the polarization state of macrophages determined by environmental factors and parameters of the biomaterials [10]. Geometry, topography, hydrophobicity or surface chemistry, as well as the mechanical properties or composition of the materials, are characteristics that influence the response of macrophages [6,11–15].

The macrophage phenotype is roughly divided into two distinct populations: M1 pro-inflammatory, classically activated macrophages and M2 anti-inflammatory, alternatively-activated macrophages. The classification is done according to in vitro stimulation, released cytokines, receptors expressed on the cell surface, as well as enzyme activity [2,16–21]. Thus, the M1 type can be activated in vitro by the stimulation of lipopolysaccharide (LPS), and secretes pro-inflammatory cytokines like tumor necrosis factor alpha (TNF-α) and exhibits predominantly C-C chemokine receptor type 7, CCR7 cellular marker. The M2 phenotype associated with wound healing and tissue repair is activated in vitro by IL-4 or IL-13 cytokines, and secretes mainly IL-10 and expresses mannose receptor CD206, Cluster of Differentiation 206 [16,19,21–23].

To optimize the response of bone biomaterials and avoid excessive inflammation or implant rejection, different immunomodulation approaches have been adopted to interfere with the immune system [24–30].

An innovative strategy to increase the efficacy of a biomaterial with medical applications could be its coating with a natural protein or a biological compound. Unlike the case of injection of a bioactive compound that is usually rapidly removed from the body, depositing the protein on the surface of the implanted material causes an increase in its concentration and results in a controlled local release. This approach may also help to prevent post-surgical infection in bone implants [31] and favor a local induction of osteogenic differentiation.

Different types of biomaterials with specific characteristics are used as drug carries for efficient delivery of a biomolecule to a specific target [32]. It is worth mentioning that new strategies have emerged to implement therapies with enhanced drug accumulation in the targeted tissue area and decreased side effects [33,34]. Biomaterials, designed for hard tissue engineering and regeneration, exploit different strategies for spatially and temporarily controlled drug delivery with osteoimmunomodulatory properties for efficient osteogenic regeneration [35].

One protein of practical interest is lactoferrin (Lf), which, besides its antimicrobial and anti-carcinogenic activities, also has an anti-inflammatory function and an osteogenic role [36,37]. Lf released at the site of injury interacts with microbial elements and with cells of the immune system, affecting them at both the cellular and molecular level [38–44]. The protein also plays an important role in the activity of bone cells, as it shows anti-apoptotic and differentiation effects on osteoblasts and an inhibitory effect on osteoclastogenesis [45–47].

It was recently shown that implants coated with different hydroxyapatite (HA) forms, crystalline or amorphous, can trigger different inflammatory responses related to surface chemistry and morphology [48,49]. Moreover, HA in combination with other growth factors, proteins or polymers has been used for studying the influence on both inflammatory and osteogenic responses. Thus, different studies have been performed with Lf-HA functionalized nanocrystals [50,51] revealing this combination as a promising system with anti-inflammatory properties and increased osteogenic capacity. It has been proved that biomimetic HA nanocrystals surface-functionalized with Lf have antibacterial activity effective against Gram-positive and Gram-negative bacteria [52–54].

There are various deposition methods which can be used for coating a surface, from click chemistry to physical simple methods, such as spin-coating, or more complex procedures, such as laser deposition [55–65]. The main disadvantages for most of the techniques are the poor control over thickness and roughness, lack of possibility to include proteins or growth factors within a wide range of synthetic polymeric matrices, or the fact that the methods imply toxic precursors, causing the destruction of the bio-compounds. There are also limited options when composite materials containing organic polymers, ceramics, and proteins are required.

In recent years, Matrix Assisted Pulsed Laser Evaporation (MAPLE) has been used for depositing single component coatings constituted of simple polymers to functional Micrococcus bacteria for a wide range of applications. MAPLE technique is derived from Pulsed Laser Evaporation, except it is using a frozen target that consists of a solution of the material of interest dissolved or suspended in a solvent (final concentration maintained below 5% (*w/v*)). The choice of solvent is based on its volatility and ability to absorb the wavelengths used, without the denaturation of the material. During the deposition process, the laser beam is scanned on the target surface, the laser beam energy being absorbed by the frozen solvent, therefore vaporizing it and transferring the target molecules toward and onto the substrate placed parallel and at a distance of few cm from the target. Depending on the number of pulses and target composition, there is a deposition/growth of a thin film on the surface of the substrate, while the solvent molecules are pumped away [57–64]. Moreover, MAPLE was shown to be an appropriate approach for obtaining hybrid coatings by embedding, in a controlled way, ceramic material, graphene, nanoparticles within a natural or synthetic polymer layer [57–64].

We recently demonstrated the potential of the MAPLE technique for single-step deposition of multiple bioactive factors as an embedding process into a biodegradable synthetic polymeric thin film (PEG-PCL-Me, Co). The major advantage of the technique was the lack of influence of solvents or specific deposition conditions on the functionality of proteins or drugs [57,58]. It was demonstrated that by entrapping the osteoconductive factors, HA and Lf, within a biodegradable copolymer matrix of PEG-PCL-Me, high performances of the multifunctional biomimetic coatings, such as enhanced proliferation, differentiation, and survival of osteoblasts, were achieved [58].

As previously mentioned, the inflammatory response can dictate the final response of the implant within the body. Therefore, a good understanding of macrophage interactions with a specific substrate could bring important information on the tailoring of material surfaces that could have an impact on further applications within biomedical devices.

Within this context, our study aimed to investigate the material-macrophage cell interactions and the inflammatory response to composites containing amorphous HA, Lf, and the polyethylene glycol-polycaprolactone (PEG-PCL) copolymer.

2. Materials and Methods

2.1. Materials

Poly (ethylene glycol)-*block*-poly(ε-caprolactone) methyl ether (570303 Aldrich) (PEG-block-PCL Me -average Mn~5,000, PCL average Mn~5,000), Lf lyophilized powder (L0520 SIGMA), and HA powder (677418 Aldrich) were obtained from Sigma-Aldrich (Saint Louis, MO, USA).

2.2. Coatings Deposition and Surface Characterization

Composite coatings were obtained by the MAPLE technique and the triple module target system, as previously described [54]. Briefly, a "Surelite II" pulsed Nd:YAG laser system (Continuum Company, Pessac, France) (266 nm, 6 ns pulse duration, 10 Hz repetition rate) was used at a fluence of 450 mJ/cm^2 (for 0.02 cm^2 laser spot size measured on the target surface) to irradiate in a single step a modular target consisting of frozen solutions of PEG-block-PCL Me copolymer (Co), Lf, HA (Sigma-Aldrich, Saint Louis, MO, USA).

Taking into consideration that the reported Lf concentration in the blood circulation is normally below 1 µg/mL, but significantly increased in inflammation/injury process, even up to 70 µg/mL [66], an intermediate value of 10 µg of Lf per sample was used. In order to ensure uniform coverage of the surfaces and sufficient polymeric layer to entrap the two bioactive components, the number of pulses/sample was chosen according to the material type: 45,000 pulses for Lf, 30,000 pulses for PEG-PCL-Me, and 30,000 pulses for HA, leading to the quantities of deposited materials of 10 µg for Lf protein and 67 µg for HA. Nevertheless, for HA, the ratio Ca/P calculated according to energy dispersive X-ray analysis (EDAX) for those conditions was close to that of bone, namely 1.58. PEG_PCL Me copolymer was used due to its ability to start degrading within the first 24 h due to PEG component, as well to maintain the coating for longer periods than 1 month when deposited as thin films [58]. Moreover, based on FTIR measurements, it was shown that the functional groups of all the elements were maintained both after MAPLE process and incorporation of LF and HA within PEG-PCL copolymer matrix, which gives an important advantage for the controlled release of functional bio-components [58].

For the single component coatings, each solution was well homogenized and rapidly frozen in a liquid nitrogen-cooled copper container. The modular target system for embedding HA and Lf in the copolymer coatings included one or two Teflon rings (concentric). Each material was frozen separately and removed after freezing, so no interaction occurred between Teflon and the laser beam during experiments [58]. Glass substrates were cleaned in acetone, ethyl alcohol, and finally deionized water and dried before being placed in the deposition chamber. The target obtained was then mounted on a copper holder inside the deposition chamber and maintained in a frozen state by liquid nitrogen. In order to avoid local damages due to overheating and drilling as a result of the laser irradiation, the target was rotated during the deposition time. The distance between the substrates and the target was kept at 3 cm. The vapors of the solvents were extracted from the chamber by a vacuum pump.

The nano- and micro-topographical characteristics were analyzed by Atomic Force Microscopy AFM (XE 100 AFM setup from Park System, Suwon, Korea) and Scanning electron microscopy (SEM) (JEOL Ltd, Tokyo, Japan). Surface and roughness measurements were performed in a non-contact mode. For each coating type, 5 areas were randomly selected and measured. SEM was performed using a JSM-531 microscope (5 kV). The contact angles were measured by the sessile drop method, and the reported values were obtained upon averaging 5 measurements performed on different areas of the sample, at 60 s time interval to obtain a steady-state value.

2.3. Cell Culture Model

Human THP-1 cells (ATCC, CRL-12424) were maintained in RPMI 1640 medium with 10% (*v/v*) inactivated fetal bovine serum (FBS) and 1% (*v/v*) streptomycin/penicillin at 37 °C in a humidified atmosphere of 5% CO_2. For in vitro biological assessment, THP-1 cells were cultured on material surfaces at a density of 4×10^5 cells/surface material in 24-well plates (NUNC). Macrophages were generated from monocytic THP-1 cells by incubation for 72 h with 100 ng/mL *phorbol* 12-myristate 13-acetate (PMA). Materials with adherent THP-1-derived macrophages were moved to a new 24-well plate and incubated for 4 h (resting phase) with glutamine-free RPMI medium supplemented with 5% (*v/v*) FBS. Cells were maintained with or without 50 ng/mL *lipopolysaccharide* LPS (*Escherichia coli* 055:B5, Sigma L4524) for a further 18 h to simulate pro-inflammatory and non-inflammatory experimental conditions.

2.4. Cell Viability and Proliferation

Before all biological experiments, the coatings were sterilized by immersion in 1% Penicillin-Streptomycin solution for 15 min.

The proliferation of THP-1 cells cultured on the material surface was evaluated by the MTS assay (CellTiter 96® AQ_ueous One Solution Cell Proliferation Assay, Promega, Fitchburg, WI, USA), which is based on reduction of a tetrazolium compound (3-(4,5-dimethylthiazol-2-yl)-5-(3-carboxymethoxyphenyl)-2-(4-sulfophenyl)-2H-tetrazolium) to formazan by a dehydrogenase present in the metabolically active cells. The amount of formazan released into the culture medium is proportional to the number of live cells. Macrophage-differentiated THP-1 cells treated or not treated with LPS were incubated with MTS solution at 37 °C. After 15 min, 100 µL of supernatant was transferred to a 96-well plate, and the optical density was measured at 450 nm using a microplate reader (Mithras Berthold LB 940, Berthold Technologies, Bad Wildbad, Germany). The viability of THP-1 cells grown on biomimetic surfaces was investigated using the LIVE/DEAD viability/cytotoxicity kit (Molecular Probes, Eugene, OR, USA). Cells were incubated for 30 min with 10 µM Calcein AM (calcein AM) and 4 µM Ethidium homodimer-1 (EthD-1) in complete RPMI medium. Subsequently, the samples were fixed with 4% paraformaldehyde (PFA) for 15 min. The nuclear labeling was carried out with Hoechst dilution 1:3000 (Life Technologies, Eugene, OR, USA) for 1 min. Negative controls (cell death) were obtained by the treatment of differentiated THP-1 cells with 70% ethanol for 5 min. The samples were mounted with ProLong Gold Antifade Reagent (Molecular Probes, Life Technologies, Eugene, OR, USA) and immediately visualized with the 10× objective using the Zeiss Axiocam ERc5s ApoTome microscope (Jena, Germany) with ApoTome.2 cursor mode. Representative images were captured with the AxioVision Rel 4.8 program that controls the camera AxioCam MRm (Jena, Germany).

2.5. Cell Adhesion and Morphology

The coating effect on macrophage morphology and spreading was investigated by fluorescence microscopy following the distribution of actin and vinculin filaments. Macrophages attached to the surfaces in the presence or absence of LPS were fixed for 15 min with 4% PFA, permeabilized with 0.2% Triton X-100, blocked for one hour with 0.5% BSA-PBS (bovine serum albumin-phosphate buffered saline), and then washed with PBS. Actin filaments were stained with Alexa Fluor 488 Phalloidin (green) (Invitrogen, Thermo Fisher Sci., CA, USA) for one hour at room temperature (RT) in 0.5% BSA-PBS solution. Vinculin labeling was performed for 30 min with a mouse anti-human antibody (Sigma) diluted at 1:50 in 0.5% BSA-PBS buffer and subsequently with Alexa Fluor 594-conjugated goat anti-mouse antibody (red) (Life Technologies) dilution 1:400 in the same buffer for 30 min. The nuclei were counterstained with Hoechst fluorescent dye (blue) (1:3000 in PBS) for 1 min at room temperature. After repeated washing with PBS, the samples were mounted on a microscope slide with ProLong Gold antifade (Molecular Probes, Life Technologies), an agent that allows the fluorescence signal to be maintained over a prolonged period. Samples were examined using 20× and 40× lens of Zeiss Axiocam ERc5s ApoTome microscope with ApoTome.2 cursor mode and AxioVision 4.8 software (Zeiss).

For scanning electron microscopy, THP-1 differentiated macrophage human cells cultured on the surface of composite materials were washed with PBS and fixed with 2.5% glutaraldehyde solution in PBS for 20 min. The samples were then subjected to gradient dehydration with 70%, 90%, and 100% ethanol solutions in two rounds of 15 min for each concentration followed by two rounds of 3 min incubation with 50%, 75% hexamethyldisilazane (HDMS, in ethanol), and then 100% HDMS solution. Evaporation of the HDMS solution was carried out in a Euroclone AURA 2000 M.C. An Inspect S Electron Scanning Microscope (FEI Company, Hillsboro, OR, USA) was used to obtain the electron microscopy images.

2.6. Cytokine Secretion Profile

The level of pro- and anti-inflammatory cytokines, TNF-α and IL-10, respectively, in cell supernatants (stored at -80 °C) was determined by the sandwich enzyme-linked immunosorbent assay (ELISA) method after 18 h of incubation with or without LPS (*E. coli*) bacterial endotoxin. The commercially available ELISA assay kit for the TNF-α and IL-10 cytokines from the R&D System, (Minneapolis, MN, USA) was used following the manufacturer's specifications. Briefly, the experiments were performed in 96-well plates (MaxiSorp, Nunc, Thermo Fisher Sci., CA, USA) treated at room temperature for 24 h with anti-TNF-α and IL-10 specific monoclonal antibodies (R&D System). After repeated washing with 0.05% Tween-PBS and blocking with 1% BSA in PBS, cell supernatants were incubated for 2 h at room temperature followed by a 2 h incubation with detection antibodies hTNF-α and hIL-10 coupled with biotin. The streptavidin-HRP conjugate and the H_2O_2: TMB enzyme substrates (BD Biosciences, Becton, Dickinson and Company, East Rutherford, NJ, USA) were used to measure the cytokine level in the cell environment. The enzyme reaction was stopped with H_2SO_4 2 N, and the optical density was measured at 450 nm (monitored by extinction reading at 450 nm) on the Mithras LB 940 DLReady spectrophotometer (BERTHOLD TECHNOLOGIES GmbH & Co. KG, Wildbad, Germany). To eliminate the variation of cell density and viability on each surface, the amount of released cytokines was normalized and expressed in relation to the results obtained in the cell viability assay against the control material. The cytokine concentration was calculated using the formula: pg/mL normalized = pg/mL measured x (DO MTS control/DO MTS sample).

2.7. Macrophage Polarization on Material Surfaces

The ability of macrophages to polarize on different coatings in the presence or absence of bacterial endotoxin was investigated by fluorescent labeling of specific markers. Thus, cells were fixed for 15 min with 4% PFA, blocked for one hour with 0.5% BSA-PBS (reagent diluent), and then washed with PBS. For chemokine type 7, macrophage M1 marker and for mannose receptor, macrophage M2 marker staining, the samples were incubated with human anti-CCR7 monoclonal antibodies and human anti CD206 antibodies (R&D systems), diluted 1:50 for 30 min at room temperature in reagent diluent. Subsequently, the samples were incubated for 30 min with anti-mouse antibodies coupled with Alexa-Fluor 594 and the anti-goat antibody coupled with Alexa-Fluor 488 (Life Technologies, 1:400 dilutions). The nuclear staining was made with the Hoechst (1: 3000 dilutions in PBS solution) for 1 min at RT, and samples were mounted on a microscope slide with ProLong Gold antifade (Molecular Probes, Life Technologies). Samples were inspected with a 40x objective (Zeiss Axiocam ERc5s ApoTome microscope with ApoTome.2 cursor mode), and the representative images were captured using AxioCam MRm camera controlled by AxioVision Rel 4.8 program.

2.8. Statistical Analysis

Data were collected from triplicate samples, and the results were expressed as mean values ± SD. Significant differences with p-value < 0.05 between results were analyzed with GraphPad Prism software (version 5; La Jolla, CA, USA) using one-way ANOVA or two-way ANOVA with Bonferroni's multiple comparison tests.

3. Results and Discussion

3.1. Surface Characterization of Composite Coatings

The chemical characterization of deposited surface used in the present work was previously described [54]. However, the reduction of a number of pulses used led to a decrease of both Lf and HA within the sample which have determined changes in the coatings morphology (Figure 1). A more detailed view and changes related to the roughness of the samples are shown in Figure 2 for a single element and for composites coatings. It can be observed that the previously observed trends, based on the composition, are maintained, with some differences in the mediated roughness of the samples containing Lf and HA. The morphology of the coatings consisted, in full coverage, of the substrates, with some random island-like structures for the Lf and Co samples, while those based on HA had the accumulation of nanoparticles onto the surface. The overview of the top morphology is shown in Figure 1.

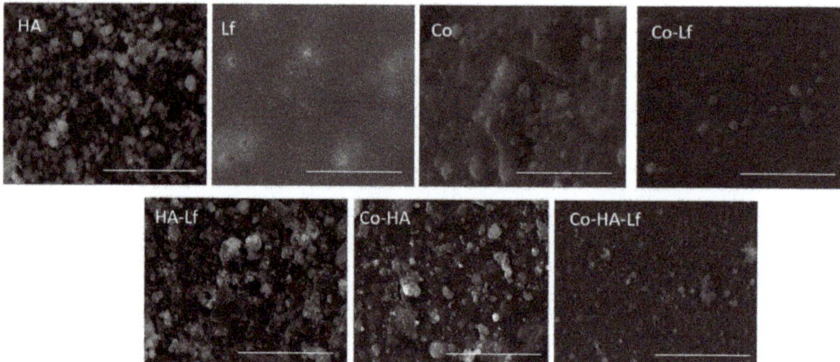

Figure 1. SEM images of the top morphology of the coatings obtained by Matrix Assisted Pulsed Laser Evaporation (MAPLE). Scale bar: 10 μm. HA: hydroxyapatite; Lf: lactoferrin.

In addition, the decrease of Lf and HA within the surface of the samples did lead to roughness and hydrophilicity changes unlike in the previously reported samples. In fact, the corresponding AFM images also showed distinct features for the composite, with the surface root mean square roughness results (obtained from the AFM measurements), revealing increased roughness, clearly indicating that surface morphology and microstructure can change depending on composition.

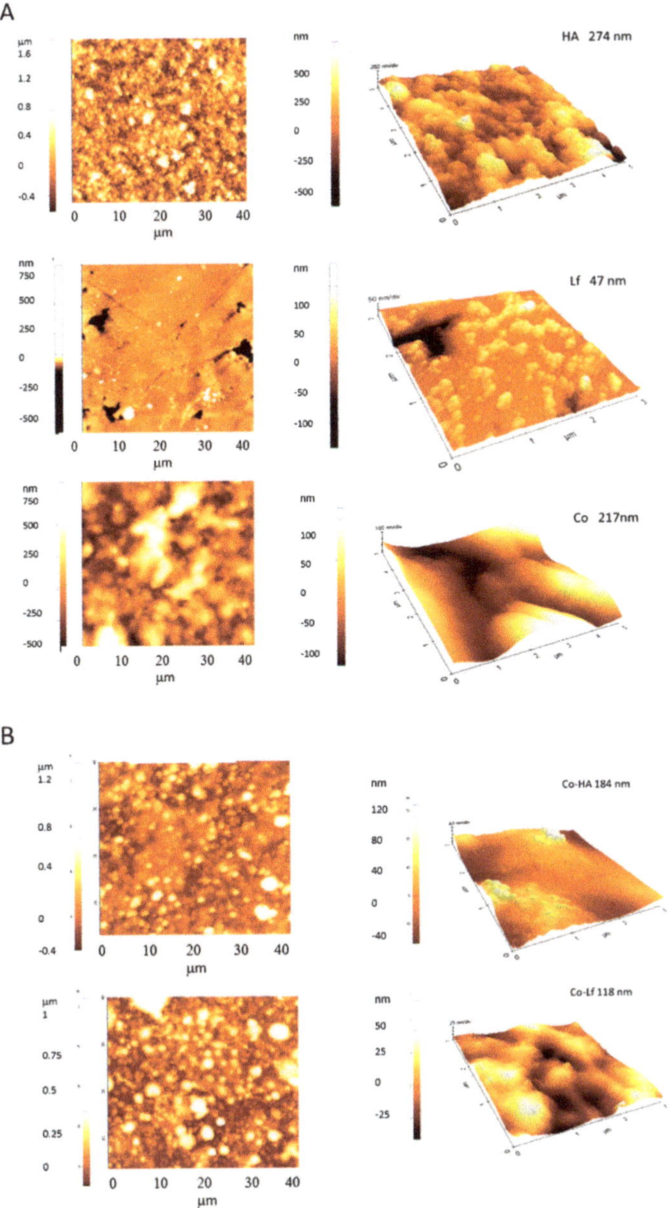

Figure 2. *Cont.*

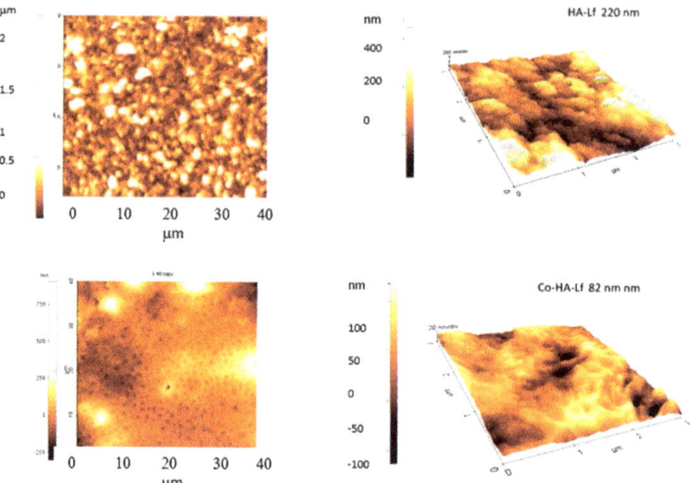

Figure 2. AFM images (40 μm × 40 μm) of single component coatings (**A**) and composite coatings (**B**). The right column depicts the detailed image of material organization for single component coatings and composite coatings with 5 μm × 5 μm and 3 μm × 3 μm, respectively. HA: hydroxyapatite; Lf: lactoferrin.

The roughness for single components ranged from 217–270 nm, except for Lf which was characterized by a 47 nm roughness, whereas by embedding Lf, HA or both within the polymeric matrix, there was a decrease of up to 118 nm for Co-Lf and only up to 220 nm for HA-Lf. The same tendency of the decrease in roughness of the coating containing the three components was observed (82 nm).

In spite of the differences observed in the surfaces roughness values, the water contact angle measurements were consistent with our previous studies, following the same trend but with slightly small differences in the measured values [58]. Thus, the hydrophilic character of the composite layers Co-Lf, Co-HA, LF-HA, and Co-Lf-HA, as well as of the LF layer, was evidenced by the measured contact angle values of 53°, 64°, 42°, 45°, and 24°, respectively. In the case of single component layer HA, the contact angle was 80°, while for the Co layer, a value of 68° was measured.

3.2. The Behavior of Human THP-1 Cells on Modified Material Surface

During inflammation, monocytes are recruited to the place of implantation where they differentiate toward macrophages. Different culture systems are used to study monocyte/macrophage interaction with biomaterials [67]. In this study, experiments were performed with THP-1 pre-monocytic human cells, a widely used cell line model for inflammation studies [66–70]. THP-1 cells were differentiated to macrophages [71], grown on modified biomimetic surfaces, and their in vitro behavior was analyzed.

3.2.1. Viability and Proliferation of Differentiated Macrophage THP-1 Cells on Modified Material Surface

The proliferation and metabolic activity of the cells attached to and differentiated on the supports were analyzed in the presence and absence of the inflammatory stimulus using the quantitative colorimetric MTS assay. The analysis reflected a directly proportional correlation between the measured absorbance and the number of viable cells. The test revealed a statistically relevant decrease ($p < 0.0001$) in viability in the presence of bacterial endotoxin after 18 h (50 ng/mL LPS) irrespective of the type of surface (Figure 3A). In the absence of lipopolysaccharide, all surfaces analyzed allow cell attachment, with the highest metabolic activity being recorded for cells incubated on the Lf-coated surface. Significantly elevated levels compared to control (uncovered surfaces) were found for THP-1 differentiated cells grown on HA-Lf, Co, Co-HA, Co-HA-Lf coatings ($p < 0.0001$). In the case of LPS treatment, a decrease in metabolic activity compared with the untreated cell was observed.

The viability of the cells attached on the analyzed surfaces was similar, except for those on the Lf and Co-coated surfaces ($p < 0.0001$) and HA-Lf and Co-HA ($p < 0.05$), which promoted survival and an increased cellular proliferation compared to control.

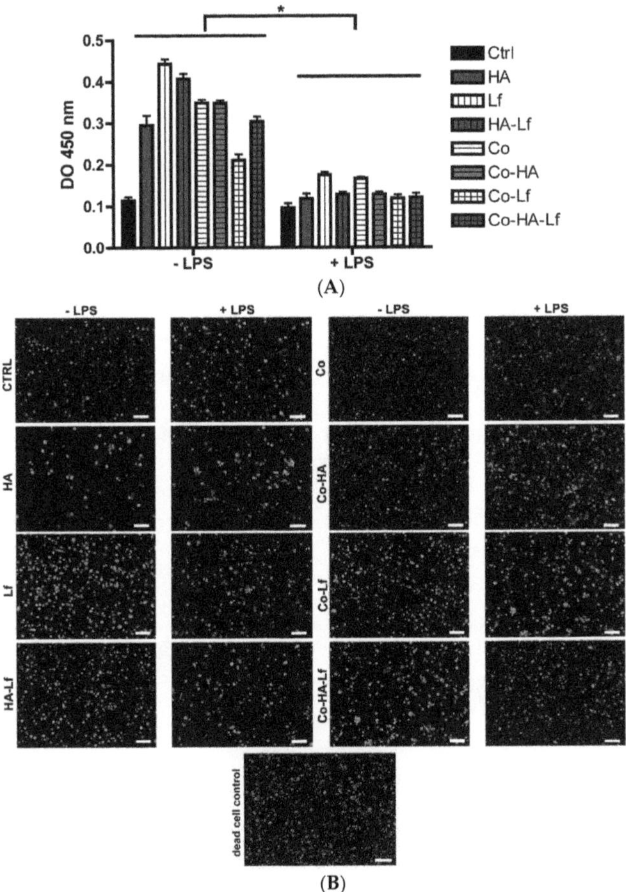

Figure 3. (**A**) Viability of THP-1 cells attached to surfaces coated with complex hybrid biomimetic components (Co, hydroxyapatite (HA), and lactoferrin (Lf)) in the absence or presence of LPS. Data are presented as mean values ± SD, and significance was determined at * $p < 0.05$; (**B**) Fluorescence microscopy images of macrophage-differentiated THP-1 cells on biomimetic modified surfaces (LIVE-DEAD method). The live cells are marked by green fluorescent calcein, and dead cells are labeled with the ethidium-1 homodimer in red. Images of fluorescence microscopy of THP-1 cells adhered for 72 h on the surface of biomaterials and treated or untreated with lipopolysaccharide (LPS) (10×). Scale bar 100 μm.

The LIVE/DEAD assay was subsequently carried out to further verify cell viability using Calcein AM/ethidium-1 homodimer dyes. The method is based on the simultaneous measurement of two parameters, the activity of intracellular esterase and the integrity of the plasma membrane. The ethidium homodimer penetrates the dead cells and binds to the DNA, thus serving as a control for cell death (red cells), while non-fluorescence Calcein AM penetrates through the live cell membrane and, by enzymatic conversion, is transformed into fluorescent calcein, marking the cells green. The results obtained by microscopic investigation of the cells were consistent with those obtained by the MTS method. Fluorescence microscopy images (Figure 3B) showed a decrease in cell number and an increase in cell death after the treatment with LPS, for almost all types of coatings. The negative effect of LPS stimulation on THP-1 cells survival has also been reported by others [72,73].

3.2.2. Adhesion and Morphology of THP-1 Cells on Modified Material Surface

The cells attach to the surfaces of the materials through a variety of micro extensions, such as filopodia or lamellipodia. In the particular case of macrophages, the interaction with different surfaces involves dynamic cellular adhesion structures [74–76]. Podosomes have an important role in the adhesion and degradation of the cell matrix, as well as in cellular motility [76,77]; these formations being dependent on nature and the characteristics of the substrate. In this study, cellular interaction and morphology were highlighted by fluorescence labeling of actin (green) and vinculin (red) filaments, two proteins of the cellular cytoskeleton involved in cell adhesion. Morphological examination of differentiated macrophage THP-1 cells under standard conditions (without LPS stimulation) indicated a predominantly rounded morphology with smooth edges (Figure 4A) with defined actin and vinculin filaments appearing at the cell peripheral. In the case of Lf-coated surfaces and of those with single or complex Co-coatings, the presence of podosomes suggested an adaptation of morphology to the substrate contact areas. When stimulated with LPS, THP-1 cells showed an increase in cell surface contact with the material. The morphology of the cell was changed and mixed morphologies, spherical, elongated or enlarged surfaces with irregular contours, could be observed (Figure 4B). This morphological behavior might be associated with a cellular activation to the pro-inflammatory phenotype (M1). The cells expressed filopodia and podosome structures that are involved in adhesion and cell motility and might be an adaptation to an external stimulus (LPS), morphology, and surface composition, as it is well known that topographical cues and surface chemistry can influence cellular spreading and the formation of focal adhesion [12,20,78].

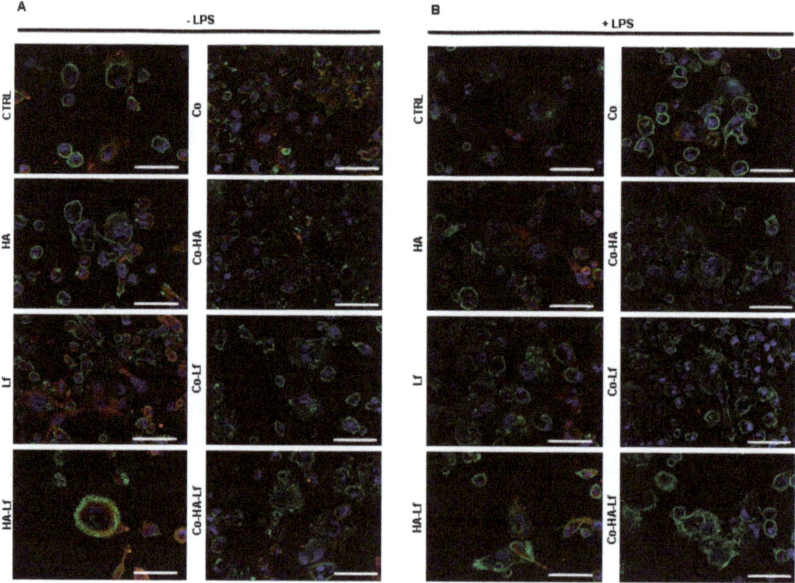

Figure 4. Adhesion and distribution of differentiated THP-1 cells on materials covered with biomimetically modified surfaces in the absence of (**A**) or in the presence of (**B**) lipopolysaccharide (LPS). Representative images of the attached cells 40× obtained by fluorescence microscopy by marking actin (green), vinculin (red), and nucleus (blue). Scale bar is 50 μm. HA: hydroxyapatite; Lf: lactoferrin.

SEM images (Figure 5) taken after macrophage differentiation revealed the presence of spherical cells in the absence of LPS and mixed morphology, an enhancement of cell spread and cell protrusions that allow cell-substrate interactions, in the case of LPS stimulation.

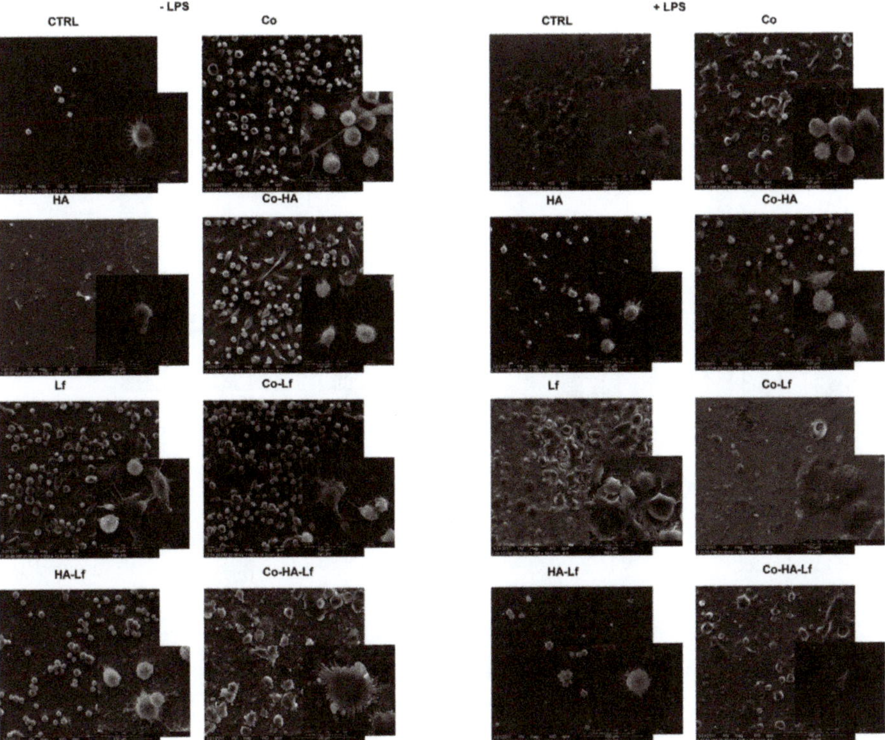

Figure 5. SEM images of THP-1-differentiated macrophages attached to the surface of biomimetically modified surfaces in the absence and presence of bacterial endotoxin. Scale bar 100 µm for 1000× images and 20 µm for the inserted image with 5000× magnification order. HA: hydroxyapatite; Lf: lactoferrin.

3.2.3. Inflammatory Response of THP-1 Cells Grown on Modified Material Surface

To assess the pro-(TNF-α) and anti-inflammatory (IL-10) cytokine secretion, experiments were performed, both under standard culture and inflammatory simulation conditions (LPS treatment), using an ELISA method. The cytokine secretion profile was detected after 18 h, in the absence or presence of the inflammatory stimulus.

TNF-α and IL-10 are two important cytokines, with an immunoregulatory role, expressed by macrophages; TNF-α is associated with the macrophage pro-inflammatory M1 phenotype, while IL-10 cytokine is associated with the M2 anti-inflammatory phenotype [16,21].

No detectable levels of cytokines released by THP-1 cells regardless of the type of surfaces were recorded, suggesting the absence of macrophage activation without LPS treatment. In contrast, treatment with endotoxin led to an increased secretion level of pro-inflammatory and anti-inflammatory cytokines (Figure 6). These findings indicate different cellular activation depending on the physical-chemical characteristics of the coating. The release of TNF-α from macrophages cultured on different surfaces revealed that Lf alone triggered a reduced release of pro-inflammatory cytokine compared to control. However, when the HA component is combined with Lf, the level of cytokine was elevated. The level of TNF-α was also increased, but to a lesser degree, when HA or Lf alone were incorporated into the polymeric matrix (Co). Lf addition within the polymeric matrix together with the HA component led to a significant reduction ($p < 0.05$) in the release of TNF-α under inflammatory conditions compared to Lf-HA. The capacity of Lf to affect the TNF-α release could be explained by either its ability to directly bind endotoxin, and thus blocking LPS interaction with macrophages, or its capacity to enter macrophages and inhibit pro-inflammatory cytokine production [72,79–81]. In the case of Co-HA-Lf, possible copolymer degradation, previously reported [58], could lead to a reduced level of pro-inflammatory cytokine secreted by macrophages, due to the controlled release of the biological component of the coating.

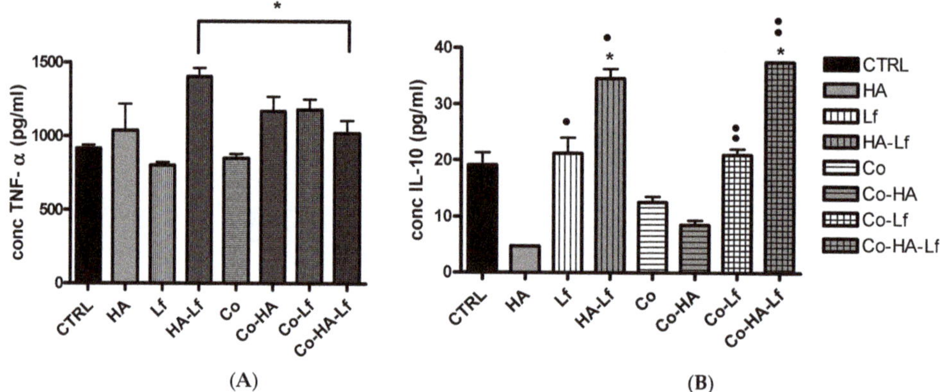

Figure 6. Secreted levels of tumor necrosis factor alpha (TNF-α) (**A**) and interleukin 10 (IL-10) cytokines (**B**) normalized to cell activity in lipopolysaccharide (LPS) treated condition. Data are presented as mean values ± SD, and significance was determined at * $p < 0.05$ vs. CTRL, • $p < 0.05$ vs. HA •• $p < 0.05$ vs. Co-HA. HA: hydroxyapatite; Lf: lactoferrin.

The level of anti-inflammatory IL-10 cytokine released from THP-1 cells seeded on HA-Lf and Co-HA-Lf coatings was notably higher than that released from macrophages cultured on the HA, Co-HA, and Co coatings. Lf and Co-Lf surfaces also induced an increased release of IL-10 from macrophages but to a lesser degree compared to coatings with combined bioactive components. Coatings without Lf seemed to induce the lowest level of anti-inflammatory cytokine release from cells. This behavior could be related to the well-known immune modulatory, anti-inflammatory activity of Lf [82], which counteracts the pro-inflammatory state of macrophages induced by LPS. Also, the increased release of IL-10 from THP-1 cells cultured on HA-Lf and Co-HA-Lf coatings compared with the (rest of) other surfaces could be partially explained by the hydrophobic/hydrophilic character of the coatings. Thus, the IL-10 level measured by ELISA after 18 h of incubation with endotoxin was found to increase with the increase in surface hydrophilicity.

Similarly, Nocerino et al. [54] observed that Lf-HA nanocomposites had an immunomodulatory activity on THP-1 cells stimulated with LPS, decreased the pro-inflammatory cytokines levels, and increased the secretion of anti-inflammatory cytokines.

Knowing that a prolonged release of pro-inflammatory TNF-α cytokine is associated with an inflammatory environment, an osteoclast activation, and alteration of implant integration [83,84], Lf is an important bioactive component that maintains a low inflammatory profile important for proper cellular activation and osteogenic proliferation and differentiation.

Taken together our results suggest that Co-HA-Lf coatings determine a cytokine secretion profile associated with a tissue regeneration-favorable immune response, in agreement with previous data [58], which clearly revealed that the incorporation of HA and Lf into polymeric coating enhanced osteoblastic differentiation.

3.2.4. Surface Effect on Macrophage Polarization

Multiple properties of biomaterial, including physical (topography and roughness) modifications or surface chemistry, can influence immune response and play a significant role in macrophage polarization towards inflammatory M1 or pro-regenerative M2 phenotypes [85–87]. This behavior of macrophages has a direct impact on bone regeneration and biological performance of the biomaterial [88–90]. Different surfaces with diverse chemistry and characteristics control macrophage reaction and phenotype shift [15,91–93], with molecular events and cellular signaling pathways involved in macrophage activation [94]. The cell response differs from the biomaterial surface and in vitro cell models used in the investigation [6,67].

Macrophage polarization induced by surface chemistry and topography correlates with changes in the profile response of cytokine [86]. For studying further the effect of hybrid surfaces on macrophage polarization to the M1 pro-inflammatory phenotype or M2 anti-inflammatory phenotype, the samples were incubated with antibodies specific for membrane markers of classical and alternative activated macrophages. In this case, the presence of a pro-inflammatory phenotypic marker, CCR7, and the anti-inflammatory phenotype specific marker, CD206 [17,95–97], was followed. LPS treatment is a classic polarization protocol to the pro-inflammatory phenotype, but the presence of lactoferrin on the analyzed surfaces, a protein with well-known anti-inflammatory effects, may cause a decrease in the endotoxin effect. Immunomarking of macrophages untreated with inflammatory stimuli revealed a particular activation. The cells displayed characteristics of both M1 and M2 polarization states, especially on coatings with Lf. The cells presented both CCR7 (macrophage M1) and CD206 (macrophage M2) markers on the membrane surface, suggesting that they might be in a state of variable continuing activation between the M1 and M2 state creating a heterogeneous population of macrophages. Representative fluorescent images of the THP-1 cell cultured on different coatings are shown in Figure 7.

This type of behavior was observed as well by Zhang et al. [50,98], who showed that material-activated macrophages do not polarize specifically into M1 or M2 phenotypes but rather are activated into a mixed phenotype with both M1/M2 characteristics.

During the inflammatory process, the macrophages suffer complex activation, expressing the features of both M1 and M2 phenotypes. This activation could explain the simultaneous expression of M1/M2 specific markers, the secretion of both pro and anti-inflammatory cytokines, as well as the presence of mixed cell morphologies on composite surfaces. In the case of endotoxin stimulation, it seemed that Lf counteracted the inflammatory effect induced by LPS, while the cells cultured on coatings with Lf alone or in combination with HA or copolymer displayed an enhanced anti-inflammatory M2 phenotype expression marker CD206 on the cell surface. The results could be correlated with the ability of Lf to modulate the polarization of macrophages through its anti-inflammatory properties [82] and also with the hydrophilic character of the hybrid coating [14]. In addition, M1-to-M2 transition on hybrid coatings of THP-1 cells stimulated with LPS could also be explained by the interference of the hydroxyapatite component [99,100] and the polymeric coating, which might be involved in a controlled release of the anti-inflammatory component [58].

Besides chemical-structural characteristics of our materials with anti-inflammatory effects (Lf, HA), the roughness could have also an impact on macrophage polarization. Generally, surface roughness maintained in the nanometer scale induces a decrease in M1 development of macrophages and an increase in M2 phenotype [101]. Our films generated by each condition exhibited a submicron roughness, with lowest values for coatings containing Lf alone or embedded within the polymeric matrix. These films exhibited increased levels of anti-inflammatory IL-10 and the increased polarization of THP-1 cells towards an M2 phenotype.

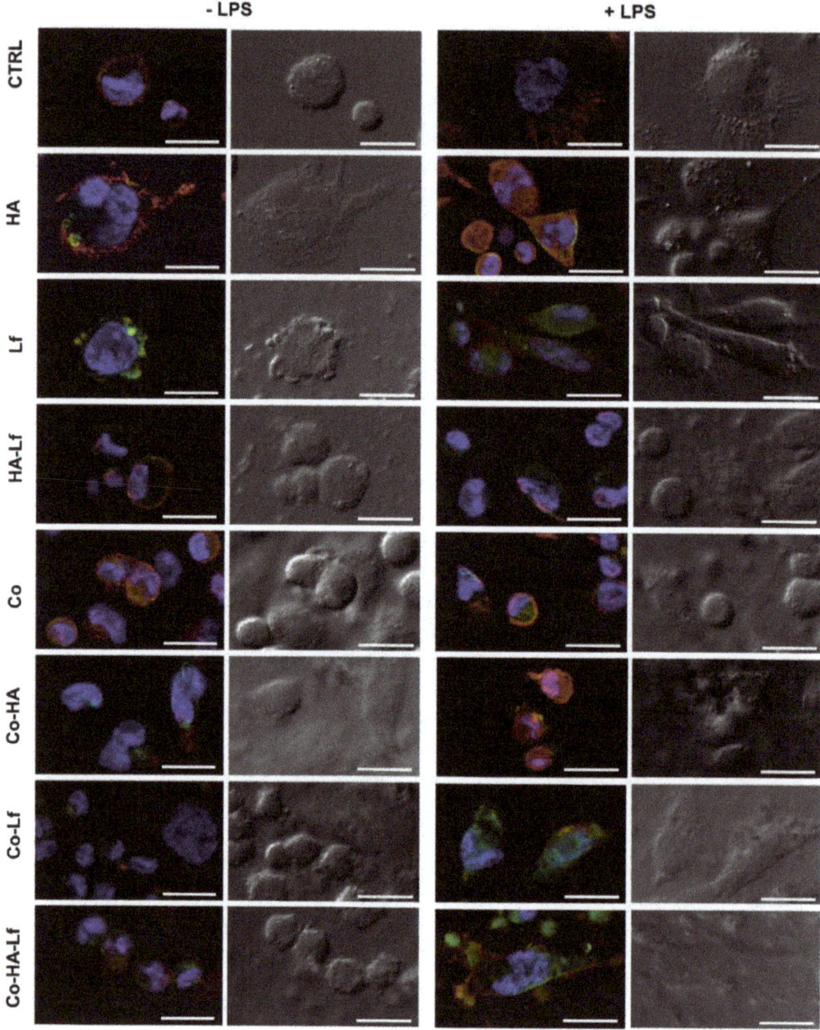

Figure 7. The polarization of differentiated THP-1 cells on biomimetic coatings. Immunofluorescence images (40×) of CCR7 M1 markers (red), CD206 M2 (green) on the cell surface, and nucleus (blue) in the absence (left) and in the presence of (right) lipopolysaccharide (LPS) (Columns 1 and 3). Associated images of optical microscopy (Columns 2 and 4). Scale bar 20 µm. HA: hydroxyapatite; Lf: lactoferrin.

4. Conclusions

The results revealed that the use of a Copolymer-HA-Lf composite could be a promising approach to the creation of coatings for bone implant materials. In the presence of an LPS stimulus, surfaces exhibited low levels of pro-inflammatory TNF-α, increased levels of anti-inflammatory IL-10, and the increased polarization of THP-1 cells towards an M2 pro-reparative phenotype. Controlled release of the components by polymeric coating allowed the combination of the properties of the two biological compounds, namely the osteogenic capacity and bone mechanical stability of HA with the anti-inflammatory and osteogenic effect of the Lf component, making the composition excellent support for bone regeneration.

Author Contributions: Conceptualization, M.I., A.R., and V.D.; Methodology, M.I., L.R., S.B., A.B., A.R., and V.D.; Validation, M.I., L.R.; Formal Analysis, M.I., V.D.; Investigation, M.I., L.R., S.B., A.B., and V.D.; Resources, V.D., M.I., and A.R.; Writing—Original Draft Preparation, M.I., V.D.; Writing—Review and Editing, A.R., R.W.E., M.D., and A.C.; Supervision, A.R., R.W.E., and A.C.

Funding: This research received funding from Romanian National Authority for Scientific Research (CNCS–UEFISCDI), under the projects TERAMED 63PCCDI/2018, PN-III-P1-1.2-PCCDI-2017-072.and Nucleu 16N/08.02.2019, Structural and Functional Proteomics Research Program of the Institute of Biochemistry of the Romanian Academy, and by the University of Bucharest-Biology Doctoral School.

Conflicts of Interest: The authors declare no conflict of interest.

References

1. Gordon, S.; Taylor, P.R. Monocyte and macrophage heterogeneity. *Nat. Rev. Immunol.* **2005**, *5*, 953–964. [CrossRef] [PubMed]
2. Gordon, S.; Martinez, F.O. Alternative Activation of Macrophages: Mechanism and Functions. *Immunity* **2010**, *32*, 593–604. [CrossRef] [PubMed]
3. Koh, T.J.; DiPietro, L.A. Inflammation and wound healing: the role of the macrophage. *Expert. Rev. Mol. Med.* **2011**, *13*, e23. [CrossRef]
4. Sica, A.; Mantovani, A. Macrophage plasticity and polarization: in vivo veritas. *J. Clin. Invest.* **2012**, *122*, 787–795. [CrossRef] [PubMed]
5. Gaffney, L.; Warren, P.; Wrona, E.A.; Fisher, M.B.; Freytes, D.O. Macrophages' Role in Tissue Disease and Regeneration. *Results Probl. Cell. D* **2017**, *62*, 245–271.
6. Rayahin, J.E.; Gemeinhart, R.A. Activation of Macrophages in Response to Biomaterials. *Results Probl. Cell. D* **2017**, *62*, 317–351.
7. Luttikhuizen, D.T.; Harmsen, M.C.; Van Luyn, M.J.A. Cellular and molecular dynamics in the foreign body reaction. *Tissue. Eng.* **2006**, *12*, 1955–1970. [CrossRef] [PubMed]
8. Anderson, J.M.; Rodriguez, A.; Chang, D.T. Foreign body reaction to biomaterials. *Semin. Immunol.* **2008**, *20*, 86–100. [CrossRef] [PubMed]
9. Franz, S.; Rammelt, S.; Scharnweber, D.; Simon, J.C. Immune responses to implants–A review of the implications for the design of immunomodulatory biomaterials. *Biomaterials* **2011**, *32*, 6692–6709. [CrossRef]
10. Ogle, M.E.; Segar, C.E.; Sridhar, S.; Botchwey, E.A. Monocytes and macrophages in tissue repair: Implications for immunoregenerative biomaterial design. *Exp. Biol. Med.* **2016**, *241*, 1084–1097. [CrossRef] [PubMed]
11. Bartneck, M.; Schulte, V.A.; Paul, N.E.; Diez, M.; Lensen, M.C.; Zwadlo-Klarwasser, G. Induction of specific macrophage subtypes by defined micro-patterned structures. *Acta Biomaterialia* **2010**, *6*, 3864–3872. [CrossRef] [PubMed]
12. McWhorter, F.Y.; Wang, T.T.; Nguyen, P.; Chung, T.; Liu, W.F. Modulation of macrophage phenotype by cell shape. *P Natl. Acad. Sci. USA* **2013**, *110*, 17253–17258. [CrossRef]
13. Singh, S.; Awuah, D.; Rostam, H.M.; Emes, R.D.; Kandola, N.K.; Onion, D.; Htwe, S.S.; Rajchagool, B.; Cha, B.H.; Kim, D.; et al. Unbiased Analysis of the Impact of Micropatterned Biomaterials on Macrophage Behavior Provides Insights beyond Predefined Polarization States. *Acs. Biomater. Sci. Eng.* **2017**, *3*, 969–978. [CrossRef]
14. Lv, L.; Xie, Y.T.; Li, K.; Hu, T.; Lu, X.; Cao, Y.Z.; Zheng, X.B. Unveiling the Mechanism of Surface Hydrophilicity-Modulated Macrophage Polarization. *Adv. Healthc. Mater.* **2018**, *7*, 1800675. [CrossRef] [PubMed]
15. Hotchkiss, K.M.; Reddy, G.B.; Hyzy, S.L.; Schwartz, Z.; Boyan, B.D.; Olivares-Navarrete, R. Titanium surface characteristics, including topography and wettability, alter macrophage activation. *Acta. Biomaterialia* **2016**, *31*, 425–434. [CrossRef] [PubMed]
16. Mosser, D.M.; Edwards, J.P. Exploring the full spectrum of macrophage activation. *Nat. Rev. Immunol.* **2008**, *8*, 958–969. [CrossRef] [PubMed]
17. Martinez, F.O.; Helming, L.; Gordon, S. Alternative Activation of Macrophages: An Immunologic Functional Perspective. *Annu. Rev. Immunol.* **2009**, *27*, 451–483. [CrossRef]
18. Brown, B.N.; Goodman, S.B.; Amar, S.; Badylak, S.F. Macrophage polarization: an opportunity for improved outcomes in biomaterials and regenerative medicine. *Biomaterials* **2012**, *33*, 3792–3802. [CrossRef]

19. Brown, B.N.; Badylak, S.F. Expanded applications, shifting paradigms and an improved understanding of host–biomaterial interactions. *Acta. Biomaterialia* **2013**, *9*, 4948–4955. [CrossRef]

20. Lee, H.S.; Stachelek, S.J.; Tomczyk, N.; Finley, M.J.; Composto, R.J.; Eckmann, D.M. Correlating macrophage morphology and cytokine production resulting from biomaterial contact. *J. Biomed. Mater. Res. A* **2013**, *101*, 203–212. [CrossRef]

21. Murray, P.J.; Allen, J.E.; Biswas, S.K.; Fisher, E.A.; Gilroy, D.W.; Goerdt, S.; Gordon, S.; Hamilton, J.A.; Ivashkiv, L.B.; Lawrence, T.; et al. Macrophage Activation and Polarization: Nomenclature and Experimental Guidelines. *Immunity* **2014**, *41*, 339–340. [CrossRef]

22. Martinez, F.O.; Gordon, S. The M1 and M2 paradigm of macrophage activation: Time for reassessment. *F1000Prime Rep.* **2014**, *6*, 13. [CrossRef]

23. Italiani, P.; Boraschi, D. From monocytes to M1/M2 macrophages: Phenotypical vs. functional differentiation. *Front. Immunol.* **2014**. [CrossRef]

24. He, X.-T.; Li, X.; Xia, Y.; Yin, Y.; Wu, R.-X.; Sun, H.-H.; Chen, F.-M. Building capacity for macrophage modulation and stem cell recruitment in high-stiffness hydrogels for complex periodontal regeneration: Experimental studies in vitro and in rats. *Acta. Biomater.* **2019**, *88*, 162–180. [CrossRef]

25. Wu, R.X.; Xu, X.Y.; Wang, J.; He, X.T.; Sun, H.H.; Chen, F.M. Biomaterials for endogenous regenerative medicine: Coaxing stem cell homing and beyond. *Appl. Mater. Today* **2018**, *11*, 144–165. [CrossRef]

26. Yu, Y.; Wu, R.X.; Yin, Y.; Chen, F.M. Directing immunomodulation using biomaterials for endogenous regeneration. *J. Mater. Chem. B* **2016**, *4*, 569–584. [CrossRef]

27. Lee, J.; Byun, H.; Madhurakkat Perikamana, S.K.; Lee, S.; Shin, H. Current Advances in Immunomodulatory Biomaterials for Bone Regeneration. *Adv. Healthc. Mater.* **2018**, *8*, 1801106. [CrossRef]

28. Chen, Z.T.; Klein, T.; Murray, R.Z.; Crawford, R.; Chang, J.; Wu, C.T.; Xiao, Y. Osteoimmunomodulation for the development of advanced bone biomaterials. *Mater. Today* **2016**, *19*, 304–321. [CrossRef]

29. Zhou, G.Y.; Groth, T. Host Responses to Biomaterials and Anti-Inflammatory Design–A Brief Review. *Macromol. Biosci.* **2018**, *18*, 1800112. [CrossRef]

30. Andorko, J.I.; Jewell, C.M. Designing biomaterials with immunomodulatory properties for tissue engineering and regenerative medicine. *Bioeng. Transl. Med.* **2017**, *2*, 139–155. [CrossRef]

31. Nagano-Takebe, F.; Miyakawa, H.; Nakazawa, F.; Endo, K. Inhibition of initial bacterial adhesion on titanium surfaces by lactoferrin coating. *Biointerphases* **2014**, *9*, 029006. [CrossRef]

32. Etheridge, M.L.; Campbell, S.A.; Erdman, A.G.; Haynes, C.L.; Wolf, S.M.; McCullough, J. The big picture on nanomedicine: the state of investigational and approved nanomedicine products. *Nanomed. Nanotechnol. Biol. Med.* **2013**, *9*, 1–14. [CrossRef] [PubMed]

33. Tang, H.; Zhao, W.; Yu, J.; Li, Y.; Zhao, C. Recent Development of pH-Responsive Polymers for Cancer Nanomedicine. *Molecules* **2018**, *24*, 4. [CrossRef] [PubMed]

34. Grossen, P.; Witzigmann, D.; Sieber, S.; Huwyler, J. PEG-PCL-based nanomedicines: A biodegradable drug delivery system and its application. *J. Control. Release* **2017**, *260*, 46–60. [CrossRef] [PubMed]

35. Zhang, K.; Wang, S.; Zhou, C.; Cheng, L.; Gao, X.; Xie, X.; Sun, J.; Wang, H.; Weir, M.D.; Reynolds, M.A.; et al. Advanced smart biomaterials and constructs for hard tissue engineering and regeneration. *Bone Res.* **2018**, *6*, 31. [CrossRef] [PubMed]

36. Moreno-Exposito, L.; Illescas-Montes, R.; Melguizo-Rodriguez, L.; Ruiz, C.; Ramos-Torrecillas, J.; de Luna-Bertos, E. Multifunctional capacity and therapeutic potential of lactoferrin. *Life Sci.* **2018**, *195*, 61–64. [CrossRef] [PubMed]

37. Garcia-Montoya, I.A.; Cendon, T.S.; Arevalo-Gallegos, S.; Rascon-Cruz, Q. Lactoferrin a multiple bioactive protein: An overview. *Bba-Gen Subjects* **2012**, *1820*, 226–236. [CrossRef]

38. Legrand, D.; Elass, E.; Pierce, A.; Mazurier, J. Lactoferrin and host defence: An overview of its immuno-modulating and anti-inflammatory properties. *Biometals* **2004**, *17*, 225–229. [CrossRef]

39. Legrand, D.; Elass, E.; Carpentier, M.; Mazurier, J. Lactoferrin: A modulator of immune and inflammatory responses. *Cell. Mol. Life Sci.* **2005**, *62*, 2549. [CrossRef]

40. Legrand, D.; Elass, E.; Carpentier, M.; Mazurier, J. Interactions of lactoferrin with cells involved in immune function. *Biochem. Cell Biol.* **2006**, *84*, 282–290. [CrossRef]

41. Suzuki, Y.A.; Lopez, V.; Lonnerdal, B. Mammalian lactoferrin receptors: Structure and function. *Cell. Mol. Life Sci.* **2005**, *62*, 2560–2575. [CrossRef] [PubMed]

42. Legrand, D.; Mazurier, J. A critical review of the roles of host lactoferrin in immunity. *BioMetals* **2010**, *23*, 365–376. [CrossRef] [PubMed]

43. Legrand, D. Lactoferrin, a key molecule in immune and inflammatory processes. *Biochem. Cell Biol.* **2011**, *90*, 252–268. [CrossRef] [PubMed]

44. Legrand, D. Overview of Lactoferrin as a Natural Immune Modulator. *J. Pediatrics* **2016**, *173*, S10–S15. [CrossRef] [PubMed]

45. Cornish, J.; Callon, K.E.; Naot, D.; Palmano, K.P.; Banovic, T.; Bava, U.; Watson, M.; Lin, J.M.; Tong, P.C.; Chen, Q.; et al. Lactoferrin is a potent regulator of bone cell activity and increases bone formation in vivo. *Endocrinology* **2004**, *145*, 4366–4374. [CrossRef]

46. Cornish, J.; Palmano, K.; Callon, K.E.; Watson, M.; Lin, J.M.; Valenti, P.; Naot, D.; Grey, A.B.; Reid, I.R. Lactoferrin and bone; structure–activity relationships. *Biochem. Cell Biol.* **2006**, *84*, 297–302. [CrossRef]

47. Cornish, J.; Naot, D. Lactoferrin as an effector molecule in the skeleton. *Biometals* **2010**, *23*, 425–430. [CrossRef]

48. Rydén, L.; Omar, O.; Johansson, A.; Jimbo, R.; Palmquist, A.; Thomsen, P. Inflammatory cell response to ultra-thin amorphous and crystalline hydroxyapatite surfaces. *J. Mater. Sci.: Mater. Med.* **2016**, *28*, 9.

49. Mestres, G.; Espanol, M.; Xia, W.; Persson, C.; Ginebra, M.P.; Ott, M.K. Inflammatory Response to Nano- and Microstructured Hydroxyapatite. *Plos One* **2015**, *10*, e0120381. [CrossRef]

50. Zhang, Y.; Cheng, X.; Jansen, J.A.; Yang, F.; van den Beucken, J.J.J.P. Titanium surfaces characteristics modulate macrophage polarization. *Mat. Sci. Eng. C–Mater.* **2019**, *95*, 143–151. [CrossRef]

51. Montesi, M.; Panseri, S.; Iafisco, M.; Adamiano, A.; Tampieri, A. Coupling Hydroxyapatite Nanocrystals with Lactoferrin as a Promising Strategy to Fine Regulate Bone Homeostasis. *PloS ONE* **2015**, *10*, e0132633. [CrossRef] [PubMed]

52. Montesi, M.; Panseri, S.; Iafisco, M.; Adamiano, A.; Tampieri, A. Effect of hydroxyapatite nanocrystals functionalized with lactoferrin in osteogenic differentiation of mesenchymal stem cells. *J. Biomed. Mater. Res. A* **2015**, *103*, 224–234. [CrossRef] [PubMed]

53. Fulgione, A.; Nocerino, N.; Iannaccone, M.; Roperto, S.; Capuano, F.; Roveri, N.; Lelli, M.; Crasto, A.; Calogero, A.; Pilloni, A.P.; et al. Lactoferrin Adsorbed onto Biomimetic Hydroxyapatite Nanocrystals Controlling—In Vivo—the Helicobacter pylori Infection. *Plos ONE* **2016**, *11*, e0158646. [CrossRef]

54. Nocerino, N.; Fulgione, A.; Iannaccone, M.; Tomasetta, L.; Ianniello, F.; Martora, F.; Lelli, M.; Roveri, N.; Capuano, F.; Capparelli, R. Biological activity of lactoferrin- functionalized biomimetic hydroxyapatite nanocrystals. *Int. J. Nanomed.* **2014**, *9*, 1175–1184.

55. Bhushan, B.; Schricker, S.R. A review of block copolymer-based biomaterials that control protein and cell interactions. *J. Biomed. Mater. Res. A* **2014**, *102*, 2467–2480. [CrossRef] [PubMed]

56. Fan, Y.Q.; Li, X.; Yang, R.J. The Surface Modification Methods for Constructing Polymer-Coated Stents. *Int. J. Polym. Sci.* **2018**. [CrossRef]

57. Dinca, V.; Florian, P.E.; Sima, L.E.; Rusen, L.; Constantinescu, C.; Evans, R.W.; Dinescu, M.; Roseanu, A. MAPLE-based method to obtain biodegradable hybrid polymeric thin films with embedded antitumoral agents. *Biomed. Microdevices* **2014**, *16*, 11–21. [CrossRef]

58. Rusen, L.; Brajnicov, S.; Neacsu, P.; Marascu, V.; Bonciu, A.; Dinescu, M.; Dinca, V.; Cimpean, A. Novel degradable biointerfacing nanocomposite coatings for modulating the osteoblast response. *Surf. Coat. Tech.* **2017**, *325*, 397–409. [CrossRef]

59. Dinca, V.; Viespe, C.; Brajnicov, S.; Constantinoiu, I.; Moldovan, A.; Bonciu, A.; Toader, C.N.; Ginghina, R.E.; Grigoriu, N.; Dinescu, M.; et al. MAPLE Assembled Acetylcholinesterase-Polyethylenimine Hybrid and Multilayered Interfaces for Toxic Gases Detection. *Sensors* **2018**, *18*, 4265. [CrossRef]

60. Pique, A.; Chrisey, D.B.; Spargo, B.J.; Bucaro, M.A.; Vachet, R.W.; Callahan, J.H.; McGill, R.A.; Leonhardt, D.; Mlsna, T.E. Use of matrix assisted pulsed laser evaporation (MAPLE) for the growth of organic thin films. *Mater. Res. Soc. Symp. P* **1998**, *526*, 421–426. [CrossRef]

61. Sima, F.; Davidson, P.; Pauthe, E.; Sima, L.E.; Gallet, O.; Mihailescu, I.N.; Anselme, K. Fibronectin layers by matrix-assisted pulsed laser evaporation from saline buffer-based cryogenic targets. *Acta. Biomater.* **2011**, *7*, 3780–3788. [CrossRef]

62. Caricato, A.P.; Luches, A. Applications of the matrix-assisted pulsed laser evaporation method for the deposition of organic, biological and nanoparticle thin films: A review. *Appl. Phys. Mater.* **2011**, *105*, 565–582. [CrossRef]

63. McGill, R.A.; Chrisey, D.B.; Pique, A.; Mlsna, T.E. Matrix-assisted pulsed-laser evaporation (MAPLE) of functionalized polymers: Applications with chemical sensors. *SPIE* **1998**, *3274*, 12.

64. Constantinescu, C.; Matei, A.; Ionita, I.; Ion, V.; Marascu, V.; Dinescu, M.; Vasiliu, I.C.; Emandi, A. Azo-derivatives thin films grown by matrix-assisted pulsed laser evaporation for nonlinear optical applications. *Appl. Surf. Sci.* **2014**, *302*, 69–73. [CrossRef]

65. Ion, V.; Matei, A.; Constantinescu, C.; Ionita, I.; Marinescu, A.; Dinescu, M.; Emandi, A. Octahydroacridine thin films grown by matrix-assisted pulsed laser evaporation for non linear optical applications. *Mat. Sci. Semicon. Proc.* **2015**, *36*, 78. [CrossRef]

66. Ward, P.P.; Uribe-Luna, S.; Conneely, O.M. Lactoferrin and host defense. *Biochem Cell Biol* **2002**, *80*, 95–102. [CrossRef]

67. Boersema, G.S.A.; Grotenhuis, N.; Bayon, Y.; Lange, J.F.; Bastiaansen-Jenniskens, Y.M. The Effect of Biomaterials Used for Tissue Regeneration Purposes on Polarization of Macrophages. *Bioresearch Open Acc* **2016**, *5*, 6–14. [CrossRef]

68. Auwerx, J. The Human Leukemia-Cell Line, Thp-1–A Multifaceted Model for the Study of Monocyte-Macrophage Differentiation. *Experientia* **1991**, *47*, 22–31. [CrossRef]

69. Park, E.K.; Jung, H.S.; Yang, H.I.; Yoo, M.C.; Kim, C.; Kim, K.S. Optimized THP-1 differentiation is required for the detection of responses to weak stimuli. *Inflamm. Res.* **2007**, *56*, 45–50. [CrossRef]

70. Chanput, W.; Mes, J.J.; Savelkoul HF, J.; Wichers, H.J. Characterization of polarized THP-1 macrophages and polarizing ability of LPS and food compounds. *Food Funct.* **2013**, *4*, 266–276. [CrossRef]

71. Lund, M.E.; To, J.; O'Brien, B.A.; Donnelly, S. The choice of phorbol 12-myristate 13-acetate differentiation protocol influences the response of THP-1 macrophages to a pro-inflammatory stimulus. *J. Immunol. Methods* **2016**, *430*, 64–70. [CrossRef]

72. Choe, Y.H.; Lee, S.W. Effect of lactoferrin on the production of tumor necrosis factor-alpha and nitric oxide. *J. Cell Biochem.* **2000**, *76*, 30–36. [CrossRef]

73. Starr, T.; Bauler, T.J.; Malik-Kale, P.; Steele-Mortimer, O. The phorbol 12-myristate-13-acetate differentiation protocol is critical to the interaction of THP-1 macrophages with Salmonella Typhimurium. *PloS ONE* **2018**, *13*, e0193601. [CrossRef]

74. Linder, S.; Aepfelbacher, M. Podosomes: Adhesion hot-spots of invasive cells. *Trends Cell Biol.* **2003**, *13*, 376–385. [CrossRef]

75. Linder, S.; Wiesner, C. Tools of the trade: Podosomes as multipurpose organelles of monocytic cells. *Cell. Mol. Life Sci.* **2015**, *72*, 121–135. [CrossRef]

76. Gimona, M.; Buccione, R.; Courtneidge, S.A.; Linder, S. Assembly and biological role of podosomes and invadopodia. *Curr. Opin. Cell Biol.* **2008**, *20*, 235–241. [CrossRef]

77. Linder, S. The matrix corroded: Podosomes and invadopodia in extracellular matrix degradation. *Trends. Cell Biol.* **2007**, *17*, 107–117. [CrossRef]

78. Rammal, H.; Bour, C.; Dubus, M.; Entz, L.; Aubert, L.; Gangloff, S.C.; Audonnet, S.; Bercu, N.B.; Boulmedais, F.; Mauprivez, C.; et al. Combining Calcium Phosphates with Polysaccharides: A Bone-Inspired Material Modulating Monocyte/Macrophage Early Inflammatory Response. *Int. J. Mol. Sci.* **2018**, *19*, 3458. [CrossRef]

79. Puddu, P.; Latorre, D.; Valenti, P.; Gessani, S. Immunoregulatory role of lactoferrin-lipopolysaccharide interactions. *Biometals* **2010**, *23*, 387–397. [CrossRef]

80. Elass-Rochard, E.; Legrand, D.; Salmon, V.; Roseanu, A.; Trif, M.; Tobias, P.S.; Mazurier, J.; Spik, G. Lactoferrin inhibits the endotoxin interaction with CD14 by competition with the lipopolysaccharide-binding protein. *Infect. Immun.* **1998**, *66*, 486–491.

81. Haversen, L.; Ohlsson, B.G.; Hahn-Zoric, M.; Hanson, L.A.; Mattsby-Baltzer, I. Lactoferrin down-regulates the LPS-induced cytokine production in monocytic cells via NF-kappa B. *Cell Immunol.* **2002**, *220*, 83–95. [CrossRef]

82. Cutone, A.; Rosa, L.; Lepanto, M.S.; Scotti, M.J.; Berlutti, F.; di Patti, M.C.B.; Musci, G.; Valenti, P. Lactoferrin Efficiently Counteracts the Inflammation-Induced Changes of the Iron Homeostasis System in Macrophages. *Front. Immunol.* **2017**, *8*, 705. [CrossRef]

83. Boyce, B.E.; Li, P.; Yao, Z.; Zhang, Q.; Badell, I.R.; Schwarz, E.M.; Keefe, R.J.; Xing, L. TNFα and pathologic bone resorption. *Keio J. Med.* **2005**, *54*, 127–131. [CrossRef]

84. Zhao, B.H.; Grimes, S.N.; Li, S.S.; Hu, X.Y.; Ivashkiv, L.B. TNF-induced osteoclastogenesis and inflammatory bone resorption are inhibited by transcription factor RBP-J. *J. Exp. Med.* **2012**, *209*, 319–334. [CrossRef]

85. Rostam, H.M.; Singh, S.; Salazar, F.; Magennis, P.; Hook, A.; Singh, T.; Vrana, N.E.; Alexander, M.R.; Ghaemmaghami, A.M. The impact of surface chemistry modification on macrophage polarisation. *Immunobiology* **2016**, *221*, 1237–1246. [CrossRef]

86. Sridharan, R.; Cameron, A.R.; Kelly, D.J.; Kearney, C.J.; O'Brien, F.J. Biomaterial based modulation of macrophage polarization: A review and suggested design principles. *Mater. Today* **2015**, *18*, 313–325. [CrossRef]

87. Luu, T.U.; Gott, S.C.; Woo BW, K.; Rao, M.P.; Liu, W.F. Micro- and Nanopatterned Topographical Cues for Regulating Macrophage Cell Shape and Phenotype. *ACS Appl. Mater. Interfaces* **2015**, *7*, 28665–28672. [CrossRef]

88. Wang, J.; Meng, F.; Song, W.; Jin, J.; Ma, Q.; Fei, D.; Fang, L.; Chen, L.; Wang, Q.; Zhang, Y. Nanostructured titanium regulates osseointegration via influencing macrophage polarization in the osteogenic environment. *Int. J. Nanomed.* **2018**, *13*, 4029–4043. [CrossRef]

89. Nassiri, S.; Graney, P.; Spiller, K.L. Manipulation of Macrophages to Enhance Bone Repair and Regeneration. In *A Tissue Regeneration Approach to Bone and Cartilage Repair*; Zreiqat, H., Dunstan, C.R., Rosen, V., Eds.; Springer International Publishing: Cham, Switzerland, 2015; pp. 65–84.

90. Ma, Q.-L.; Zhao, L.-Z.; Liu, R.-R.; Jin, B.-Q.; Song, W.; Wang, Y.; Zhang, Y.-S.; Chen, L.-H.; Zhang, Y.-M. Improved implant osseointegration of a nanostructured titanium surface via mediation of macrophage polarization. *Biomaterials* **2014**, *35*, 9853–9867. [CrossRef]

91. Chung, L.; Maestas, D.R.; Housseau, F.; Elisseeff, J.H. Key players in the immune response to biomaterial scaffolds for regenerative medicine. *Adv. Drug. Deliv. Rev.* **2017**, *114*, 184–192. [CrossRef]

92. Fernandes, K.R.; Zhang, Y.; Magri, A.M.P.; Renno, A.C.M.; van den Beucken, J.J.J.P. Biomaterial Property Effects on Platelets and Macrophages: An *in Vitro* Study. *ACS Biomater. Sci. Eng.* **2017**, *3*, 3318–3327. [CrossRef]

93. Chen, Z.; Bachhuka, A.; Han, S.; Wei, F.; Lu, S.; Visalakshan, R.M.; Vasilev, K.; Xiao, Y. Tuning Chemistry and Topography of Nanoengineered Surfaces to Manipulate Immune Response for Bone Regeneration Applications. *ACS Nano.* **2017**, *11*, 4494–4506. [CrossRef]

94. Moore, L.B.; Kyriakides, T.R. Immune Responses to Biosurfaces. In *Molecular Characterization of Macrophage-Biomaterial Interactions*; Lambris, J.D., Ekdahl, K.N., Ricklin, D., Nilsson, B., Eds.; Springer International Publishing: Cham, Switzerland, 2015; pp. 109–122.

95. Van Ginderachter, J.A.; Movahedi, K.; Ghassabeh, G.H.; Meerschaut, S.; Beschin, A.; Raes, G.; De Baetselier, P. Classical and alternative activation of mononuclear phagocytes: Picking the best of both worlds for tumor promotion. *Immunobiology* **2006**, *211*, 487–501. [CrossRef]

96. Joshi, A.D.; Oak, S.R.; Hartigan, A.J.; Finn, W.G.; Kunkel, S.L.; Duffy, K.E.; Das, A.; Hogaboam, C.M. Interleukin-33 contributes to both M1 and M2 chemokine marker expression in human macrophages. *Bmc. Immunol* **2010**, *11*, 52. [CrossRef]

97. Miao, X.C.; Wang, D.H.; Xu, L.Y.; Wang, J.; Zeng, D.L.; Lin, S.X.; Huang, C.; Liu, X.Y.; Jiang, X.Q. The response of human osteoblasts, epithelial cells, fibroblasts, macrophages and oral bacteria to nanostructured titanium surfaces: a systematic study. *Int. J. Nanomed.* **2017**, *12*, 1415–1430. [CrossRef]

98. Zhang, Y.; Bose, T.; Unger, R.E.; Jansen, J.A.; Kirkpatrick, C.J.; van den Beucken, J.J.J.P. Macrophage type modulates osteogenic differentiation of adipose tissue MSCs. *Cell Tissue. Res.* **2017**, *369*, 273–286. [CrossRef]

99. Linares, J.; Fernandez, A.B.; Feito, M.J.; Matesanz, M.C.; Sanchez-Salcedo, S.; Arcos, D.; Vallet-Regi, M.; Rojo, J.M.; Portoles, M.T. Effects of nanocrystalline hydroxyapatites on macrophage polarization. *J. Mater. Chem. B* **2016**, *4*, 1951–1959. [CrossRef]

100. Alhamdi, J.R.; Peng, T.; Al-Naggar, I.M.; Hawley, K.L.; Spiller, K.L.; Kuhn, L.T. Controlled M1-to-M2 transition of aged macrophages by calcium phosphate coatings. *Biomaterials* **2019**, *196*, 90–99. [CrossRef]

101. Li, X.; Huang, Q.; Elkhooly, T.A.; Liu, Y.; Wu, H.; Feng, Q.; Liu, L.; Fang, Y.; Zhu, W.; Hu, T. Effects of titanium surface roughness on the mediation of osteogenesis via modulating the immune response of macrophages. *Biomed. Mater.* **2018**, *13*, 045013. [CrossRef]

© 2019 by the authors. Licensee MDPI, Basel, Switzerland. This article is an open access article distributed under the terms and conditions of the Creative Commons Attribution (CC BY) license (http://creativecommons.org/licenses/by/4.0/).

 *coatings*

Review

Current Status on Pulsed Laser Deposition of Coatings from Animal-Origin Calcium Phosphate Sources

Liviu Duta [1,*] and Andrei C. Popescu [2,*]

[1]   Lasers Department, National Institute for Lasers, Plasma, and Radiation Physics, 409 Atomistilor Street, 077125 Magurele, Romania

[2]   Center for Advanced Laser Technologies (CETAL), National Institute for Lasers, Plasma and Radiation Physics, 077125 Magurele, Romania

*   Correspondence: liviu.duta@inflpr.ro (L.D.); andrei.popescu@inflpr.ro (A.C.P.)

Received: 30 April 2019; Accepted: 22 May 2019; Published: 24 May 2019

Abstract: The aim of this paper is to present the current status on animal-origin hydroxyapatite (HA) coatings synthesized by Pulsed Laser Deposition (PLD) technique for medical implant applications. PLD as a thin film synthesis method, although limited in terms of surface covered area, still gathers interest among researchers due to its advantages such as stoichiometric transfer, thickness control, film adherence, and relatively simple experimental set-up. While animal-origin HA synthesized by bacteria or extracted from animal bones, eggshells, and clams was tested in the form of thin films or scaffolds as a bioactive agent before, the reported results on PLD coatings from HA materials extracted from natural sources were not gathered and compared until the present study. Since natural apatite contains trace elements and new functional groups, such as CO_3^{2-} and HPO_4^{2-} in its complex molecules, physical-chemical results on the transfer of animal-origin HA by PLD are extremely interesting due to the stoichiometric transfer possibilities of this technique. The points of interest of this paper are the origin of HA from various sustainable resources, the extraction methods employed, the supplemental functional groups, and ions present in animal-origin HA targets and coatings as compared to synthetic HA, the coatings' morphology function of the type of HA, and the structure and crystalline status after deposition (where properties were superior to synthetic HA), and the influence of various dopants on these properties. The most interesting studies published in the last decade in scientific literature were compared and morphological, elemental, structural, and mechanical data were compiled and interpreted. The biological response of different types of animal-origin apatites on a variety of cell types was qualitatively assessed by comparing MTS assay data of various studies, where the testing conditions were possible. Antibacterial and antifungal activity of some doped animal-origin HA coatings was also discussed.

Keywords: animal-origin calcium phosphate coatings; natural hydroxyapatite; doping; high adherence; pulsed laser deposition technique; biomimetic applications

---

1. Introduction

In the last few decades, the field of bone tissue engineering has been widely studied and expanded for addressing bone-related traumas. By combining biomaterials and cells for bone tissue ingrowth, an efficient and viable alternative to allografts or autografts could be delivered.

Bioactive materials represent a vast bioengineering research field with tremendous interest for the production of durable implants and bone substitutes able to bypass rejection difficulties.

Calcium phosphates (CaP) are bio-ceramic materials used especially for orthopedic and dental medical applications [1,2]. Synthetic hydroxyapatite (HA), with the complex chemical formula

$Ca_{10}(PO_4)_6(OH)_2$, is the most well-known CaP material, and is frequently used in implantology due to its close chemical composition and crystallographic structure resemblance with the mineral phase of vertebrate bones (50% in mass and 70% in volume) [3–5]. We stress upon that the mineral constituent of vertebrate skeletal systems mainly consists of a calcium-deficient HA, doped with various ions [6]. According to the biomimetic approach, a material designed to repair the skeletal system must be similar to the biological one in terms of composition, stoichiometry, the crystallinity degree, morphology, and functionality. It was reported that HA can promote new bone in-growth through the osteoconduction mechanism without eliciting local or systemic toxicity, inflammation, or foreign body response [7–9]. When an HA-based ceramic is implanted, a fibrous tissue-free layer containing carbonated apatite forms on its surfaces and contributes to the implant bonding to the bone. This results in earlier implant stabilization and superior fixation of the implant to the surrounding tissues [8–10]. Moreover, the development of microbial biofilms onto the surface of medical devices or human tissues represents a worrying health problem, which can lead to a high diversity of biofilm-associated infections, with increasing incidence [11]. Therefore, the antimicrobial properties of HA are also of key importance [12,13]. Taking into account all these demands, one could, therefore, explain the large interest for synthesis and deposition of apatites enriched with biologically-active ions or molecules, as well as more resorbable (soluble) CaPs.

Despite its excellent bone regeneration properties, HA has also some important disadvantages: HA-based ceramics are very brittle in bulk [14] and are characterized by poor mechanical properties, especially in liquid media. Therefore, HA-based materials cannot be used in bulk for orthopedic devices, which must withstand the application of high loads during their lifetime [15]. To overcome these drawbacks, HA can be applied as a coating onto the surface of metallic or polymeric implants, which aim to significantly improve implants' overall performances, by successfully combining the excellent bioactivity of the ceramic with the mechanical advantages of the substrate implants [14,16].

Various techniques have been and are continuously developed to obtain HA. In this respect, they can be categorized in two main paths of producing HA: (*i*) the first one implies the use of chemical routes, and (*ii*) the second one involves extracting it naturally, from biogenic, mammalian, or fish bone sustainable, low-cost resources (further denoted as BioHA).

For depositing CaP coatings onto metallic implants, the industrial technique of choice is plasma-spraying, due to the synthesis speed, large area of deposition, and work in an ambient atmosphere [17]. However, HA coatings produced using this technique are prone to cracking and delamination and, because of high-processing temperatures, could contain residual decomposition phases. In this respect, current interests are quickly advancing toward two focused research directions: (*i*) increasing the biomimicry of HA-based coatings with respect to the composition and structure of bone apatite, and (*ii*) improving or even discovering alternative deposition techniques, which can allow for the achievement of novel HA (doped) coatings with increased mechanical and biological characteristics.

When compared to other physical vapor deposition techniques, i.e. thermal evaporation or sputtering, the pulsed laser deposition (PLD) technique stands as a simple, versatile, rapid, and cost-effective method, which can enable precise control of thickness and morphology for the fabrication of high-quality films [18,19]. The main advantage of the PLD technique applied for HA-based bio-ceramics is represented by its capacity to grow stoichiometric films with a controlled degree of crystallinity and thickness. By this method, one can assure the flexibility to control the morphology, phase, crystallinity, and chemical composition of obtained CaPs. These characteristics have a special influence over bio-resorption or dissolution, which are directly involved in the process of films' osseointegration.

Taking into account all these aspects, the aim of the current review is to emphasize the advantages of using animal-origin HA coatings as viable substitutes of synthetic HA ones, which are fabricated by the PLD technique for implantable applications. Conclusions will be drawn, future perspectives will be advanced, and a series of recommendations will be highlighted.

2. Review of Literature

A digital search was performed using Web of Science (http://apps.webofknowledge.com), following the criteria described below.

2.1. Inclusion Criteria

1. Articles written in English. 2. Use of animal-origin HA materials for physical-chemical, in vitro and in vivo studies. 3. Pulsed laser deposition as the technique used for the synthesis of animal-origin HA coatings. 4. Articles written starting from the year 2001.

2.2. Exclusion Criteria

1. Non-English articles. 2. Nanoparticles. 3. Other deposition techniques (such as Plasma spraying, Magnetron sputtering, etc.) used for the synthesis of animal-origin HA coatings.

After the review process, a total of 49 articles met the inclusion criteria and were further assessed in detail and parts of the reported results were considered and discussed in this review.

3. BioHA *vs.* Synthetic HA

Various methods have been used to obtain HA due to its attractive biological properties and resemblance to the mineral part of the bone. There are two main categories of methods used to produce HA.

(*i*) The first one, which is generally utilized due its reliability, implies the use of inorganic synthesis by different methods, such as hydrothermal [20,21], co-precipitation [22,23], or sol-gel [24]. However, these approaches are, on one hand, polluting and time-consuming, and, on the other hand, quite expensive [25,26]. Moreover, the resulting synthetic HA, which is a stoichiometric material with Ca/P ratio of 1.67, does not completely match the chemical composition of bone [27].

(*ii*) Thus, researchers managed to find alternative, low-cost methods to produce HA, such as obtaining it from biogenic, mammalian, or natural fish bone (sustainable/renewable) sources. BioHA is a carbonated, non-stoichiometric Ca-deficient material with a reduced degree of crystallinity. Therefore, it differs from synthetic HA in terms of composition, stoichiometry, crystal size/morphology, crystallinity degree, degradation rate, and overall biological performance. One should also note that there might exist some concerns about the use of natural HA because of the potential risk of dangerous diseases transmission, when the material is not well-prepared [27]. However, when all health security issues are handled [28,29], then HA derived directly from sustainable, low production cost materials (such as animal bones) has a composition that closely matches the morphological and structural architecture of the inorganic components of the human bone. In fact, it has been reported that HA obtained from biowaste such as eggshells, bovine bones, fish-scales, and fish bones can lead to overall properties and behavior comparable or even better than synthetic ones due to the similarities of bone apatites [2,30,31]. However, in order to completely benefit from the previously mentioned advantages against synthetic HA, BioHA should be extracted in a controlled manner, i.e., from resources for which one knows the exact alimentation and way of life of animals, since this could directly influence the overall quality of the final product, HA.

Functional groups from biological apatites can be substituted with trace amounts of cations (Na^+, Mg^{2+}, and K^+), anions (F^-, Cl^-, SiO_4^{4-} and CO_3^{2-}), or, in some cases, by both [2,32,33], either adsorbed on the surface of the crystal or incorporated in the lattice structure [34–37]. Specifically, in the apatite structure, the carbonate ions can replace either the hydroxyl or the phosphate ion sites, which leads to a type-A or type-B structure [38]. If these substitutions take place simultaneously, a type-AB substitution occurs, as in the case of the bone mineral [32,39,40]. One should note that the mineral bone substitutions with various trace elements are considered directly responsible for the modifications in crystallinity, solubility, and biological response [41] and, therefore, they play a key role in the performance of hard tissues and the overall osseointegration process. As a consequence, alternative methods used to modify the HA structure by incorporating different ions to improve the osteoconductive properties of synthetic HA is currently of interest to the scientific community [2].

Apatite crystals produced in the biological system are different in many ways from the crystals obtained using synthetic precursors. The apatite crystals grown in the living system bear smaller crystallite size. Therefore, they have a large surface area, which further allows them to absorb an extra number of ions. In short, biological minerals tend to attain an organized structure in a very short time [42].

Table 1 introduces the sample codes, which will be further used in the text.

Table 1. Sample acronyms related to different materials used throughout the review and their explanation.

Sample Code	Sample Description
Ti	Titanium (control specimen or deposition substrate)
HA_{syn}	Synthetic hydroxyapatite
DHA	Dentine hydroxyapatite
BHA	Bovine hydroxyapatite
SHA	Ovine (sheep) hydroxyapatite
BHA:Li	Bovine hydroxyapatite doped with Li_2O
BHA:CIG	Bovine hydroxyapatite doped with commercial inert glass
SHA:Ti	Ovine (sheep) hydroxyapatite doped with titanium
BHA:MgF	Bovine hydroxyapatite doped with MgF_2
BHA:MgO	Bovine hydroxyapatite doped with MgO
BHA:LiC	Bovine hydroxyapatite doped with Li_2CO_3
BHA:LiP	Bovine hydroxyapatite doped with Li_3PO_4

The use of BioHA comes from the ambition and continuous efforts of researchers to attain biomimetism. While the chemically-synthesized HA is similar to the mineral part of the bone, it lacks trace elements and functional groups that modify the chemical formula of the natural HA in bone. Because of the fact that it is too laborious and probably impossible to chemically synthesize HA with the right amounts of trace elements and functional groups, an easy method is to isolate the biological HA from bones of various animals, to transform it into powder, and to press it into pellets that will be further used as PLD targets. Mammalian bones contain a higher source of ions and trace elements [43–46], with Mg^{2+} and Na^+ being the most frequently found ones. One notes that the presence of Na^+ and Mg^{2+} ions alongside HA play an important role in the development of teeth and bone, whereas their absence could cause bone loss and fragility [2]. Other trace elements such as K^+, Sr^{2+}, Zn^{2+}, Si^{4+}, Ba^{2+}, F^-, and Al^- have been identified. However, the overall concentration of these elements can vary and this is due to differences in the diet of the animals [47].

BioHA mainly differentiates from chemically synthesized HA by the presence of carbonate ions. The $Ca_{10}(PO_4)_6(OH)_2$ complex formula of pure HA changes into:

$$Ca_{10-x}(PO_4)_{6-x}(CO_3)_x(OH)_{2-x}$$

where $0 \leq x \leq 2$ [48].

Another possibility can be that in $(PO_4)^{3-}$ sites to find hydrogen phosphate $(HPO_4)^{2-}$ ions [49]. Winand et al. [50] proposed a general formula for hydrogen phosphate-containing apatites:

$$Ca_{10-x}(PO_4)_{6-x}(HPO_4)_x(OH)_{2-x}$$

where $0 \leq x \leq 2$.

However, it is highly possible that BioHA can concomitantly have substitutions with CO_3^{2-} and HPO_4^{2-} groups. Hence, Combes et al. [37] advanced the following formula:

$$Ca_{10-x}(PO_4)_{6-x}(CO_3 \text{ or } HPO_4)_x(OH \text{ or } 1/2CO_3)^{2-x}$$

where $0 \leq x \leq 2$.

It should be mentioned that various studies showed a deviation of OH^- groups in the composition of bone-extracted apatites as compared to the standard HA formula. A possible explanation could be that PO_4^{3-} groups are reactive with water molecules trapped in the lattice or from the surrounding medium, which results in the formation of hydrogen phosphate:

$$H_2O + PO_4^{3-} \rightarrow HPO_4^{2-} + OH^-.$$

Trace elements in the composition of BioHA materials are commonly determined qualitatively by spectroscopic techniques. It should be noted that there are chemical methods that involve dissolution of the mineral part of the bone and identification of constituents, but they are susceptible to post-decomposition chemical reactions. For metallic ions such as Na, K, Al, and Sr, it is believed that they do not influence the apatite properties, as long as they remain in the normal concentrations known for the human body. However, by accumulation following diet intake, some elements' increased content might cause biological reactions such as bone weakening or induction of osteoporosis.

A trace element with a clear effect on bones is fluorine, which forms in bone apatites fluoride ions. In addition, 99% of fluorine in most organisms is accumulated in bones in the form of fluorides (as e.g. NaF) [51]. In the human body, the fluorine content is of 0.05% to 0.1%. Its presence favorizes an increase of bone density, but, in high content, causes a disease called fluorosis, in which bones become hard and brittle. Studies suggest that fluoride also has anti-plaque properties. It was reported that amine fluoride and stannous fluoride possess bactericidal properties against oral bacteria [52].

Dopants may have an important role in the structure and properties of BioHA coatings. For the crystallites size function of dopant materials in the case of BioHA coatings synthesized by PLD, the reader is guided to consult Table 2. Duta et al. [25] observed that a percentage of 1.5 wt.% Ti in composition of PLD targets from SHA causes synthesis of layers with a reduced degree of crystallinity. Deposited SHA coatings without Ti display a crystallinity degree of ~30%, while, for the ones doped with Ti, deposited in the same experimental conditions, the crystallinity degree dropped to 7% only.

Table 2. Crystalline coherence length in the case of BioHA coatings synthesized by PLD, function of doping materials. Determinations were performed along the *c*-axis [(002) crystal plane] and *a*-axis [(300) crystal plane] by applying the Scherrer Equation.

Sample Material	D002 (nm)	D300 (nm)	D002/D300	Reference
HA$_{syn}$	59.2	47.4	1.25	[53]
SHA	50.1	48.5	1.03	[25]
BHA	152.8	103.8	~1.47	[54]
SHA:Ti	41.6	32.1	1.30	[25]
BHA:Li	83.5	56	1.49	[53]
BHA:CIG	48.9	23.4	2.1	
BHA:MgF	100.0	89.6	~1.11	[54]
BHA:MgO	169.7	141.2	~1.20	

BHA was doped with various elements in order to identify their role on the crystallinity of the obtained coatings. Thus, it was shown that by inserting 1 wt.% lithium into the BHA target, a film with a higher degree of crystallinity could be obtained, as compared to undoped BHA structures. A mix of 10% commercial bioinert glass and 90% BHA could produce a film consisting mainly of HA with small crystallites (48.9 nm in length vs. 83.5 nm for BHA:Li and 71 nm for BHA coatings) and a high degree of anisotropy (D002/D300 = 2.1 vs. 1.49 in the case of BHA:Li and 1.34 in the case of BHA coatings) [53].

In FTIR investigations, bands characteristic to carbonate groups were identified both in targets and films made of BHA. The lines peaking at (1419, 1457, and 1544 cm^{-1}, respectively) were characteristic to C–O asymmetric stretching and were present in both spectra of targets and films.

In the case of bovine HA doped with MgF_2 (BHA:MgF) structures [54], the dopant indirectly induced the presence of MgO traces caused during sintering of the targets by partial oxidation of MgF_2. The amount of MgF_2 in targets was of 2% and the amount transformed in MgO was estimated to be of 0.4 wt.%. However, in films, peaks of MgF_2 were not identified, but only supplemental weak MgO peaks were observed. This can hint toward the possible structural substitution of the Mg^{2+} cations (in the Ca^{2+} sites) and F^- anions (in the OH^- group sites) into the HA lattice during the deposition process. By incorporating MgF_2 in the BHA structure, the HA crystallites' length was reduced from 152 to 100 nm, while the ratio between length and width, D300/D002, dropped from 1.47 to 1.11. Mihailescu et al. [54] also inserted 5 wt.% MgO in BHA targets and observed that, in this case, crystallites length increased from 152 to 170 nm. The D300/D002 dropped from 1.47 to 1.20.

One notes that, in the case of MgF_2 doping, FTIR investigations revealed the disappearance of the $(OH)^-$ bands from the synthesized coatings. Moreover, supplemental vibration bands have been recorded at ~876 to 873 cm^{-1}, which is ascribed to the overlapping of the characteristic peaks of $(HPO_4)^{2-}$ and $(CO3)^{2-}$ groups.

4. Preparation of Materials

Powder Preparation

It is important to mention that all the experimental procedures used for the fabrication of animal-origin HA materials are conducted in accordance with the European Regulations [28] and ISO 22442-1 standard [29].

4.0.1. Extraction of HA from Mammalian Bones

BioHA, is usually extracted from bones and teeth of various animals. To obtain the raw materials, cortical bones originating from bovines or swine are the choices generally reported in various studies tackling the extraction of natural/biological HA [2,41]. Two possible explanations for this preference might be advanced: (*i*) the immediate availability of cow and sheep bones in commercial shops and abattoirs to the difference of more exotic sources such as marine animals, and (*ii*) the extracellular matrix of bovine bones is mainly composed of HA nanocrystals and collagen fibers. The bones represent 11% of pork, 15% of beef, and 16% of sheep carcasses, respectively. Due to human consumption of these animals' meat, billions of tons of such by-products are generated every year. Moreover, millions of tons of eggshells are thrown out as waste material yearly. Part of these by-products are repurposed as ingredients for pet food, fertilizers, or in pharmaceutic and cosmetic industries. Their applicability can be increased to the biomedical sector, where bones can be used as grafts, scaffolds, or as raw materials for the production of HA for coating of implants [2,55].

BioHA powders are usually obtained from the cortical part of the femoral bones of freshly slaughtered animals (received from slaughter houses, which use the other animal parts for general consumption) [11]. Bones are generally delivered on ice to laboratories and, before use, they are submitted to a veterinary control. HA derived from animal bones is generally prepared following a three-step procedure: (*i*) mechanical cleaning of soft tissue, (*ii*) deproteinization in the alkali media, and (*iii*) calcination at temperatures sufficiently high to remove any remnant organic and biological hazardous components. Additionally, all bones are thoroughly washed and/or boiled with distilled water. After this preliminary step of washing and drying, the bones are cut off, gently checked for any remnant ligaments and soft tissues and further processed. Next, bone marrows are extracted, and all other unwanted soft tissue residues or macroscopic adhering impurities and substances are removed from shafts. Then, the shafts are cut off into slices, cleaned, and washed with distilled water. Cleaned parts are deproteinized in either sodium hypochlorite, hydrogen peroxide, diammonium hydrogen phosphate, acetone, or NaOH solutions [56–59]. After washing and drying, for the calcination process, the bone pieces are heated in a furnace (at various temperatures ranging from 350 °C to 1400 °C, for 1 to 18 h in air (Table 3), in order to completely remove the organic matter from the bones. One should emphasize that the as-prepared powders are biologically safe due to the high temperature fabrication route, which not only favors the crystallization of the material, but also completely prevents any risk of diseases transmission, since no pathogen can survive to such extreme conditions [25,53,54,60]. The resulting calcined bone specimens are cooled to room temperature (RT) by slow furnace cooling. Then, they are generally crushed inside an agate crucible, using a mortar and pestle, prior to undergoing ball-milling to fine powders (i.e., with particles of submicron size).

Figueiredo et al. [61] performed a comparative study of the structural and chemical properties of both human and animal bone-derived HA, submitted to different calcination temperatures (600, 900, and 1200 °C, respectively). The obtained results indicated that the calcination temperature strongly influenced the properties of the bone samples. At higher temperatures, pure HA could be obtained, which presents higher crystallinity degrees, larger crystallite sizes, and a less porous structure. Figueiredo et al. concluded that the mammalian bone samples calcined at 600 °C exhibited the most promising combination of chemical composition and structure that could be exploited to provide good alternatives to synthetic apatite and/or allogeneic bone.

The viability of producing biogenic HA from bio-waste animal bones (bovine, caprine, and galline), by heat treatments in air atmosphere at different temperatures (600–1000 °C) was demonstrated by Ramesh et al. [62]. Among the three types of investigated animal bones, it was demonstrated that the bovine-derived HA was stable for all investigated temperatures, while those produced from both caprine and galline bones exhibited signs of decomposition, with the appearance of the β-TCP phase at temperatures beyond 700 °C. In addition, the bovine and caprine bones presented hardness values comparable with the ones of the human cortical bone, while, for galline bone samples, higher porosity levels and low hardness values were inferred.

Akyurt et al. [63] produced HA from sheep teeth, by calcination and sintering at different temperatures (1000–1300 °C). Compression strength and microhardness measurements were performed and the obtained results were the best in the case of samples sintered at 1300 °C. This behavior was attributed to the F-content presence in the enamel structure.

Gheisari et al. [64] prepared natural HA-Hardystonite (HT) ceramics with different percentages of HT (5, 10, and 15 wt.%, respectively). Their results showed that the mechanical properties of HA increased and the bioactivity behavior improved with the addition of HT to natural HA. The maximum value of the density was inferred for the 10 wt.% HT samples, which was attributed to the formation of HT silicate phases between the matrix particles and formation of glass bonds. Moreover, the 15 wt.% HT nanocomposite samples had the lower cold crushing strength as compared to the other ones investigated in the paper, which was due to the overlapping of glass bonds. Simulated body fluid test results indicated that the Si content increased in the samples with a higher amount of HT. Consequently, the number of Si-OH nucleation sites expands and the formation of apatite layers takes place along with an increase in the ceramics' bioactivity. In a similar study, Khandan et al. [65] reported on natural HA-diopside (Di) bio-ceramics, with different Di content (10, 20, and 30 wt.%, respectively). The obtained results indicated that an elevated content of Di in the samples determined an increase of both the density on the surface and the adhesion values. In order to maintain a surface free of cracks, Khandan et al. indicated the optimum values for voltage, time, and temperature as being 50 V, 10 min, and 850 °C, respectively.

Barakat et al. [30] proposed in their study three different methods to extract HA from bovine bones, which include thermal decomposition, subcritical water, and alkaline hydrothermal processes. One advantage of the first method was to produce an HA nano-rode rather than the nanoparticles obtained by using the other two proposed methods. However, the morphological investigations illustrated smaller-size particles in the case of subcritical and alkaline hydrothermal processes, as compared to thermal decomposition. Barakat et al. concluded that the proposed methods are simple and cheap, which are two important advantages that might advance their use for large scale applications.

Giraldo-Betancur et al. [66] reported on a physical-chemical comparison of synthetic and biological HA. The bio-HA samples originated from bovine bones and were prepared by using three different treatments (defat, alkaline, and calcination). It was indicated that the calcination and alkaline processes delivered crystalline HA, which presented a comparable quality to the synthetic one. In addition, it was illustrated that the calcination process was appropriate to obtain HA with the minor and major elements that appear in the natural bone tissue. After applying the alkaline process, a crystallinity degree greater than 62% was detected. Moreover, the surface of the alkaline sample presented a transition behavior between dense and porous morphology. In a similar study, Rincón-López et al. [67] performed a comparative study on the physical-chemical properties and biological behavior of sintered-bovine-derived HA and commercial HA. It was demonstrated that highly crystalline HA could be obtained from bovine bones. One should note that the Na^+ and Mg^{2+} ions intrinsically presented in HA of bovine origin seemed to influence the sintering behavior evolving to ceramics with lower porosity and coarser microstructure compared to those obtained with synthetic HA. Both these studies indicated the possibility to use HA derived from bovine bones as a viable alternative to synthetic HA for biomedical applications.

In Table 3, the details for HA powder preparation using mammalian sources are summarized.

The colour of the resulting powders (Table 3) is either milky-white or green.

For more information on BioHA extraction methods from mammalian bones, the reader can consult additional information contained in references from Table 3.

Table 3. Preparation of HA powders derived from mammalian bones, presented in the order of the calcination temperature.

Source Material	Pre-Treatment	Calcination		Dopants	Reference
		[°C]	Time [min]		
Bovine bones (femur)	Immersion in 2.6 wt.% sodium hypochlorite solution for 14 days	–	–	–	[68]
Bovine bones (femur)	Boiling in distilled water for 2.5 h, ultrasonication with acetone for 5 min, dried at 120 °C for 12 h in an oven	350–900	180	–	[69]
Bovine bones	Boiling in deionized water for 30 min, petroleum ether with constant agitation at 30 °C and sodium hydroxide solution, drying in a vacuum oven at 1.33 Pa and 70 °C for 5 h	400–900	180	–	[66]
Bovine bones (femur)	Boiling water followed by sun drying	500–1400	120–240	–	[70]
Bovine, caprine, and galline bones	Autoclave at 100 °C for 1 h, rinsing with water and drying for 3 h at 70 °C in a box oven	600–1000	120	–	[62]
Bovine bones	Applying direct flame from a gas torch to the cleaned bones	600–1100	180	–	[71]
Bovine bones	Boiling with water for 2 to 3 h, drying in an oven at 80 °C for 72 h	600–1100	180	–	[72]
Human, bovine, porcine bones	Boiling in distilled water for 30 min, immersion in ethanol, hydrogen peroxide, formaldehyde solution, drying in a vacuum oven at 50 °C for three days	600–1200	1080	–	[61]
Bovine bones	Immersion for 14 days in an alkali solution of 1% sodium hypochlorite	700	240	1 wt.% Li_2CO_3, 1 wt.% Li_3PO_4	[60]
Bovine bones	Washing by water and acetone, drying at 160 °C for 48 h	750	360	–	[30]
Bovine bones	Boiling in distilled water for 8 h, drying overnight at 200 °C	800	180	–	[73]
Bovine bones (femur)	Keeping in boiling distilled water for 2 h, heated at 60 °C for 24 h	800	120	–	[74]
Bovine bones	Boiling for 15 min, filtrating, washing with distilled water for several times and drying at 100 °C in a vacuum oven for 48 h	800–1100	180	–	[75]
Ovine and bovine bones	Immersion for 14 days in an alkali solution of 1% sodium hypochlorite	850	240	1.5 wt.% Ti	[53]
Veal bones	Sodium hydroxide solution in a beaker, neutralization with distilled water, drying in an oven	850	180	–	[64]

Table 3. *Cont.*

Source Material	Pre-Treatment	Calcination		Dopants	Reference
		[°C]	Time [min]		
Calf bones	Sodium hydroxide solution in a beaker, neutralization with distilled water, drying in an oven	850	180	–	[65]
Bovine bones (femur)	Cleaning and washing with distilled water, immersion for 14 days in an alkali solution of 1% sodium hypochlorite	850	240	1 wt.% of Li_2CO_3 and Li_3PO_4	[11]
Bovine bones	Washing by water and acetone, drying at 160 °C for 48 h	850	60	–	[76]
Bovine bones	Boiling and sun drying	900	120	–	[77]
Bovine bones	Boiling in distilled water for 3 h, drying in an oven at 100 °C for 24 h	900	120	–	[78]
Bovine bones (femur)	Immersion for 14 days in an alkali solution of 1% sodium hypochlorite	1000	240	5 wt.% MgO, 2 wt.% MgF_2	[54]
Ovine dentine bones	Cleaning and washing, drying at 750 °C for 5–6 h	1000	240	1.5 wt.% Ti	[25]
Camelus dromedarius bones (femur)	Boiling in distilled water for 1 h, washing with a strong water jet, drying at 100 °C, for 60 min, drying at RT for 7 days, immersion in acetone for 1 h, washing with distilled water	1000	180	–	[79]
Sheep teeth	Cleaning and washing, drying at 750 °C for 5 to 6 h	1000–1300	240	–	[63]
Human dentine, ovine, and bovine bones	Immersion for 14 days in an alkali solution of 1% sodium hypochlorite	1100	240	–	[80]
Bovine bones	Boiling in deionized water for 30 min, petroleum ether with constant agitation at 30 °C and sodium hydroxide solution, drying in a vacuum oven at 1.33 Pa and 70 °C for 5 h	1200	120–240	–	[67]
Bovine bones	Boiling in distilled water for 2 h, heating in an electrical furnace at 500 °C for 6 h	1200	240	–	[81]

4.0.2. Extraction of HA from Fish Sources

Even though fish bones represent a rich source of calcium, phosphate, and carbonate [2], there are scarce reports on HA synthesis from these natural sources for biomedical applications [82]. In general, fish bones originate from fisheries, which capture fish and use it mainly for obtaining meat, oil, or some low-value fertilizers [2]. The fish bones are collected after gently removing the flesh parts from the entire fish. The bones are washed with a hot water jet or steam to remove all types of proteins and other organic impurities. The cleaned bony parts are further treated with reagents such as NaOH [83]. Similar to the case of mammalian bones, after washing and drying, the bone pieces are subjected to heating at various temperatures ranging from 100 to 1200 °C, for 1 to 12 h in air (see Table 4), to completely eliminate any organic matter from the bones. The resulting calcined fish bone specimens are cooled to RT by slow furnace cooling. Afterward, the specimens are either hand-crushed with a mortar and pestle or ball-milled to fine powders (Table 3). One notes that the color of the resulting powders (Table 4) is white.

Table 4. Preparation of HA powders derived from fish sources, presented in order of the calcination temperature.

Source Material	Pre-Treatment	Calcination [°C]	Time [min]	Reference
Cuttlefish bones	Heating at 200 °C for 6, 12, and 24 h and drying	100–1200	60	[84]
Sword fish and tuna bones	Boiling in water for 1 h and washing by a strong water jet, drying at RT in air for 24 h	600–950	720	[85]
Fish bone wastes	Incineration at 300 °C for 3 h	750	300	[88]
Fish scale	Hydrolyzation under 1% protease N for 2.5 h, and 0.5% flavourzyme for 0.5 h, stirring and heating in boiling water for 10 min, drying by hot air, and storing at –20 °C	800	240	[86]
Cod fish bones	Immersion in $CaCl_2 \cdot 2H_2O$, $Ca(C_2H_3O_2)_2$ and NaF solutions, for different time intervals, stirring at 75 °C	900–1200	60	[87]
Tuna bones	Washing with hot water for two days, mixing with 1.0% sodium hydroxide and acetone, drying at 60 °C for 24 h	900	300	[83]
Shark tooth enameloid	Keeping in boiling water for 3 h, drying in a laboratory oven at 60 °C for 24 h	950	720	[89]
Dentine and enameloid of shark teeth and deer antlers	Boiling water for 3 h and drying in a laboratory oven at 60 °C for 24 h	950	720	[90–92]

Venkatesan et al. [83] isolated HA from fish sources by applying alkaline hydrolysis and thermal calcination methods. An increase in the particle size from 0.3 to 1.0 μm was observed when using the thermal calcination method, which is due to the particle agglomeration. HA obtained by the second method are of a nanorod shape, with 17–71 nm in length and 5–10 nm in width. Thermal calcination, in comparison to alkaline hydrolysis, produced HA with a higher crystallinity degree.

Kannan et al. [84] used hydrothermal transformation of aragonitic cuttlefish bones at 200 °C and calcination at temperatures up to 1200 °C to produce porous HA scaffolds with different levels of fluorine substitution (46% and 85%) on the OH sites. The F incorporation in the HA lattice determined a lowering of the unit cell volume because of the reduction of the *a*-axis length. The crystallites formed were close in size to bone-like apatite and were orientated along the *a*-axis rather than the *c*-axis. An type-AB carbonated apatite was also detected.

Boutinguiza et al. [85] reported on BioHA obtained from sword and tuna fish bones. The prepared powders consisted of a B-type carbonate HA with a Ca/P ratio higher than that of stoichiometric HA due to carbonate ions substituting phosphates. The presence of minor elements such as Na, K, Mg, and Sr substituting Ca was also indicated. The calcination treatment performed at 600 °C delivered a B-type carbonated HA. When reaching 950 °C, β-TCP was present in a minor amount and the carbonate content decreased. This indicated a partial decomposition of the material. It was concluded that these fish bone-derived materials originate from sustainable and cheap sources and could, therefore, represent a promising future alternative to synthetic HA for medical applications.

Huang et al. [86] compared the composition and biological properties of bio-waste HA derived from fish scale with those of synthetic HA. The experimental results indicated that the fish HA materials presented nano-sized particles with a high Ca/P ratio. Compared to synthetic HA, the sintering process increased porosity and surface roughness for fish-derived HA. The alkaline phosphate assay and von Kossa staining demonstrated that the fish-HA particles promoted osteogenic differentiation and mineralization of MG63 cells. Taking into account these results, one should recommend fish scales as a cost-effective and environmentally-friendly source for producing HA.

Piccirillo et al. [87] produced apatite-based and tricalcium phosphate-based materials from codfish bones, by annealing at temperatures between 900 °C and 1200 °C. Single phase HA, chlorapatite, and fluorapatite were obtained using $CaCl_2$ and NaF solutions, respectively. One should note that this was the first study to report on compositional modifications of natural origin materials used to tailor the relative concentrations of elements. The obtained results indicated that, by using a simple and effective valorization technique, the conversion of this by-product into a viable compound for biomedical applications can be attained.

In Table 4, the details for HA powder preparation using fish resources are summarized.

For more information on BioHA extraction methods from fish sources, the reader can consult additional information contained in references from Table 4.

4.0.3. Extraction of HA from Biogenic Sources

Every year, impressive quantities of biogenic resources such as eggshells, sea shells, and other calcite materials are disposed of as waste by restaurants, hatcheries, bakeries, or homes [2]. Due to the fact that they are very cheap and accessible, eggshells represent an important source of calcium precursor required for the synthesis of HA.

In general, the preparation procedure to extract CaO from shells consists first in washing them with boiling water or steam to completely remove all impurities. In the next step, the shells are crushed to fine powders, which is followed by heating at elevated temperatures. The resulting CaO reacts with phosphorous precursors to prepare HA by following a protocol detailed in Reference [93].

Starting from eggshells, Elizondo-Villarreal et al. [94] synthesized HA using a simple hydrothermal approach. The obtained product was a mixture of HA and $CaHPO_4$ in a 3:1 ratio, with HA morphology in the form of whiskers. It was concluded that the obtained physical-chemical characteristics of HA should advance this type of material for dental prosthesis applications.

HA was prepared by Chaudhuri et al. [95] from eggshells and a solvent such as dipotassium phosphate. An appropriate amount of eggshells-derived CaO was immersed in K_2HPO_4 solution, for different soaking times, to obtain nanocrystalline HA. Both grain size and pH indicated a slight decreasing tendency with increasing soaking time. The lattice strain also showed similar behavior with increasing soaking time. The proposed procedure for the low-temperature synthesis of large-scale HA is rather simple, low cost, and eco-friendly.

Tamasan et al. [96] reported on the preparation and characterization of powders consisting of the different phases of CaPs obtained from marine-origin raw materials of sea-shells (*Sputnik sea urchins* and *Trochidae I. concavus*) in reaction with H_3PO_4. In the developing CaP powders, hot-plate, and ultrasound methods were used. In the Sputnik sea urchins samples, brushite was found to be predominant, while calcite was shown to exist as a small secondary phase. For the second analyzed material, monetite and HA phases were identified. A thermal treatment performed at 850 °C resulted in flat-plate whitlockite crystals (β-MgTCP) for both samples, regardless of the fabrication method.

Gunduz et al. prepared biphasic bioceramic nano-powders of HA and β-TCP [97] from shells of a sea snail, by using a novel mechano-chemical method. When compared to the conventional hydrothermal method, this chemical method is simple and economic, due to inexpensive and safe equipment. The characteristics of the as-produced powders, along with their biological origin, should advance these materials for further consideration and experimentation to fabricate nanoceramic biomaterials.

In Table 5, the details for HA powder preparation using biogenic resources are summarized.

Table 5. Preparation of HA powders derived from biogenic sources, presented in the order of the calcination temperature.

Source Material	Pre-Treatment	Calcination		Reference
		[°C]	Time [min]	
Shell of sea snail	Cleaning thoroughly from sand particles and other foreign materials, drying, solution of H_3PO_4, hot-plate stirring at 80 °C for 8 h, filtration, and drying at 100 °C overnight in an incubator.	400–800	240	[97]
Sputnik Sea urchin and sea snail shells	Heating on a hotplate at 80 °C for 15 min	450–850	240	[96]
Chicken eggshells	Cleaning with distilled water and keeping into 1 M H_2O_2 solution for a week, drying at 90 °C	700	300	[94]
Egg shell	Cleaning and washing with flowing distilled water, drying at 300 °C for 1 h	850–900	180	[95]
Eggshells	Cleaning in boiling water	900	120	[98]
Hen eggshell	Stripping the membrane off the eggshell, rinsing with water, drying	1000	180	[99]

For more information on BioHA extraction methods from biogenic sources, the reader can consult additional information contained in references from Table 5.

5. Pulsed Laser Deposition Method

5.1. Method Overview

To the difference of wet synthesis methods, plasma-assisted techniques offer important advantages, such as: (*i*) a much faster process for surface deposition, (*ii*) industrial scaling, (*iii*) a stoichiometric transfer of the target composition in the synthesized structures, (*iv*) a better uniformity in terms of morphology and composition, (*v*) a lower porosity, and (*vi*) a decreased tendency of the deposited structures to crack or delaminate [100,101]. In the biomedical domain, for the fabrication of CaP coatings for bone implant applications, the most applied plasma-assisted techniques are radio-frequency magnetron sputtering and PLD [102].

PLD is a thin films synthesis technique consisting of the ablation by a focused, high power, pulsed laser beam of a solid target in a vacuum environment and the condensation of the resulting vapor phase on a deposition substrate placed parallel to the holder [19]. The most basic set-up consists of a vacuum chamber, a laser source, a focusing lens, a target holder rotated by a motor, and a substrate usually placed on a heater with the role to increase film adherence and promote crystallinity.

The wavelengths of choice for the laser sources used for ablation are in the UV range due to the higher penetration depth of this type of beam in the target material as compared to visible or IR laser sources and higher photon energy that allows for a more efficient vaporization of the target [103]. Popular laser sources used in PLD experiments are excimer lasers such as ArF [104], KrF [105], or XeCl [106] emitting at 193, 248, or 308 nm, respectively, or a solid-state laser such as a Nd:YAG [107], which emits at 266 nm.

To increase the amount of evaporated target material to the detriment of expulsed liquid or solid phases, lasers emitting in a pulsed regime, with pulse durations in the nanoseconds-picoseconds range, are generally used [108]. In these regimes, the absorption process occurs on a much shorter time-scale as compared to the thermal diffusion process. The surface layers of the target reach temperatures of tens of thousands of K and are instantly vaporized. A plasma plume consisting of atoms, ions, electrons, molecules, free radicals, and condensed particles expands at supersonic speed and is accelerated toward the substrate (on a direction perpendicular on the target), where the deposition takes place [109].

If a solid collector is placed in front of the target at a distance approximately equal to the plasma plume's length, the material condenses and, pulse-by-pulse, a thin film is grown [110]. If the placement of the substrate is closer to the target than the plasma plume's length, the material will be deposited and, subsequently, washed by the plasma plume, with the deposition being inefficient and uncontrollable. Similarly, if the distance is larger than the plume's length, the deposition will also be inefficient, since the film thickness is inversely proportional to the square of the target–to–substrate distance. Femtosecond laser sources are not considered optimal for thin film synthesis. Due to the high energy of the ablated species, the growing film is heavily bombarded by high speed particles, which makes the process inefficient and results in films with defects, vacancies, or made of small solid particles ripped from the target surface [111].

Due to the high temperatures generated on the target surface, any type of material can be ablated and, therefore, a wide variety of targets can be used for thin film deposition.

The technique is versatile in terms of process output. By independent control of the deposition temperature, background pressure, and substrate positioning, one can change the film's crystallinity, composition, adherence, surface roughness, and/or thickness [18]. The use of multiple targets irradiated subsequently allows for the deposition of multi-layered structures [112].

Moreover, the PLD technique can be easily adapted to increase the range of applications: by limiting the number of pulses, one can deposit nanoparticles instead of thin films [113]. Two targets of different materials can be irradiated simultaneously and the intermixing plasma plumes can create a compositional library on a substrate [114]. By irradiating a frozen target made from an organic compound dissolved into a matrix material, an organic thin film can be laser deposited without degradation of the target material in a variation of the method, denominated MAPLE, i.e. Matrix Assisted Pulsed Laser Evaporation [115].

The film growth by PLD also offers different types of advantages over other deposition techniques, such as:

- The irradiation source is situated outside the deposition chamber, which offers a high degree of flexibility in using the material, set-up, and adjustment of deposition parameters;
- most solid materials can be laser ablated and deposited as films;
- due to the laser operating in a pulsed regime, the film growth rate can be controlled with a highly precise degree (10^{-2} –10^{-1} nm/pulse);
- in optimal conditions, the stoichiometry of the deposited layer coincides with the one of the targets, even for very complex materials with a high degree of instability;
- the high energy of ablated species determines the synthesis of extremely adherent layers;
- one can obtain species with electronic states different from the equilibrium ones and new, metastable phases of the material;
- even though their thickness might have very low values, the films uniformly cover the substrate and prevent the release of ions from the implant into the body;
- when reducing the thickness, the risk of delamination decreases.

Like all the other synthesis techniques, PLD presents some disadvantages too, such as a low deposition rate, an applicability domain limited to compounds that are not sensible to thermal decomposition and degradation that appears during processing by prolonged exposure to UV laser radiation, the presence of droplets (of a different composition and dimensions), and a low deposition area (no bigger than a few cm^2). The last two drawbacks can be diminished or even eliminated by additional set-ups [18,116].

5.2. BioHA Targets Preparation

The powders obtained following the protocols mentioned in Section 4 are generally pressed and the resulting pellets are heat-treated in a furnace, to reach compactness by eliminating air bubbles and water vapors. The as-sintered pellets are compact and tough (Figure 1) and can be used as targets in the PLD experiments [11,25,53,54,60,80].

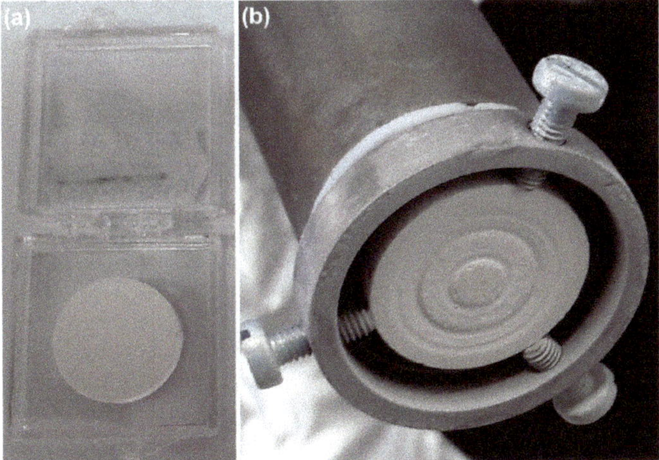

Figure 1. Photos of sintered PLD targets: (**a**) before and (**b**) after the action of the beam generated by a KrF* excimer laser source.

The Importance of Thermal Treatments in the Case of Targets

Even though CaP bio-ceramics possess excellent biomedical characteristics, their poor mechanical behavior determined focused research to find viable solutions to this important drawback.

The properties, efficiency, phase purity, and size distribution of HA extracted from natural resources, especially from bones, depend upon factors like the extraction technique, the calcination temperature, and the nature of bones [2]. HA derived from natural resources (such as animal bones) is available in unlimited supply and can be produced by calcination methods [27,117]. Ozyegin et al. [117] reported that sintered BHA is safe from potentially transmitted diseases such as bovine spongiform encephalopathy ("mad cow disease"). The manufacturing of these powders is easy to carry out and economically viable [2,117]. Prions, which are the smallest proteinaceous infectious particles, can resist inactivation by procedures that modify nucleic acids and are able to transmit alone an infectious disease, which causes harm to the host tissue [118]. Thus, the temperature generally applied for calcination is around 850 °C and it was demonstrated that no disease transmission pathogens (including prions) can survive to such high temperatures [2,27,117]. However, a special calcination regime has to be carefully selected in order to remove all organic matter and enhance the crystallinity of HA, while avoiding thermal decomposition of the final product, HA [2].

5.3. Substrates Used as Pulsed Laser Deposition Collectors

In general, for the fabrication of bio-implants and bio–devices, there are five classes of materials used: (i) metallic materials, (ii) polymers, (iii) ceramics, (iv) composites, and (v) natural materials.

The most commonly used metals and alloys for medical device applications include stainless steels, cobalt-based alloys, and commercially-pure titanium and its alloys [119]. The last ones are generally preferred for bone-replacement applications mainly due to their intrinsic resistance. These materials have been widely used as endosseous implant materials in dentistry and orthopedics due to their high strength, good corrosion resistance, no allergic problems, low density, and excellent biocompatibility [120]. Moreover, the thin oxide layer, which is naturally formed on their surface, is acting as a protective barrier, which confers its well-known resistance among metals to corrosion in physiological conditions. However, it is far from being an ideal material from a medical point of view because of its insufficient bioactivity when implanted in the human body [121].

Polymers represent the largest class of biomaterials, which have gained greater attention due to some important advantages such as (i) ease of fabrication to various complex shapes and structures, (ii) wide range of bulk compositions and physical properties, and (iii) easy tailoring of surface properties [119].

Ceramics are inorganic compounds of metallic or non-metallic materials. When used for skeletal or hard tissue repairing, they are referred to as bio-ceramics. These bio-ceramics can be classified as bioinert (e.g. alumina, zirconia), bioresorbable (e.g. TCP), bioactive (e.g. HA, bioactive glasses, glass ceramics), and porous (e.g. HA coatings on metallic substrates for tissue in-growth).

A composite consists of two or more materials, characterized by a combination of the best physical and chemical characteristics of each component materials. In the biomedical field, composite materials are generally designed for superior mechanical and biological properties. They can be classified as a function of the (i) matrix material and (ii) bioactivity of the composite. In the first case, there are three types of composite materials: (i) polymer matrix composites, (ii) metal matrix composites, and (iii) ceramic matrix composites. For the second case, there are: (i) bioinert composites, (ii) bioactive composites, and (iii) bioresorbable composites.

Collagen and glycosaminoglycans are the most commonly used natural materials for clinical applications [122] due to a series of important advantages, such as: (i) easy recognition by the biological environment due to their similarity with the macromolecular substances, (ii) possibility to by-pass toxicity-related issues, chronic inflammation, or lack of recognition by cells (which generally occurs when dealing with synthetic materials), and (iii) biodegradable nature of the materials, which permits their use for focused applications, where a specific function is needed for a limited period of time [123].

In general, prior to the coating process and immediately after it, special attention needs to be paid to the surface preparation of various types of used substrates. In this respect, different substrate preparation techniques have been reported, from simple to complex ones, with their own peculiarities. Thus, simple substrate preparation procedures include cleaning or degreasing to remove any remnant surface contamination because of manufacturing conditions or improper storage. Solvents like acetone, ethylic alcohol, and/or distilled water might be used in this case, always keeping in mind the nature and properties of the used substrates. By contrast, more complex substrate preparation techniques include (i) mechanical modification of the surface, by sand-blasting [124], grit-blasting [125,126], polishing [127,128], or grinding [129], and (ii) chemical treatments, modifications, or functionalization, using chemical activation [130], alkaline treatments [131], anodization [124,125], acid-etching [124–126], oxidizing [132], and many other actions [124,133]. Additional physical treatments, resulting in various types of surface modifications, were briefly discussed in Reference [133]. After CaP synthesis, the applied post-deposition treatments are intended to (i) provide crystallization/recrystallization of various phases, (ii) improve their fixation to the substrate, and (iii) evaporate traces of solvent(s) that might remain trapped within the deposited structures.

5.4. Pulsed Laser Deposition Experimental Set-Up

A PLD experimental set-up is typically composed of a UV laser source, a vacuum stainless-steel deposition chamber (Figure 2), equipped with a rotating target, a fixed substrate holder, and pumping systems. Following target irradiation, a plasma cloud expands, either in vacuum or in different gas atmospheres, and the vapors are collected on substrates (generally placed parallel to the targets and heated up to a temperature in the range of 350–600 °C), in the form of thin, adherent films [18]. In this respect, amorphous or crystalline, extremely adherent, stoichiometric, and dense or porous structures from various complex materials can be synthesized, even at relatively low deposition temperatures, by varying the experimental parameters that are mainly related to the (i) laser (fluence, wavelength, pulse-duration, and repetition rate), and (ii) deposition conditions (target-to-substrate distance, substrate temperature, nature, and pressure of the environment) [19,134,135].

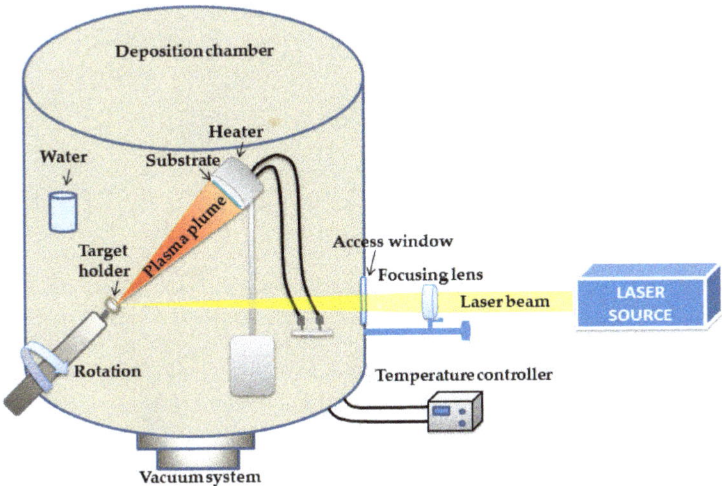

Figure 2. Schematic of an experimental set-up used for PLD synthesis of animal-origin HA coatings.

Due to the possibility to independently vary a large number of parameters (such as the wavelength, fluence, pulse repetition frequency, energy, target preparation, target–to–substrate distance, substrate temperature, area of the laser spot, deposition geometry, nature, and pressure of the gas in the deposition chamber), PLD represents a versatile technique to produce films with a high diversity of morphological and structural characteristics, which are superior to the ones appertaining to conventional deposition techniques (fast processing, safety in functioning, and low production costs) [18,19,116,136].

The substrate temperature during film synthesis by PLD is usually situated in the range of 350 to 600 °C. This way, one can assure the fabrication of highly crystalline and phase-pure films [137]. Moreover, lower or higher values for substrate temperature can be chosen to obtain layers with fine textures or different roughness, with improved adherence to the substrate, depending on the intended applications.

Various experimental conditions used for PLD synthesis of BioHA coatings have been reported in the literature. They are summarized in Table 6. The results from these papers will be thoroughly analyzed and discussed in Section 6.

Table 6. Different experimental conditions used for PLD synthesis of BioHA coatings.

HA Source Material/Used Substrate	Type of Laser Used	Target-to-Substrate Separation Distance [cm]	Energy [mJ]	Temperature during Deposition [°C]	Water Vapor Atmosphere [mbar]	Pulse Frequency [Hz]	Number of Applied Laser Pulses	Reference
Enameloid of shark teeth/Ti6Al4V disc	ArF* (193 nm)	–	320	460	0.15–0.45	10	–	[89]
Ovine and bovine bone/Ti disc and Si wafer			330					[53]
Human dentine, ovine, and bovine bones/Ti disc and Si wafer			350					[80]
Bovine bone/Ti disc and Si wafer	KrF* (248 nm)	5		500	0.50	10	15,000	[54]
Sheep dentine/Ti disc and Si wafer			330					[25]
Bovine bone/Ti disc and Si wafer			360					[60]
Bovine bone/Ti disc								[11]

5.5. Thermal Treatments Applied to Pulsed Laser Deposited Coatings

BioHA coatings are generally amorphous when deposited by PLD. HA loses (OH)$^-$ molecules and on the deposition substrate a mix of CaPs is formed. The most prominent of them are tricalcium phosphate (TCP) and octacalcium phosphate. The process is reversible and a post-deposition thermal treatment around 500 °C in water vapors leads to conversion of the deposited amorphous and non-apatite phases into more stable compounds, like HA. In addition, their crystallinity and corrosion resistance increase, and the residual stress is considerably reduced [138–141]. It is also important to note that, the presence of water during the post-deposition heat treatment also plays an important role in this conversion [142,143].

Dinda et al. [144] have reported that a post-deposition thermal treatment at 300 °C can produce pure, adherent, and crystalline HA coatings, which will not dissolve in simulated body fluids.

Post-deposition thermal treatments, generally performed in air or enriched water-vapor atmospheres, are of key importance. Their role is, on one hand, to restore the stoichiometry of the synthesized compounds and, on the other hand, to improve the overall crystalline status of the coatings [53].

The sintering temperature and atmosphere are considered as important parameters able to tailor the strength and toughness of HA [145]. For example, sintering at elevated temperatures has the tendency to eliminate the functional group OH in the HA matrix and this would result in the decomposition of the HA phase to form α-TCP, β-TCP, and tetra-calcium phosphate [145].

6. Characterization of Pulsed Laser Deposited Coatings of Undoped and Doped Animal-Origin Hydroxyapatite

The purpose of CaP coatings that cover metallic medical prostheses is to ensure biomimetism, i.e. to "mislead" the human body that the newly implanted device is very similar to its bones and, thus, start osteoblasts proliferation and osseous matrix synthesis. There is no perfect coating technique and the existence of shortcomings for each of them determined the implementation of others, among which PLD is worth mentioning due to its advantages such as stoichiometric transfer and precise thickness control.

For characterization of the PLD synthesized coatings, the most common investigation techniques are: Scanning Electron Microscopy (SEM) for surface morphology and cross-sectional investigations, Energy-dispersive X-ray spectroscopy (EDS) for Ca/P ratio determinations, profilometry or Atomic Force Microscopy (AFM) for surface parameters valuation, X-ray diffraction (XRD) for crystalline status, stress, crystallites size, and orientation investigations, Fourier Transform InfraRed (FTIR) spectroscopy for identification of chemical bonds, pull-out bonding strength or scratch tests for assessment of films' adherence to the substrate, and in vitro and in vivo tests. Therefore, we will concentrate in the following sections on these techniques for comparing the morphological, structural, and functional properties of PLD coatings synthesized from targets made of CaPs obtained from animal sources.

6.1. Morphological and Compositional Analyses

PLD coatings have general surfaces covered by round particles, which are further denominated as "droplets." Their presence on the surface can be explained like this: following target surface laser irradiation, the surface top layers are instantly vaporized. Part of the pulse energy is transformed into heat, which is concentrated in a subsurface layer. Therefore, due to heat accumulation, a mix of liquid and gaseous phases is generated. A bubble is formed, which expands until the top solid layer breaks and a plasma plume is released. The shockwave of the plume splashes the liquid phase on the target surface and liquid droplets are released with supersonic speeds reaching the substrate. Their number and diameter depend on the type of deposited material as well as pressure in the deposition chamber and on the laser parameters [18]. Droplets can be trapped using ingenious particle captors or off-axis geometries in order to obtain particle-free films. However, in the case of biological applications, a rough morphology could be beneficial for osteoblast proliferation and droplet-containing films are considered acceptable.

Sheep derived HA (SHA) coatings synthesized by PLD using a KrF* excimer laser source [25] also had an irregular morphology composed of spherical particles whose distribution could be approximated by a Gaussian fitting centered around a mean diameter of 1 μm. The particles size varied between 0.35 to 3.5 μm. An interesting observation was made for SHA doped with Ti (SHA:Ti, 1.5 wt.%) coatings. The particle number on the surface almost doubled as compared to undoped material, which produced rougher surfaces. When depositing coatings with the same number of laser pulses from SHA and SHA:Ti targets, Duta et al. [25] obtained thicker SHA coatings as compared to SHA:Ti ones (1.8 vs. 1.2 μm, respectively).

In Reference [53], Duta et al. conducted a comparison between film morphology in the case of SHA and BHA doped with Li_2O or a commercial inert glass. In all cases, the coatings displayed the morphology with round micronic droplets covering the surface. There were no big differences between particles diameter. However, a variation in the number of particles/cm^2 was statistically significant. Table 7 summarizes these differences.

Table 7. Droplets density in the case of synthetic HA and different BioHA coatings synthesized by PLD, using the same experimental conditions.

Sample Material	Density (Particles/cm^2)
SHA	$(5.5 \pm 0.8) \times 10^7$
BHA:Li	$(11.2 \pm 0.7) \times 10^7$
BHA:CIG	$(13.3 \pm 1.1) \times 10^7$
HA$_{syn}$	$(1.4 \pm 0.5) \times 10^7$

The coatings' thickness decreased in the following order: HA$_{syn}$, BHA:Li, SHA, BHA:CIG (3.71 ± 0.15 μm → 3.52 ± 0.16 μm → 2.95 ± 0.21 μm → 2.79 ± 0.02 μm). The roughness of the films decreased in the same order with values of 0.76 ± 0.1 for HA$_{syn}$, 0.71 ± 0.06 for BHA: Li, 0.61 ± 0.05 for SHA and 0.77 ± 0.03 μm for BHA:CIG, respectively [53].

To the difference, Mihailescu et al. [54] obtained smooth surfaces of BHA coatings on which a low number of spherical droplets appeared as sputtered. Their diameter ranged from 1.6 to 2.5 μm. The beginning of an organization of some droplets into clusters could be assumed from the SEM images. For BHA doped with MgF_2 and MgO, deposited in the same experimental conditions, a clear increase of the droplets' density was observed. They tend to accumulate in a majority in clusters of particles, with only a few isolated particles being identified. BHA doped with MgO (BHA:MgO) displayed the largest area of film surface covered by particle clusters. Apparently, the mean size of the droplets was very similar between undoped and doped materials.

In the case of BHA doped with Li_2CO_3 and Li_3PO_4 structures (further denoted as BHA:LiC and BHA:LiP) [60], the films' surface was flat, with isolated round droplets, sputtered on the surface. Most droplets were more than half embedded in the film mass, which is an indicative for their high velocity at the moment of impact with the surface. The droplets' number quantification was performed on four randomly chosen SEM fields and the obtained values were of $(2.2 \pm 0.2) \times 10^4$ particulates/mm^2 (for BHA), $(2.8 \pm 0.2) \times 10^4$ particulates/mm^2 (for BHA:LiC) and $(1.5 \pm 0.1) \times 10^4$ particulates/mm^2 (for BHA:LiP) [60].

From the cross-sectional SEM images, thicknesses of ~8.5, ~7.2, and ~6 μm, respectively, were inferred for BHA, BHA:LiC, and BHA:LiP coatings [60].

Comparative SEM micrographies showing different surface morphologies, which correspond to titanium and various types of BioHA coatings synthesized by PLD, using the same deposition conditions, are presented in Figure 3.

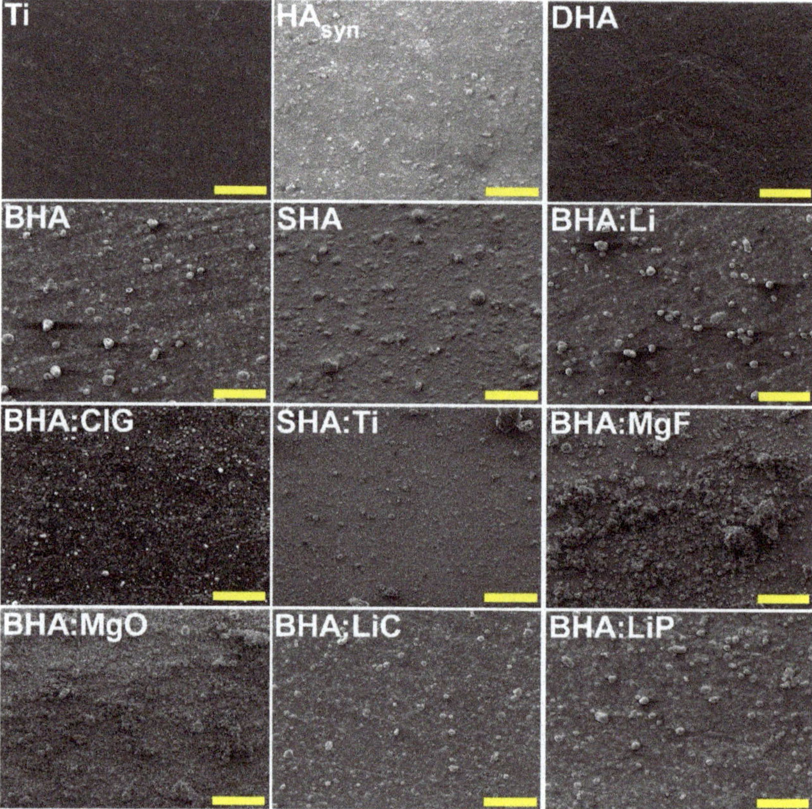

Figure 3. SEM micrographies corresponding to the surfaces of titanium and different types of BioHA coatings synthesized by PLD. Magnification bar: 50 μm.

Thickness of the PLD synthesized coatings can play an important role on their morphology, especially in the case of rough substrates. In Reference [89], films deposited in a water-vapor atmosphere onto Ti substrates were significantly thinner as compared to the ones deposited in vacuum. The calculated deposition rates were 3.5 nm/min for depositions conducted in water vapors, and 18.1 nm/min for depositions performed in vacuum. Therefore, films deposited in water vapors were thin and mimicked the Ti surface morphology. For deposition times under 30 min, surface features were not evident. For deposition times between 30 and 120 min, surface features became prominent. Aggregates of globular shape became visible at magnifications of 5000×, while at 20,000×, the droplets appeared to be made of elongated acicular crystalline structures of < 1 μm in length.

In general, the EDS quantitative measurements are performed during SEM investigations to determine the elemental distribution in the synthesized films and to estimate the Ca/P ratio.

A cross-sectional SEM image and its corresponding EDS spectra in the case of a BHA:CIG coating synthesized by PLD is presented in Figure 4. A compact structure of the coating can be observed. Moreover, the presence of trace elements (Na, Mg, Si, etc.) is also emphasized. A quasi-stoichiometric target-to-substrate transfer is reported in which the inferred Ca/P molar ratio is ~1.70. This value is typical to natural bone due to multiple doping with trace elements and functional groups.

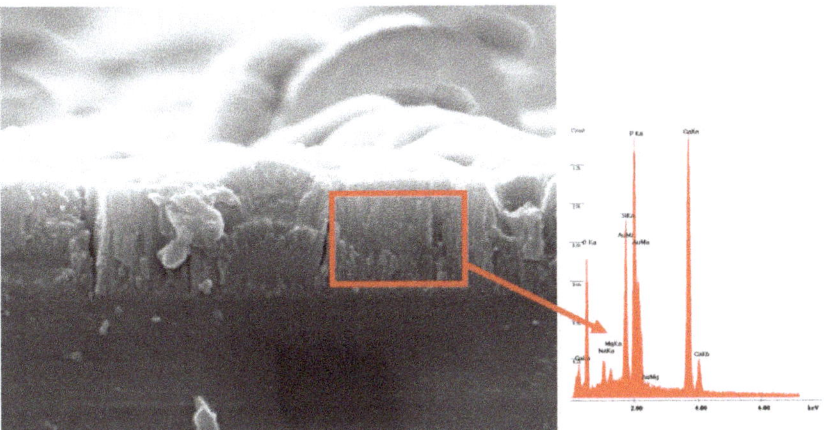

Figure 4. Cross-sectional SEM image and the corresponding EDS spectrum of a bovine hydroxyapatite doped with commercial inert glass coating synthesized by PLD.

6.2. Structural Investigations

GIXRD patterns of post-deposition treated BioHA coatings synthesized by PLD should display the characteristic peaks of HA [hexagonal, P63/m (176) space group, ICDD: 00-009-0432] [25]. It is quite possible that the GIXRD patterns contain, besides the HA peaks, some pronounced humps (centered at $2\theta \approx 30°$), which are proof of an amorphous phase in the deposited films. Duta et al. [25] reported for SHA films deposited by PLD a crystallinity degree of ~35%. The starting powders for PLD targets displayed in the XRD patterns some peaks that could be associated to $MgCO_3$ (ICDD: 01-086-2345) and $CaMg_2$ (ICDD: 00-013-0450) phases. However, after the deposition, these crystalline peaks were not present in the GIXRD patterns [25].

Some typical XRD patterns of HA_{syn} and BioHA coatings undoped or doped with Li_2O (further denoted as BHA:Li) synthesized by PLD are presented comparatively in Figure 5. One can observe that the most intensive peaks of the patterns in Figure 5 appertain to the Ti peaks of the substrate (ICDD: 44-1294). The fabricated coatings consisted of an HA phase with different degrees of crystallinity, as observed from the diffracted intensity variation and the mean crystallite sizes estimated from the FWHM of (002) and (300) XRD lines [53]. It is visible in Figure 5 that a more pronounced crystallinity corresponds to BHA:Li films. In addition, several weak lines were also identified whose origin might be due to a small percent of oxygen-deficient titanium oxides. Their formation might be due to the substrate oxidation during the post-deposition treatments [53].

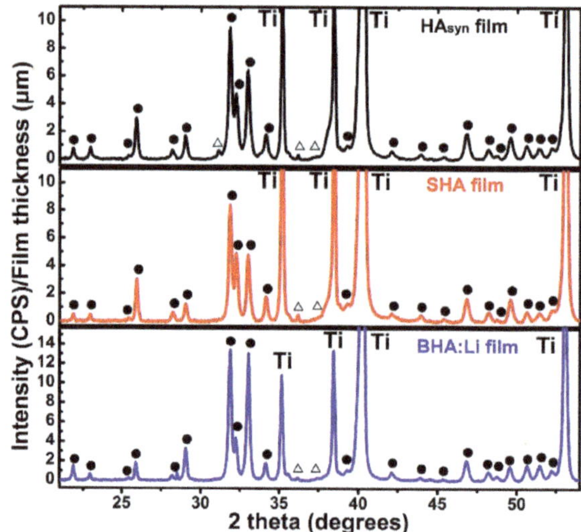

Figure 5. Comparative XRD patterns of synthetic and BioHA coatings synthesized by PLD onto titanium substrates (●, HA; Δ, TiO, and Ti_2O sub-oxides). Reprinted from Reference [53], with permission from Elsevier.

Another way to identify BioHA is by means of characteristic functional groups that can be detected by FTIR. Besides the characteristic HA bands of ν3 asymmetric stretching mode of $(PO_4)^{3-}$ groups (1010–982, 1063–1023, and 1091–1090 cm^{-1}, respectively), the ν1 symmetric stretching mode of $(PO_4)^{3-}$ groups (961–945 cm^{-1}), ν4 bending mode of $(PO_4)^{3-}$ groups (558–553 and 600–594 cm^{-1}, respectively), and the peak at ~630 cm^{-1} characteristic to $(OH)^-$ groups, in the case of BioHA, which is a band centered at 874–870 cm^{-1}, could be present in the spectra, assigned to $(HPO_4)^{2-}$ groups.

In the case of BHA targets and films, Duta et al. [53] identified small traces of hydrogen phosphorus fluoride hydrate (ICDD:77-0136) and brushite in the PLD targets [$CaPO_3(OH)\cdot 2H_2O$, ICDD:11-0293]. These traces were not present in the films obtained from irradiating these targets. Both targets and films displayed low amounts of fluorine besides the main HA phase. Popescu et al. [60] identified in BHA targets and films only FTIR bands corresponding to carbonate groups besides the characteristic orthophosphate bands of HA.

In the case of HA obtained from shark teeth, Hidalgo-Robatto et al. [89] reported on depositions conducted in water vapors. Depending on the pressure, different degrees of crystallinity were obtained. Under 0.25 mbar, spectra revealed a singular broad peak centered at ~31°, while, for a pressure of 0.45 mbar, the characteristic peaks for apatites located at 29.4, 32.1, and 33.4°, respectively, were present. In the FTIR spectra, besides the characteristic groups of CaPs, a peak associated with carbonate groups was present at 870 cm^{-1}. Depending on the deposition conditions, the importance of this peak varied. As the vapor pressure increased, the intensity of carbonate bands also increased. The CO_3^{2-}/PO_4^{3-} ratio was 30% higher in the BioHA coating deposited at 0.45 mbar H_2O vapors as compared to the case of depositions conducted without water vapors. F1s transition peaks were also shown by XPS in the PLD targets and in deposited films. Based on FTIR, XRD, and XPS results, one could conclude that films deposited by PLD from shark-derived HA contained a hydrated and fluorinated carbonated HA.

6.3. Bonding Strength Tests

The bonding strength at the bio-functional coating–substrate interface is considered a critical parameter for the fabrication, successful implantation, and long-term stability of implant-type structures, since it governs both the initial stability and long-term functioning of the medical devices [16,146].

The bounding strength values inferred for Ti, BHA, and doped BHA coatings synthesized by PLD are presented in Figure 6. The raw data of the obtained results reported in References [25,53,54,60,80] were plotted together for comparison reasons only.

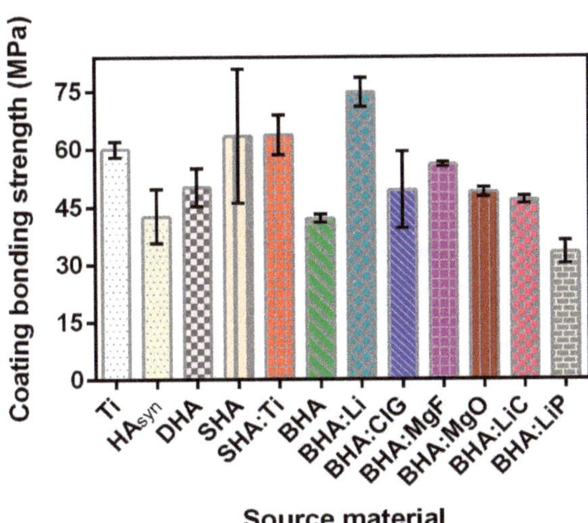

Figure 6. Bonding strength values in the case of titanium and different BioHA coatings synthesized by PLD using the same experimental conditions.

First, in order to verify the quality of the bonding adhesive, control tests are carried out on bare Ti substrates. The adherence values determined at the stainless-steel dolly/Ti substrate interface were of ~60 MPa. We note that this value is in accordance with the specifications provided by the manufacturer [147].

Duta et al. [53] recorded for the case of HA_{syn} coatings, adherence values generally similar to the ones reported in literature for HA films synthesized by PLD [148,149]. Significantly higher values were measured in the case of SHA and BHA:Li coatings. The decreased value of adherence registered for the BHA:CIG structure was attributed to the intrinsic friability of the glassy doping phase under mechanical stress [53].

Duta et al. [25] performed comparative pull-out adherence tests between SHA and SHA:Ti structures. The values recorded for SHA coatings were of ~55 MPa, while, for the doped ones, the values were of ~64 MPa. The increase in bonding strength of SHA:Ti coatings was due to the presence of the Ti-Ca segregation at the interface with the substrate. One notes that, during post-deposition annealing treatments, atomic inter-diffusion processes are known to take place at the coating–Ti substrate interface [150–152]. In general, this leads to the strengthening of the films' adherence by smoothing the interface between the two different materials. Therefore, the existence at the interfacial region of a segregated Ti-Ca inter-layer could accommodate an easier and more homogeneous inter-diffusion during the progress of the heat-treatments [25].

Mihailescu et al. [54] investigated the bonding strength for BHA and doped BHA structures. Compared to BHA coatings, the addition of MgF_2 and MgO seemed to result in an improvement of the films' adherence to the Ti substrate with mean bonding strength values of ~49 and ~56 MPa, respectively, being measured [54].

Popescu et al. [60] reported on similar coating adherence values of (42–46) MPa, in the case of BHA and BHA:LiC coatings. These values are in agreement with the ones inferred by Duta et al. [80]. A decrease of the bonding strength down to ~33 MPa was recorded in the case of BHA:LiP structures, which was attributed to the existence of a larger content of the amorphous counterpart [60].

It is important to emphasize that the bonding strength response measured for all coatings synthesized by PLD (Figure 6) was superior to the minimum mandatory value of 15 MPa imposed by international standard 13779-2 [153] in the case of implant-type coatings used for load-bearing applications.

6.4. In vitro Biological Observations

6.4.1. Bioactivity Effect

The most common method to qualitatively assess BioHA coatings' biocompatibility and cells' proliferation is the MTS assay, while cells' toxicity can be determined by using a lactate dehydrogenase (LDH) activity test. The MTS assay is based on the use of a tetrazolium compound (3-(4,5-dimethylthiazol-2-yl)-5-(3-carboxymethoxyphenyl)-2-(4-sulfophenyl)-2H-tetrazolium, inner salt – MTS) that is chemically reduced by viable cells into formazan, which is soluble in tissue culture medium. Since the production of formazan is proportional to the number of living cells, the intensity of the produced color can be used as an indicator of cell proliferation.

LDH (or LD) is an enzyme involved in energy production that is found in almost all of the body's cells. Elevated levels of LD usually indicate some type of cell damage. LD levels will typically peak as the cellular destruction begins. Thus, if a substrate is cytotoxic, the detected LD levels will be high. In Reference [25], tests were conducted according to the ISO standard 10993-5 [154].

In literature, the biocompatibility of synthesized coatings was assessed by the MTS tests, by applying different protocols. In this respect, human mesenchymal stem cells (hMSC) were seeded on Ti and various BioHA surfaces and the obtained results are plotted in Figure 7a,b and reported in References [25,60].

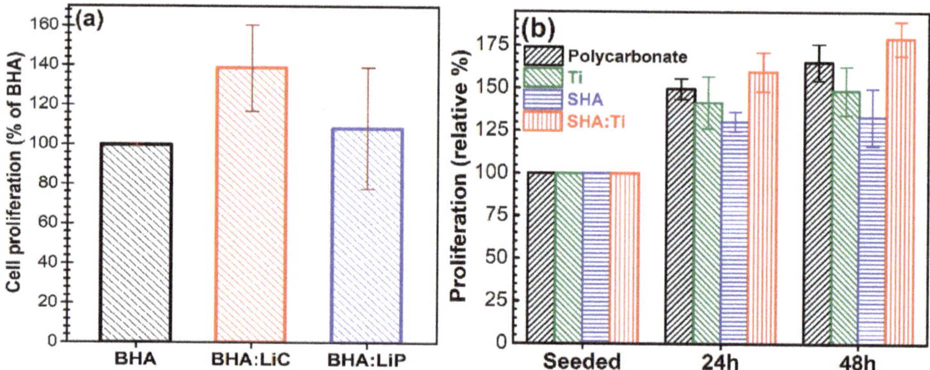

Figure 7. Proliferation of human mesenchymal stem cells assessed by MTS assay in the case of (**a**) bovine hydroxyapatite (BHA) and doped BHA coatings and (**b**) Ti, sheep hydroxyapatite (SHA), and doped SHA coatings synthesized by PLD. Reprinted from References [25,60], with permission from Elsevier. Slight modifications of the acronyms from the original figures were done to be in agreement with the sample codes used in this review.

In Figure 7a, one can observe that the cells' proliferation was better in the case of doped BHA coatings with respect to undoped ones. The proliferation increase was demonstrated to be in a direct relation with the augmentation of the polar component of the surface-free energy. Therefore, the beneficial effect of Li_2CO_3-doping could be emphasized [60].

After 24 and 48 h, the proliferation capacity of cells grown on the surface of SHA and SHA:Ti coatings was improved as compared to Ti surfaces for the same time intervals (Figure 7b). In addition, cell death at 24 h of growth was tested by using the LDH method [25].

Lithium is a trace metal in organisms. In pharmacy, it is used in composition of various drugs for the treatment of bipolar affective disorder and depression [155]. Lithium was also administered in drugs for treating osteoporosis in the last 20 years [156,157] due to its proven ability to enhance osteogenesis [158]. Despite its proven efficacy [159], there are reports when lithium administration determined side effects such as thirst and excessive urination, nausea, diarrhea, tremor, weight gain, sexual dysfunction, dermatological effects, and long-term effects on the thyroid gland, kidneys, and parathyroid glands [160]. However, recent studies challenge this negative perception of lithium administration, stating that intoxication occurs only when the concentration is higher than 1.5 mmol/L, and its main adverse effects can be properly monitored and managed [11,161–163].

Hidalgo-Robatto et al. [89] tested BioHA coatings (obtained from shark teeth) synthesized by PLD using a pre-osteoblastic cell line MC3T3-E1. The BioHA was compared to synthetic HA and a control of tissue culture polystyrene. Cell morphology was similar in the case of synthetic HA and BioHA. Both coatings were able to promote healthy morphology on the cells. MC3T3-E1 cells made contact with neighboring cells and covered the entire surface of their respective coatings. The MTT assay showed after seven days from seeding, an intense proliferation of cells on the biological apatite coatings, which is significantly higher than the proliferation on the synthetic HA coatings or the control by the MTS assay. After 21 days, all tested materials had higher values of cell proliferation than on day 7 and displayed similar absorbance values. Alkaline phosphatase (ALP) activity was tested as a measure of osteoblastic activity and bone formation. Statistically significant higher ALP activity was quantified on the BioHA coatings compared to both the synthetic HA and TCP, which presented lower values in the same range. After 21 days of incubation, both types of apatite coatings apparently manifested higher levels of ALP activity as compared to TCP values. Overall, BioHA seemed to accelerate osteogenic activity as compared to synthetic HA due to its fluoride content and its trace elements such as Na and Mg.

Doping of BioHA with various elements can drastically increase the cellular response toward PLD synthesized coatings. In such a study, the surfaces of medical-grade Ti, SHA, and SHA:Ti coatings (the precursor material for PLD targets was a mix of SHA powder and 1.5 wt.% Ti powder) were seeded with human mesenchymal stem cells (hMSC) [25]. On all types of samples, cells displayed the same typical morphology: small dimensions, well-spread, and having a polygonal shape. They have established liaisons with neighbouring cells by thin cellular extensions. The lowest proliferation value was reached for the SHA coatings, with respect to both SHA:Ti ones and controls (bare Ti and polycarbonate) [25]. The experiments also showed an important cell number growth (of ~60%) in the case of SHA:Ti coated samples. The proliferation increase of SHA:Ti *vs.* SHA was found statistically significant (p < 0.05). The LDH results showed that both SHA and SHA:Ti coatings fostered a good cells' cytocompatibility. However, the amount of cells that underwent apoptosis and death was lower in the case of the polycarbonate control and SHA:Ti coatings. SHA structures presented almost double cell apoptosis death as compared to SHA:Ti ones. The dead cells' index (percent of dead cells from the total number of cells) was lower than 4.5% for all cases, which constitutes another proof of the good cytocompatibility of the PLD coatings. The beneficial effect of Ti doping in HA has not been yet completely elucidated. Duta et al. [25] advanced its lower degree of crystallinity as a possible explanation for the superior biocompatibility of SHA:Ti, which generated the premises for a better interaction with the organic component of the mesenchymal cell culture media. Ca^{2+} ions released by dissolution may act as binders for organic moieties from culture media with roles in the healthy development of cells. A similar interpretation of enhanced protein binding by HA:Ti has been advanced by Kandori et al. [164] and Wei et al. [165], respectively.

In Reference [60], the influence of lithium doping of BHA on the biological response of the same type of cells was tested. The cells were stained with calcein/EthD-1 in order to evaluate the viable ones and the morphology of hMSC was observed three days after seeding on the surface of undoped and doped coatings. BHA:LiC and BHA:LiP samples produced a decrease in cell spreading without significantly modifying cell viability. Cells were adherent on the tested materials and exhibited typical fibroblast-like characteristics, with dendrites and prominent nuclei. There was no visible difference between lithium-doped BHA coatings and control materials (bare Ti and undoped BHA samples), as shown in Figure 8.

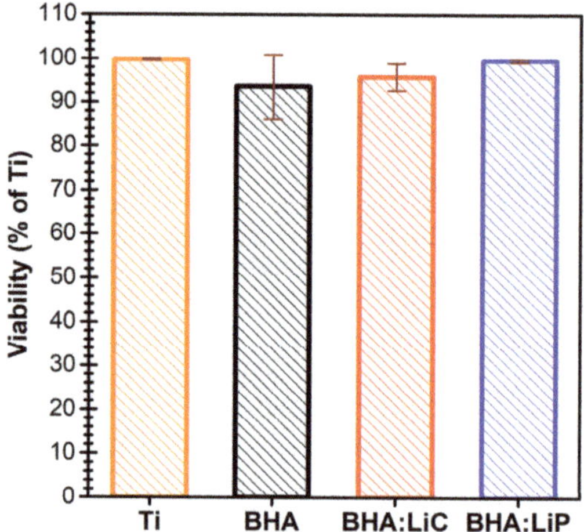

Figure 8. Cellular viability of human mesenchymal stem cells after 72 h of growth on the surface of Ti, bovine hydroxyapatite (BHA), and doped BHA coatings synthesized by PLD using the same experimental conditions. Reprinted from Reference [60], with permission from Elsevier. Slight modifications of the acronyms from the original figure were done for uniformization with the sample codes used in this review.

The in vitro cellular viability of hMSC grown for 72 h on Ti, BHA, and doped BHA coatings was assessed using the calcein AM/EthD-1 [60] method and the results are presented in Figure 8.

Based on relative intensity of calcein and EthD-1 signals on resulting scatter-grams, positive from negative cell populations could be distinguished. Compared to BHA samples, lithium-doped and MgF$_2$-doped surfaces led to a slight improved growth of hMSC cells, whilst, for BHA:CIG and BHA:MgO surfaces, a reduced cellular viability was inferred (Figure 8).

The high biocompatibility of BHA:LiC coatings was assumed to be favored by the highest wettability and overall surface energy, as well as by an augmented contribution of the polar component of the surface energy. Other studies report in favor of this hypothesis that, in order to enhance the cellular response (including here adhesion, spreading, and cytoskeletal organization), an optimal window for the contact angle in the range of 60° to 80° between surface and water molecules should exist [166–168]. In this case, the cell proliferation results are in good agreement with this rationale, with the best results being obtained for coatings to elicit a water contact angle of ~72° and the highest polar component.

Dos Santos et al. [169] and Redey et al. [170] reported that the higher polar component of carbonated HA and natural calcium carbonated surfaces induced an increased osteoblast and osteoclast adhesion, respectively. Furthermore, Zhao et al. [171] indicated that a higher surface free energy is connected with the increased production of bone ALP and osteocalcin, in the case of MG63 osteoblastic cells grown on hydroxylated Ti surfaces.

6.4.2. Antibacterial Effect

Mihailescu et al. [54] tested BHA and BHA:MgF or BHA:MgO against gram-positive *Micrococcus sp.* and gram-negative *Enterobacter sp.* for 72 h. The BHA coatings and Ti substrate used as controls had no antibacterial effect. When BHA was doped with MgF$_2$, and pulsed laser deposited on a substrate, it exhibited a slight anti-biofilm activity (1 log reduction of viable cell counts as compared to bare Ti) toward *Enterobacter sp.* The highest anti-biofilm activity was observed for the BHA:MgO coatings, which inhibited the *Enterobacter sp.* biofilm in all development stages, *Micrococcus sp.* biofilm at 24 and 72 h, and *C. albicans* biofilm at 48 h.

In Reference [172], it was shown that MgO exhibits stronger bactericidal activity as compared to TiO_2. Two hypotheses explaining the decrease of the number of biofilm-embedded cells could be considered, i.e. (*i*) the surface coatings prevent the microbial/bacteria adhesion and (*ii*) the coatings exhibit a microbicidal activity, which kills the cells before or after contact with the coated surfaces. The second hypothesis is supported by the results of the minimal inhibitory concentration assay showing an increased microbicidal activity of BHA:MgO and BHA:MgF coatings as compared to BHA against the tested strains. No cytotoxic activity of deposited coatings against human cells was revealed during tests on Hep2 cells. It has been reported in literature that pure Mg is highly effective against *Staphylococcus aureus (S. aureus)* [173]. Numerous studies tested the antibacterial efficiency of MgO against bacteria. Sawai et al. [174] demonstrated the efficiency of MgO powders against *S. aureus* and *Escherichia coli*, while Gayathri et al. [175] created a doped HA containing Mg in the form of nanoparticles that had antibacterial efficiency, but also insecticide action, which killed the larvae of three mosquito species. Photoluminescence studies conducted by Sawai et al. [174] showed that MgO slurries generate active oxygen species such as O^{2-} and this seems to be the most probable cause for bacterial apoptosis. These oxygen species form in wet media hydroxyl radicals and hydrogen peroxide. Hydroxyl radical is the most reactive oxygen radical, able to interact with almost every type of molecule of the living cells of bacteria and fungi, which causes irreversible damage to cellular components and, eventually, apoptosis [176].

Mg is not toxic for the human organism in a dose of 200 to 400 mg per day. The Food and Nutrition Board of the Institute of Medicine in the United States has set an upper tolerable limit of 350 mg per day for supplemental magnesium. In larger doses, it can induce nausea and vomiting, muscle pain and weakness, confusion, and cardiac arrhythmias [177].

BHA and BHA doped with lithium compounds such as carbonate and phosphate (i.e., BHA:LiC or BHA:LiP coatings) proved antibacterial and antifungal activity against *S. aureus* and *C. albicans* strains [11], which was an unexpected discovery. After 24 h, all tested coatings manifested inhibitory activity for *S. aureus* colonies. After 48 and 72 h from inoculation, the antibacterial effect was much more visible. Therefore, a drastic decrease in the number of CFU/mL of 2 to 4 logs at these two-time intervals was recorded (Figure 9).

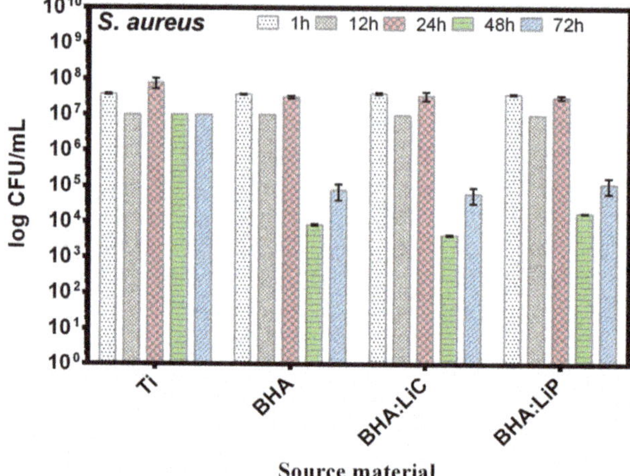

Figure 9. Graphic representation of the logarithmic values of *S. aureus* biofilm cells developed on the surfaces of Ti, bovine hydroxyapatite (BHA), and doped BHA coatings, at different time intervals (1–72 h) [11]. Slight modifications of the acronyms from the original figure were done for uniformization with the sample codes used in this review.

Representative SEM images of *S. aureus* biofilm developed on the surface of Ti and BHA:Li structures are given in Figure 10, at a magnification of 10,000×. One can observe that the density of *S. aureus* cells was significantly more reduced on BHA:Li surfaces as compared to Ti ones, which is in good agreement with the quantitative results presented in Figure 9.

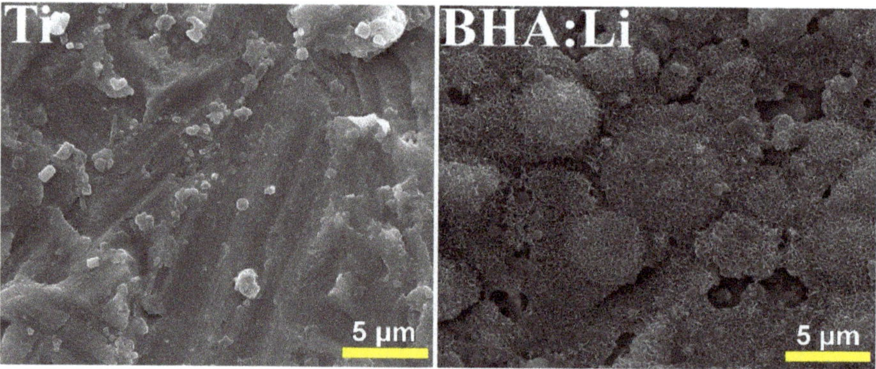

Figure 10. SEM images corresponding to *S. aureus* biofilm after 72 h of development on the surfaces of Ti and bovine hydroxyapatite doped with Li$_2$O coatings.

BHA:LiC coatings exhibited a slightly increased antimicrobial activity against *S. aureus* biofilm, as compared to BHA and BHA:LiP ones. In the case of *C. albicans*, the antifungal activity started to be pronounced after 12 h of incubation and with a drastic decrease of the fungal biofilm after 48 and 72 h, respectively (Figure 11).

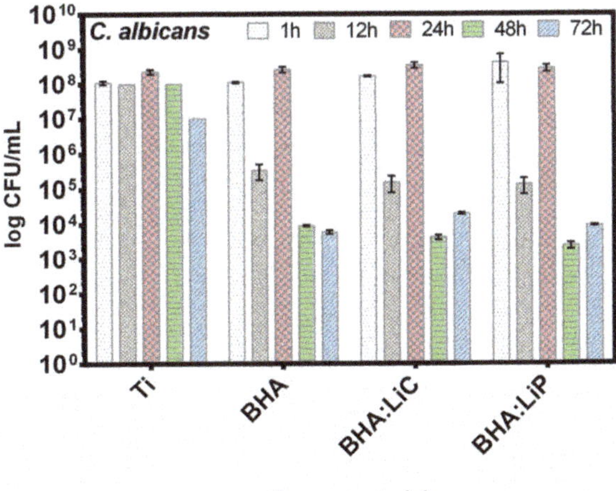

Source material

Figure 11. Graphic representation of the logarithmic values of *C. albicans* biofilm cells developed on the surfaces of Ti, bovine hydroxyapatite (BHA), and doped BHA coatings, at different time intervals (1–72 h) [11]. Slight modifications of the acronyms from the original figure were done for uniformization with the sample codes used in this review.

Again, the order of efficiency in reducing colonies was BHA:LiC > BHA:LiP > BHA. SEM images revealed that, when grown on the surface of Ti, *C. albicans* produced biofilms, with interconnected cells through filaments, while, when seeded on the surface of the synthesized BHA or doped BHA coatings, the cells were ovoid and adhered in isolated unicellular forms, with completely absent filaments (Figure 12).

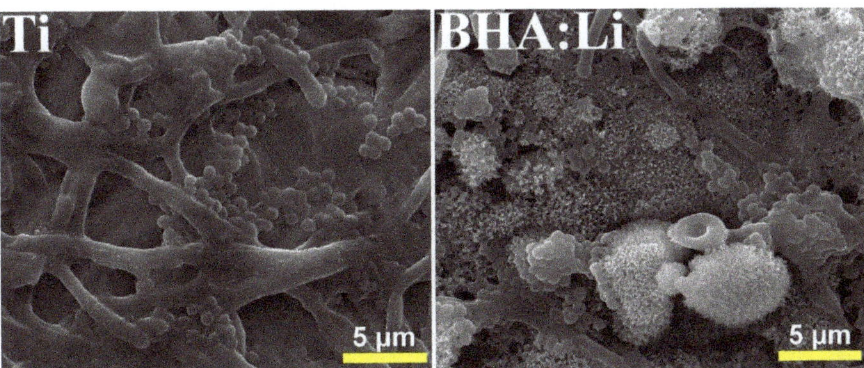

Figure 12. SEM images corresponding to *C. albicans* biofilm after 72 h of development on the surfaces of Ti and bovine hydroxyapatite doped with Li_2O coatings.

Representative SEM images of *C. albicans* biofilm developed on the surfaces of Ti and BHA:Li structures are given in Figure 12, at a magnification of 10,000×. While bare Ti samples were completely covered by a continuous layer of cells, the number of yeast cells developed on the surface of BHA:Li coatings was significantly decreased. They adhered exclusively in isolated, unicellular form, without interconnecting filaments. Duta et al. [11] assessed the antibacterial and antifungal activity of BioHA coatings to one of the following factors: (*i*) difference in polarity between negative-coated surfaces (associated with the presence of hydroxyl hydrophilic functional groups) and *S. aureus/C.*

albicans biofilms matrix, which is also negatively charged, due to its main component, represented by self-produced exopolysaccharides, (*ii*) coatings will facilitate the formation of a highly alkaline medium, which consequently leads to the destruction of lipids that are the main component of microbial cell membrane, or (*iii*) the smooth surfaces of the films, which reduce the contact area between material surface and bacterial cells, or (iv) metal accumulation inside bacterial cells, followed by the cellular wall disruption, and, eventually, cellular lysis.

6.5. In Vivo Tests

In vitro tests were demonstrated to play a key role in the biological assessment of biomaterials by offering in-sight information on their possible behavior in a biological environment. On the other hand, to understand the complex processes that might occur in a living system and to provide accurate data to completely validate the performances of a given biomaterial designed for clinical trials, in vivo tests are extremely important and should, therefore, follow in vitro ones.

In the literature, there are many in vivo reports on CaPs, but, to the best of our knowledge, there are few works only on HA coatings synthesized by PLD, and, unfortunately, no studies at all on BioHA coatings synthesized by PLD. However, the authors of this review are confident that future in vivo studies of BioHA coatings should deliver interesting results, since BioHA materials proved many times to possess important advantages over synthetic HA. In the following paragraphs, a selection of the papers dealing with PLD of synthetic HA will be considered and the obtained results will be reviewed.

The main purpose of CaP coatings is to stimulate and accelerate new bone formation around an implant [178], and most of the reported research studies revealed positive and favorable effects in this direction [179]. In general, the spectrum of animals used for testing CaP coatings synthesized by PLD is reduced, and this includes rats, rabbits, mini-pigs, dogs, goats, or sheep [14]. The advantages and disadvantages of different animal models used were discussed in Reference [139].

Chen et al. [180] synthesized using PLD fluoridated synthetic HA (HA:F) structures onto Ti implants and inserted them into rat femurs. By using microcomputed tomography, histology, and sequential polychromatic fluorescent investigations, their osteo-inductive and osseointegration activity in comparison with that of uncoated Ti implants was assessed. The results of their study demonstrated that HA:F-coated implants were characterized by superior osteogenesis and osseointegration properties (1.5 times more new bone at one week and more than four times more new bone at 4 and 8 weeks, respectively) as compared to bare (uncoated) Ti specimens, which can advance them as viable candidates for future applications in dentistry.

Mroz et al. [181] reported on Mg-doped HA coatings fabricated by PLD onto porous Ti implants, which were introduced in rabbit femoral bones for six months. The evaluation of bone–implant interaction and bone volume in the region of interest were performed. Histopathological investigations showed that all implants integrated well with the surrounding bone with ingrowth of newly formed bone into the pores of the implants. No inflammatory or foreign body reaction were noticed at the implant site. They note that, a significant increase in bone volume for Mg-doped HA implants, as compared to uncoated implants, was detected.

Peraire et al. [182] conducted a comparative study of the biological stability and the osteo-conductivity of HA coatings produced by PLD and plasma spraying. The functionalized grit-blasted titanium rods were implanted in the proximal tibia of mature New Zealand White rabbits. The obtained data suggested superior bone growth (percentage of bone contact) in the case of HA–PLD coatings (86%) as compared to the HA–PS coatings (67%) or the uncoated grit-blasted titanium (69%), after a period of 24 weeks implantation. Taking into consideration these results, the initiation of clinical evaluation of this coating method for medical applications was encouraged.

Mihailescu et al. [183] reported on synthetic HA, Mn^{2+}-doped carbonated HA and octacalcium phosphate (OCP) coatings fabricated by PLD onto Ti implants for insertion in the tibia bones of New Zealand White female rabbits. The bone–implants bonding was assessed by a tensile (pull-out) test. The measured value of the pull-out force was found to be more than double in the case of

CaP-functionalized Ti implants, as compared to the control (uncoated) ones. In addition, the pull-out force increased by 25% for the Mn^{2+}-doped carbonated HA and 10% in OCP-coated implants, in comparison with the synthetic HA-coated implants. Mihailescu et al. concluded that nanostructured CaP depositions manifested a significant potential for improving the performance of functionalized Ti implants in bone, and the composition and structure of the CaP coating have a significant influence on their biological effect.

Dostalova et al. [184] functionalized by PLD the surface of Ti6Al4V dental implants with HA coatings. Coated and uncoated implants were inserted into the lower jaw of mini-pigs. The evaluation of implants' osseointegration degree was assessed by polarized and fluorescent light microscopy with computer image processing. The results of the study proved that the osseointegration of the coated layer was superior for HA coatings (~77.2%) than the integration of uncoated Ti implants (~65.2%). With the owned advantage of delivering an inert and osseoconductive ceramic coating, the same team concluded that the PLD could be a viable method for coating metallic implants.

7. Conclusions and Perspectives

When animal-origin calcium phosphates (CaPs) are tackled, pulsed laser deposition (PLD) could be one of the most interesting choices for coating synthesis. If synthetic hydroxyapatite (HA) has a complex molecule with a large number of atoms and functional groups difficult to transfer in form of thin films by physical deposition techniques, natural apatites become even more complicated to transfer because of new functional groups and substitution ions, which further complicate the molecule. PLD is renowned for the ability to transfer stoichiometrically very complex molecules and this advantage should make it one of the premiere candidates for the transfer of such complex materials.

The most common sources for obtaining animal-origin HA for PLD targets are swine, ovine, and bovine teeth or bones. They are immersed in alkali solutions and further calcinated at ~800 °C to remove the organic parts. The powder resulting after grinding and milling can be pressed into targets or mixed with various doping agents to change mechanical or biocompatibility properties.

Coatings' morphology depends on the thickness, the origin of the natural apatite, and on the doping agents. In general, films are covered by round droplets that originate from the expulsion of the liquid phase during target irradiation. Their number and size can drastically differ depending on the experimental conditions. Usually, films in the range of hundreds of nanometers mimic the substrate morphology (especially if it is rough), with surface features that are not evident. Thickness seems to be dependent on the experimental conditions (number of pulses, laser fluence, and gas pressure), but also on the type of apatite. Studies on coatings synthesized in the same experimental conditions on sheep, cow, and cow-HA and bio-glass revealed different thicknesses for different types of materials.

Structural differences between synthetic HA and animal-origin apatites or between different types of animal origin-apatites cannot be shown by X-ray diffraction only. All these materials display the same crystalline peaks in the diffraction patterns. Differences appear in Fourier transform infrared spectroscopy that contain, in the case of animal-origin HA, bands corresponding to $(HPO_4)^{2-}$ or $(CO_3)^{2-}$ groups.

In terms of adherence to the substrate, PLD synthesized coatings from doped or undoped animal-origin HA could easily surpass the mandatory value of 15 MPa of bonding strength established by international standards for biocompatible coatings to be used in implantology.

Biocompatibility of animal-origin HA was consistently being reported as surpassing that of synthetic HA both in cells morphology, cells proliferation, and bioactivity. Moreover, when doped with Ti or Li_2CO_3, the effect was a boost in cell proliferation.

Animal-origin HA could be, therefore, a reliable, promising, safe, and, most importantly, low-cost source for coatings that reach biomimetism. For PLD, the field is still in his incipient development, with numerous possibilities to expand in terms of natural-HA sources, doping agents, or morphology and structural control of the obtained coatings. The transfer from PLD, which is essentially a laboratory technique to an industrial coating technique also belongs to the future.

Besides their often-increased bioactivity as compared to synthetic HA, BioHA coatings can display antibacterial and antifungal properties, which are not present in the case of synthetic HA. Undoped or BioHA doped with MgO, or carbonates and phosphates of lithium, were active against various microorganisms. Possible explanations for this property could be the presence of dopants, which have antibacterial and antifungal properties by themselves, the smoother surfaces of PLD deposited BioHA films, which reduce the contact area between coating and bacteria, and the metal ions accumulation inside microorganisms followed by the membrane disruption and cell apoptosis.

The authors of this review express their confidence that the future of biomimetic coatings belongs to the natural-origin CaPs source materials and their perfect transfer and delicate tuning in terms of thickness, crystallinity, and functional groups can be optimally studied using PLD as deposition method.

Author Contributions: Conceptualization, L.D. and A.C.P. Methodology, L.D. and A.C.P. Software, L.D. Validation, L.D. and A.C.P. Formal analysis, L.D. and A.C.P. Investigation, L.D. and A.C.P. Resources, L.D. and A.C.P. Data curation, L.D. and A.C.P. Writing–original draft preparation, L.D. and A.C.P. Writing–review and editing, L.D. and A.C.P. Visualization, L.D. and A.C.P. Supervision, L.D. and A.C.P. Project administration, L.D. and A.C.P.

Funding: Two grants from the Ministry of Research and Innovation, CNCS-UEFISCDI, No. PN-III-P1-1.1-PD-2016-1568 (PD 6/2018) and PN-III-P1-1.1-TE-2016-2015 (TE136/2018), within PNCDI III supported this work. The authors acknowledge and thank the partial support from the Core Programme – Contract 16N/2019.

Acknowledgments: The authors acknowledge G.-P. Pelin for the SEM images that were used in this review for exemplification purposes and for the schematic with the PLD experimental set-up, C. Chifiriuc for data on biofilm development on bovine hydroxyapatite doped with Li_2O coatings, and F.N Oktar for providing animal-origin powders that were used for numerous studies of bioactive coating synthesis.

Conflicts of Interest: The authors declare no conflict of interest.

References

1. Oladele, I.; Agbabiaka, O.; Olasunkanmi, O. Non-synthetic sources for the development of hydroxyapatite. *J. Appl. Biotechnol. Bioeng.* **2018**, *5*, 92–99. [CrossRef]

2. Akram, M.; Ahmed, R.; Shakir, I.; Ibrahim, W.A.W.; Hussain, R. Extracting hydroxyapatite and its precursors from natural resources. *J. Mater. Sci.* **2014**, *49*, 1461–1475. [CrossRef]

3. Bigi, A.; Bracci, B.; Cuisinier, F.; Elkaim, R.; Fini, M.; Mayer, I.; Mihailescu, I.N.; Socol, G.; Sturba, L.; Torricelli, P. Human osteoblast response to pulsed laser deposited calcium phosphate coatings. *Biomaterials* **2005**, *26*, 2381–2385. [CrossRef]

4. Mihailescu, I.N.; Torricelli, P.; Bigi, A.; Mayer, I.; Iliescu, M.; Werckmann, J.; Socol, G.; Miroiu, F.; Cuisinier, F.; Elkaim, R.; et al. Calcium phosphate thin films synthesized by pulsed laser deposition: Physico-chemical characterization and in vitro cell response. *Appl. Surf. Sci.* **2005**, *248*, 344–348. [CrossRef]

5. Dorozhkin, S. History of Calcium Phosphates in Regenerative Medicine. In *Advances in Calcium Phosphate Biomaterials*; Ben-Nissan, B., Ed.; Springer: Berlin/Heidelberg, Germany, 2014; Volume 2, pp. 435–483. [CrossRef]

6. Mihailescu, I.N.; Ristoscu, C.; Bigi, A.; Mayer, I. Advanced biomimetic implants based on nanostructured coatings synthesized by pulsed laser technologies. In *Laser-Surface Interactions for New Materials Production, Tailoring Structure and Properties*; Miotello, A., Ossi, M., Eds.; Springer: Berlin/Heidelberg, Germany, 2010; pp. 235–268. [CrossRef]

7. O'Hare, P.; Meenan, B.J.; Burke, G.A.; Byrne, G.; Dowling, D.; Hunt, J.A. Biological responses to hydroxyapatite surfaces deposited via a co-incident microblasting technique. *Biomaterials* **2010**, *31*, 515–522. [CrossRef]

8. Habibovic, P.; Kruyt, M.C.; Juhl, M.V.; Clyens, S.; Martinetti, R.; Dolcini, L.; Theilgaard, N.; van Blitterswijk, C.A. Comparative in vivo study of six hydroxyapatite-based bone graft substitutes. *J. Orthop. Res.* **2008**, *26*, 1363–1370. [CrossRef]

9. Marini, E.; Ballanti, P.; Silvestrini, G.; Valdinucci, F.; Bonucci, E. The presence of different growth factors does not influence bone response to hydroxyapatite: preliminary results. *J. Orthopaed. Traumatol.* **2004**, *5*, 34–43. [CrossRef]

10. Rabiee, S.; Moztarzadeh, F.; Solati-Hashjin, M. Synthesis and characterization of hydroxyapatite cement. *J. Mol. Struct.* **2010**, *969*, 172–175. [CrossRef]

11. Duta, L.; Chifiriuc, M.C.; Popescu-Pelin, G.; Bleotu, C.; (Pircalabioru) Gradisteanu, G.; Anastasescu, M.; Achim, A.; Popescu, A. Pulsed Laser Deposited Biocompatible Lithium-Doped Hydroxyapatite Coatings with Antimicrobial Activity. *Coatings* **2019**, *9*, 54. [CrossRef]

12. Predoi, D.; Iconaru, S.L.; Predoi, M.V. Bioceramic Layers with Antifungal Properties. *Coatings* **2018**, *8*, 276. [CrossRef]

13. Prosolov, K.A.; Belyavskaya, O.A.; Linders, J.; Loza, K.; Prymak, O.; Mayer, C.; Rau, J.V.; Epple, M.; Sharkeev, Y.P. Glancing Angle Deposition of Zn-Doped Calcium Phosphate Coatings by RF Magnetron Sputtering. *Coatings* **2019**, *9*, 220. [CrossRef]

14. León, B.; Jansen, J.A. *Thin Calcium Phosphate Coatings for Medical Implants*; Springer: New York, NY, USA, 2009; p. 328. [CrossRef]

15. Rodríguez-Lorenzo, L.M.; Vallet-Regí, M.; Ferreira, J.M.F.; Ginebra, M.P.; Aparicio, C.; Planell, J.A. Hydroxyapatite ceramic bodies with tailored mechanical properties for different applications. *J. Biomed. Mater. Res.* **2002**, *60*, 159–166. [CrossRef]

16. Sima, L.E.; Stan, G.E.; Morosanu, C.O.; Melinescu, A.; Ianculescu, A.; Melinte, R.; Neamtu, J.; Petrescu, S.M. Differentiation of mesenchymal stem cells onto highly adherent radio frequency-sputtered carbonated hydroxylapatite thin films. *J. Biomed. Mater. Res.* **2010**, *95*, 1203–1214. [CrossRef]

17. Liu, X.; He, D.; Zhou, Z.; Wang, G.; Wang, Z.; Wu, X.; Tan, Z. Characteristics of (002) Oriented Hydroxyapatite Coatings Deposited by Atmospheric Plasma Spraying. *Coatings* **2018**, *8*, 258. [CrossRef]

18. Chrisey, D.B.; Hubler, G.K. *Pulsed Laser Deposition of Thin Films*, 1st ed.; John Wiley & Sons: Hoboken, NJ, USA, 1994; pp. 1–649.

19. Eason, R. *Pulsed Laser Deposition of Thin Films-Applications-Led Growth of Functional Materials*; Wiley-Interscience: Hoboken, NJ, USA, 2006; pp. 1–682.

20. Yang, Y.; Wu, Q.; Wang, M.; Long, J.; Mao, Z.; Chen, X. Hydrothermal Synthesis of Hydroxyapatite with Different Morphologies: Influence of Supersaturation of the Reaction System. *Cryst. Growth Des.* **2014**, *14*, 4864–4871. [CrossRef]

21. Zhu, Y.; Xu, L.; Liu, C.; Zhang, C.; Wu, N. Nucleation and growth of hydroxyapatite nanocrystals by hydrothermal method. *AIP Adv.* **2018**, *8*, 085221. [CrossRef]

22. Pham, V.H.; Van, H.N.; Tam, P.D.; Ha, H.N.T. A novel 1540 nm light emission from erbium doped hydroxyapatite/beta-tricalcium phosphate through co-precipitation method. *Mater. Lett.* **2016**, *167*, 145–147. [CrossRef]

23. Ciobanu, C.S.; Popa, C.L.; Predoi, D. Cerium-doped hydroxyapatite nanoparticles synthesized by the co-precipitation method. *J. Serb. Chem. Soc.* **2016**, *81*, 433–446. [CrossRef]

24. Papynov, E.K.; Shichalina, O.O.; Modin, E.B.; Yu Mayorov, V.; Portnyagin, A.S.; Kobylyakov, S.P.; Golub, A.V.; Medkov, M.A.; Tananaev, I.G.; Avramenk, V.A. Wollastonite ceramics with bimodal porous structure prepared by Sol-Gel and SPS techniques. *RSC Adv.* **2016**, *6*, 34066–34073. [CrossRef]

25. Duta, L.; Mihailescu, N.; Popescu, A.C.; Luculescu, C.R.; Mihailescu, I.N.; Cetin, G.; Gunduz, O.; Oktar, F.N.; Popa, A.C.; Kuncser, A.; et al. Comparative physical, chemical and biological assessment of simple and titanium-doped ovine dentine-derived hydroxyapatite coatings fabricated by pulsed laser deposition. *Appl. Surf. Sci.* **2017**, *413*, 129–139. [CrossRef]

26. Oktar, F. Microstructure and mechanical properties of sintered enamel hydroxyapatite. *Ceram. Int.* **2007**, *33*, 1309–1314. [CrossRef]

27. Goller, G.; Oktar, F.N.; Agathopoulos, S.; Tulyaganov, D.U.; Ferreira, J.M.F.; Kayali, E.S.; Peker, I. Effect of sintering temperature on mechanical and microstructural properties of bovine hydroxyapatite (BHA). *J. Sol-Gel Sci. Techn.* **2006**, *37*, 111–115. [CrossRef]

28. Eisenhart, S. *EU Regulation 722. New EU Animal Tissue Regulations in Effect for Some Medical Devices*; Emergo: Hong Kong, China, 2013; Available online: https://www.emergobyul.com/blog/2013/09/new-eu-animaltissue-regulations-effect-some-medical-devices (accessed on 29 April 2019).

29. ISO 22442-1. *Medical Devices Utilizing Animal Tissues and Their Derivatives-Part 1: Application of Risk Management*; International Organization for Standardization: Berlin, Germany, 2015.

30. Barakat, N.A.M.; Khil, M.S.; Omran, A.M.; Sheikh, F.A.; Kim, H.Y. Extraction of pure natural hydroxyapatite from the bovine bones bio waste by three different methods. *J. Mater. Process. Technol.* **2009**, *209*, 3408–3415. [CrossRef]

31. Sadat-Shojai, M.; Khorasani, M.T.; Dinpanah-Khoshdargi, E.; Jamshidi, A. Synthesis methods for nanosized hydroxyapatite with diverse structures. *Acta Biomater.* **2013**, *9*, 7591–7621. [CrossRef] [PubMed]

32. Elliott, J.C.; Wilson, R.M.; Dowker, S.E.P. Apatite structures. *Adv. X-ray Anal.* **2002**, *45*, 172–181.

33. Wopenka, B.; Pasteris, J.D. A mineralogical perspective on the apatite in bone. *Mater. Sci. Eng. C* **2005**, *25*, 131–143. [CrossRef]

34. Ben-Nissan, B. *Advances in calcium phosphate biomaterials*; Part of the Springer Series in Biomaterials Science and Engineering book series; Springer: Berlin, Germany, 2014; Volume 2, p. 547.

35. Mucalo, M. *Hydroxyapatite (HAp) for Biomedical Applications*; Elsevier: Amsterdam, The Netherlands, 2015; p. 404. [CrossRef]

36. Šupová, M. Substituted hydroxyapatites for biomedical applications: A review. *Ceram. Int.* **2015**, *41*, 9203–9231. [CrossRef]

37. Combes, C.; Cazalbou, S.; Rey, C. Apatite biominerals. *Minerals* **2016**, *6*, 34. [CrossRef]

38. Fleet, M.E.; Liu, X. Coupled substitution of type A and B carbonate in sodium-bearing apatite. *Biomaterials* **2007**, *28*, 916–926. [CrossRef]

39. Astala, R.; Stott, M.J. First principles investigation of mineral component of bone: CO_3 substitutions in hydroxyapatite. *Chem. Mater.* **2005**, *17*, 4125–4133. [CrossRef]

40. Landi, E.; Tampieri, A.; Mattioli-Belmonte, M.; Celotti, G.; Sandri, M.; Gigante, A.; Fava, P.; Biagini, G. Biomimetic Mg- and Mg, CO3-substituted hydroxyapatites: Synthesis characterization and in vitro behaviour. *J. Eur. Ceram. Soc.* **2006**, *26*, 2593–2601. [CrossRef]

41. Tite, T.; Popa, A.-C.; Balescu, L.M.; Bogdan, I.M.; Pasuk, I.; Ferreira, J.M.F.; Stan, G.E. Cationic Substitutions in Hydroxyapatite: Current Status of the Derived Biofunctional Effects and Their In Vitro Interrogation Methods. *Materials* **2018**, *11*, 2081. [CrossRef]

42. Verdelis, K.; Lukashova, L.; Wright, J.T.; Mendelsohn, R.; Peterson, M.G.E.; Doty, S.; Boskey, A.L. Maturational changes in dentin mineral properties. *Bone* **2007**, *40*, 1399–1407. [CrossRef]

43. Sun, R.-X.; Lv, Y.; Niu, Y.-R.; Zhao, X.-H.; Cao, D.-S.; Tang, J.; Sun, X.-C.; Chen, K.-Z. Physicochemical and biological properties of bovine-derived porous hydroxyapatite/collagen composite and its hydroxyapatite powders. *Ceram. Int.* **2017**, *43*, 16792–16798. [CrossRef]

44. Ahmad, N.; Bukhari, S.A.; Akhtar, N.; Haq, I. Serum hormonal, electrolytes and trace element profiles in the rutting and non-rutting one-humped male camel (*Camelus dromedarius*). *Anim. Reprod. Sci.* **2007**, *101*, 172–178.

45. Rahavi, S.S.; Ghaderi, O.; Monshi, A.; Fathi, M.H. A comparative study on physicochemical properties of hydroxyapatite powders derived from natural and synthetic sources. *Russ. J. Non-Ferrous Metals* **2017**, *58*, 276–286. [CrossRef]

46. Ofudje, E.A.; Rajendran, A.; Adeogun, A.I.; Idowu, M.A.; Kareem, S.O.; Pattanayak, D.K. Synthesis of organic derived hydroxyapatite scaffold from pig bone waste for tissue engineering applications. *Adv. Powder Technol.* **2017**, *29*, 1–8. [CrossRef]

47. Mohd Pu'ad, N.A.S.; Koshy, P.; Abdullah, H.Z.; Idris, M.I.; Lee, T.C. Syntheses of hydroxyapatite from natural sources. *Heliyon* **2019**, *5*, e01588. [CrossRef]

48. Labarthe, J.C.; Bonel, G.; Montel, G. Structure and properties of B-type phosphocalcium carbonate apatites. *Ann. Chim.* **1973**, *8*, 289–301.

49. Wu, Y.; Glimcher, M.J.; Rey, C.; Ackerman, J.L. A unique protonated phosphate group in bone mineral not present in synthetic calcium phosphates. Identification by phosphorus-31 solid state NMR spectroscopy. *J. Mol. Biol.* **1994**, *244*, 423–435. [CrossRef]

50. Winand, L. Etude physico-chimique du phosphate tricalcique hydraté et de l'hydroxyapatite. *Ann. Chim.* **1961**, *6*, 951–967.

51. Riggs, B.L.; Hodgson, S.F.; O'Fallon, W.M.; Chao, E.Y.; Wahner, H.W.; Muhs, J.M.; Cedel, S.L.; Melton III, J. Effect of fluoride treatment on the fracture rate in postmenopausal women with osteoporosis. *N. Engl. J. Med.* **1990**, *322*, 802–809. [CrossRef]

52. Kay, H.M.; Wilson, M. The in vitro effects of amine fluorides on plaque bacteria. *J. Periodontol.* **1988**, *59*, 266–269. [CrossRef]

53. Duta, L.; Oktar, F.N.; Stan, G.E.; Popescu-Pelin, G.; Serban, N.; Luculescu, C.; Mihailescu, I.N. Novel doped hydroxyapatite thin films obtained by pulsed laser deposition. *Appl. Surf. Sci.* **2013**, *265*, 41–49. [CrossRef]

54. Mihailescu, N.; Stan, G.E.; Duta, L.; Chifiriuc, C.M.; Coralia, B.; Sopronyi, M.; Luculescu, C.; Oktar, F.N.; Mihailescu, I.N. Structural, compositional, mechanical characterization and biological assessment of bovine-derived hydroxyapatite coatings reinforced with MgF_2 or MgO for implants functionalization. *Mater. Sci. Eng. C* **2016**, *59*, 863–874. [CrossRef]

55. Jayathilakan, K.; Sultana, K.; Radhakrishna, K.; Bawa, A.S. Utilization of byproducts and waste materials from meat, poultry and fish processing industries: a review. *J Food Sci Technol.* **2012**, *49*, 278–293. [CrossRef]

56. Kim, S.-H.; Shin, J.-W.; Park, S.-A.; Kim, Y.K.; Park, M.S.; Mok, J.M.; Yang, W.I.; Lee, J.W. Chemical, structural properties, and osteoconductive effectiveness of bone block derived from porcine cancellous bone. *J. Biomed. Mater. Res. Part B* **2004**, *68*, 69–74. [CrossRef]

57. Barakat, N.A.M.; Khalil, K.A.; Sheikh, F.A.; Omran, A.M.; Gaihre, B.; Khil, S.M.; Kim, H.Y. Physiochemical characterizations of hydroxyapatite extracted from bovine bones by three different methods: extraction of biologically desirable Hap. *Mater. Sci. Eng. C-Mater.* **2008**, *28*, 1381–1387. [CrossRef]

58. Liu, Q.; Huang, S.; Matinlinna, J.P.; Chen, Z.; Pan, H. Insight into Biological Apatite: Physiochemical Properties and Preparation Approaches. *Biomed Res. Int.* **2013**, *2013*, 929748. [CrossRef]

59. Seo, D.S.; Kim, Y.G.; Lee, J.K. Sintering and dissolution of bone ash-derived hydroxyapatite. *Met. Mater.-Int.* **2010**, *16*, 687–692. [CrossRef]

60. Popescu, A.C.; Florian, P.E.; Stan, G.E.; Popescu-Pelin, G.; Zgura, I.; Enculescu, M.; Oktar, F.N.; Trusca, R.; Sima, L.E.; Roseanu, A.; et al. Physical-chemical characterization and biological assessment of simple and lithium-doped biological-derived hydroxyapatite thin films for a new generation of metallic implants. *Appl. Surf. Sci.* **2018**, *439*, 724–735. [CrossRef]

61. Figueiredo, M.; Fernando, A.; Martins, G.; Freitas, J.; Judas, F.; Figueiredo, H. Effect of the calcination temperature on the composition and microstructure of hydroxyapatite derived from human and animal bone. *Ceram. Int.* **2010**, *36*, 2383–2393. [CrossRef]

62. Ramesh, S.; Loo, Z.Z.; Tan, C.Y.; Chew, W.J.K.; Ching, Y.C.; Tarlochan, F.; Chandran, H.; Krishnasamy, S.; Bang, L.T.; Sarhan, A.A.D.M. Characterization of biogenic hydroxyapatite derived from animal bones for biomedical applications. *Ceram. Int.* **2018**, *44*, 10525–10530. [CrossRef]

63. Akyurt, N.; Yetmez, M.; Karacayli, U.; Gunduz, O.; Agathopoulos, S.; Gökçe, H.; Öveçoğlu, M.L.; Oktar, F.N. A New Natural Biomaterial: Sheep Dentine Derived Hydroxyapatite. *Key. Eng. Mater* **2012**, *493–494*, 281–286. [CrossRef]

64. Gheisari, H.; Karamian, E.; Abdellahi, M. A novel hydroxyapatite – hardstonite nanocomposite ceramic. *Ceram. Int.* **2015**, *41*, 5967–5975. [CrossRef]

65. Khandan, A.; Abdellahi, M.; Barenji, R.V.; Ozada, N.; Karamian, E. Introducing natural hydroxyapatite-diopside(NHA-Di) nano-bioceramic coating. *Ceram. Int.* **2015**, *41*, 12355–12363. [CrossRef]

66. Giraldo-Betancur, A.L.; Espinosa-Arbelaez, D.G.; del Real-López, A.; Millan-Malo, B.M.; Rivera-Muñoz, E.M.; Gutierrez-Cortez, E.; Pineda-Gomez, P.; Jimenez-Sandoval, S.; Rodriguez-García, M.E. Comparison of physicochemical properties of bio and commercial hydroxyapatite. *Curr. Appl. Phys.* **2013**, *13*, 1383–1390. [CrossRef]

67. Rincón-López, J.A.; Hermann-Muñoz, J.A.; Giraldo-Betancur, A.L.; De Vizcaya-Ruiz, A.; Alvarado-Orozco, J.M.; Muñoz-Saldaña, J. Synthesis, Characterization and In Vitro Study of Synthetic and Bovine-Derived Hydroxyapatite Ceramics: A Comparison. *Materials* **2018**, *11*, 333. [CrossRef] [PubMed]

68. Bianchi, M.; Gambardella, A.; Graziani, G.; Liscio, F.; Maltarello, M.C.; Boi, M.; Berni, M.; Bellucci, D.; Marchiori, G.; Valle, F.; et al. Plasma-assisted deposition of bone apatite-like thin films from natural apatite. *Mater. Lett.* **2017**, *199*, 32–36. [CrossRef]

69. Pramanik, S.; Hanif, A.S.M.; Pingguan-Murphy, B.; Osman, N.A.A. Morphological change of heat treated bovine bone: a comparative study. *Mater.* **2013**, *6*, 65–75. [CrossRef] [PubMed]

70. Kusrini, E.; Sontang, M. Characterization of x-ray diffraction and electron spin resonance: Effects of sintering time and temperature on bovine hydroxyapatite. *Radiat. Phys. Chem.* **2012**, *81*, 118–125. [CrossRef]

71. Bahrololoom, M.E.; Javidi, M.; Javadpour, S.; Ma, J. Characterisation of natural hydroxyapatite extracted from bovine cortical bone ash. *J. Ceram. Process. Res.* **2009**, *10*, 129–138.

72. Bano, N.; Jikan, S.S.; Basri, H.; Bakar, S.A.A.S.; Nuhu, A.H. Natural hydroxyapatite extracted from bovine bone. *J. Sci. Technol.* **2017**, *9*, 22–28.

73. Ruksudjarit, A.; Pengpat, K.; Rujijanagul, G.; Tunkasiri, T. Synthesis and characterization of nanocrystalline hydroxyapatite from natural bovine bone. *Curr. Appl. Phys.* **2008**, *8*, 270–272. [CrossRef]

74. Nasiri-Tabrizi, B.; Fahami, A.; Ebrahimi-Kahrizsangi, R. A comparative study of hydroxyapatite nanostructures produced under different milling conditions and thermal treatment of bovine bone. *J. Ind. Eng. Chem.* **2014**, *20*, 245–258. [CrossRef]

75. Rakmae, S.; Lorprayoon, C.; Ekgasit, S.; Suppakarn, N. Influence of heat-treated bovine bone-derived hydroxyapatite on physical properties and in vitro degradation behavior of poly (lactic acid) composites. *Polym. Plast. Technol.* **2013**, *52*, 1043–1053. [CrossRef]

76. Nirmala, R.; Sheikh, F.A.; Kanjwal, M.A.; Lee, J.H.; Park, S.-J.; Navamathavan, R.; Kim, H.Y. Synthesis and characterization of bovine femur bone hydroxyapatite containing silver nanoparticles for the biomedical applications. *J. Nanopart. Res.* **2011**, *13*, 1917–1927. [CrossRef]

77. Herliansyah, M.K.; Hamdi, M.; Ide-Ektessabi, A.; Wildan, M.W.; Toque, J.A. The influence of sintering temperature on the properties of compacted bovine hydroxyapatite. *Mater. Sci. Eng. C* **2009**, *29*, 1674–1680. [CrossRef]

78. Ayatollahi, M.R.; Yahya, M.Y.; Asgharzadeh Shirazi, H.; Hassan, S.A. Mechanical and tribological properties of hydroxyapatite nanoparticles extracted from natural bovine bone and the bone cement developed by nano-sized bovine hydroxyapatite filler. *Ceram. Int.* **2015**, *41*, 10818–10827. [CrossRef]

79. Jaber, H.L.; Hammood, A.S.; Parvin, N. Synthesis and characterization of hydroxyapatite powder from natural Camelus bone. *J. Aust. Ceram. Soc.* **2018**, *54*, 1–10. [CrossRef]

80. Duta, L.; Serban, N.; Oktar, F.N.; Mihailescu, I.N. Biological hydroxyapatite thin films synthesized by pulsed laser deposition. *Optoelectron. Adv. Mater.-Rapid Commun.* **2013**, *7*, 1040–1044.

81. Maidaniuc, A.; Miculescu, F.; Voicu, S.I.; Andronescu, C.; Miculescu, M.; Matei, E.; Mocanu, A.C.; Pencea, I.; Csaki, I.; Machedon-Pisu, T.; et al. Induced wettability and surface-volume correlation of composition for bovine bone derived hydroxyapatite particles. *Appl. Surf. Sci.* **2018**, *438*, 158–166. [CrossRef]

82. Ferraro, V.; Carvalho, A.P.; Piccirillo, C.; Santos, M.M.; Castro, P.M.L.; Pintado, M.E. Extraction of high added value biological compounds from sardine, sardine-type fish and mackerel canning residues-A review. *Mat. Sci. Eng. C-Mater.* **2013**, *33*, 3111–3120. [CrossRef] [PubMed]

83. Venkatesan, J.; Qian, Z.J.; Ryu, M.; Thomas, N.V.; Kim, S.K. A comparative study of thermal calcination and an alkaline hydrolysis method in the isolation of hydroxyapatite from Thunnus obesus bone. *Biomed. Mater.* **2011**, *6*, 12. [CrossRef]

84. Kannan, S.; Rocha, J.H.G.; Agathopoulos, S.; Ferreira, J.M.F. Fluorine-substituted hydroxyapatite scaffolds hydrothermally grown from aragonitic cuttlefish bones. *Acta Biomater.* **2007**, *3*, 243–249. [CrossRef] [PubMed]

85. Boutinguiza, M.; Pou, J.; Comesaña, R.; Lusquiños, F.; de Carlos, A.; León, B. Biological hydroxyapatite obtained from fish bones. *Mater. Sci. Eng. C-Mater. Biol. Appl.* **2012**, *32*, 478–486. [CrossRef]

86. Huang, Y.C.; Hsiao, P.C.; Chai, H.J. Hydroxyapatite extracted from fish scale: Effects on MG63 osteoblast-like cells. *Ceram. Int.* **2011**, *37*, 1825–1831. [CrossRef]

87. Piccirillo, C.; Silva, M.F.; Pullar, R.C.; Braga da Cruz, I.; Jorge, R.; Pintado, M.M.; Castro, P.M. Extraction and characterisation of apatite- and tricalcium phosphate-based materials from cod fish bones. *Mat. Sci. Eng. C-Mater.* **2013**, *33*, 103–110. [CrossRef] [PubMed]

88. Senthil, R.; Vedakumari, S.W.; Sastry, T.P. Hydroxyapatite and Demineralized Bone Matrix from Marine Food Waste – A Possible Bone Implant. *Am. J. Mat. Synth. Proc.* **2018**, *3*, 1–6. [CrossRef]

89. Hidalgo-Robatto, B.M.; Aguilera-Correa, J.J.; López-Álvarez, M.; Romera, D.; Esteban, J.; Gonzalez, P.; Serra, J. Fluor-carbonated hydroxyapatite coatings by pulsed laser deposition to promote cell viability and antibacterial properties. *Surf. Coat. Technol.* **2018**, *349*, 736–744. [CrossRef]

90. Aguiar, H.; Chiussi, S.; López-Álvarez, M.; González, P.; Serra, J. Structural characterization of bioceramics and mineralized tissues based on Raman and XRD techniques. *Ceram. Int.* **2018**, *44*, 495–504. [CrossRef]

91. Lopez-Alvarez, M.; Vigo, E.; Rodriguez-Valencia, C.; Outeirino-Iglesias, V.; Gonzalez, P.; Serra, J. In vivo evaluation of shark teeth-derived bioapatites. *Clin. Oral Impl. Res.* **2016**, *28*, e91–e100. [CrossRef] [PubMed]

92. Lopez-Alvarez, M.; Pérez-Davila, S.; Rodríguez-Valencia, C.; González, P.; Serra, J. The improved biological response of shark tooth bioapatites in a comparative in vitro study with synthetic and bovine bone grafts. *Biomed. Mater.* **2016**, *11*, 035011. [CrossRef]

93. Rivera, E.M.; Araiza, M.; Brostow, W.; Castano, V.M.; Estrada, J.R.D.; Hernandez, R.; Rodrigues, J.R. Synthesis of hydroxyapatite from eggshells. *Mater. Lett.* **1999**, *41*, 128–134. [CrossRef]

94. Elizondo-Villarreal, N.; Martínez-de-la-Cruz, A.; Obregón Guerra, R.; Gómez-Ortega, J.L.; Torres-Martínez, L.M.; Castaño, V.M. Biomaterials from Agricultural Waste: Eggshell-based Hydroxyapatite. *Water Air Soil Pollut.* **2012**, *223*, 3643–3646. [CrossRef]

95. Chaudhuri, B.; Mondal, B.; Modak, D.K.; Pramanik, K.; Chaudhuri, B.K. Preparation and characterization of nanocrystalline hydroxyapatite from eggshell and K2HPO4 solution. *Mater. Lett.* **2013**, *97*, 148–150. [CrossRef]

96. Tamasan, M.; Ozyegin, L.S.; Oktar, F.N.; Simon, V. Characterization of calcium phosphate powders originating from Phyllacanthus imperialis and Trochidae Infundibulum concavus marine shells. *Mat. Sci. Eng. C-Mater* **2013**, *33*, 2569–2577. [CrossRef]

97. Gunduz, O.; Sahin, Y.M.; Agathopoulos, S.; Ben-Nissan, B.; Oktar, F.N. A New Method for Fabrication of Nanohydroxyapatite and TCP from the Sea Snail Cerithium vulgatum. *J. Nanomater.* **2014**, *2014*, 382861. [CrossRef]

98. Li, S.; Wang, J.; Jing, X.; Liu, Q.; Saba, J.; Mann, T.; Zhang, M.; Wei, H.; Chen, R.; Liu, L. Conversion of calcined eggshells into flower-like hydroxyapatite agglomerates by solvothermal method using Hydrogen peroxide/N,N-dimethylformamide mixed solvents. *J. Am. Ceram. Soc.* **2012**, *95*, 3377–3379. [CrossRef]

99. Ho, W.-F.; Hsu, H.-C.; Hsu, S.-K.; Hung, C.-W.; Wu, S.-C. Calcium phosphate bioceramics synthesized from eggshell powders through a solid state reaction. *Ceram. Int.* **2013**, *39*, 6467–6473. [CrossRef]

100. Surmenev, R.A. A review of plasma-assisted methods for calcium phosphate-based coatings fabrication. *Surf. Coat. Technol.* **2012**, *206*, 2035–2056. [CrossRef]

101. Surmenev, R.A.; Surmeneva, M.A.; Ivanova, A.A. Significance of calcium phosphate coatings for the enhancement of new bone osteogenesis—A review. *Acta Biomater.* **2014**, *10*, 557–579. [CrossRef]

102. Graziani, G.; Boi, M.; Bianchi, M. A Review on Ionic Substitutions in Hydroxyapatite Thin Films: Towards Complete Biomimetism. *Coatings* **2018**, *8*, 269. [CrossRef]

103. Schneider, C.W.; Lippert, T. Laser Ablation and Thin Film Deposition. In *Laser Processing of Materials*; Schaaf, P., Ed.; Springer Series in Materials Science; Springer: Berlin/Heidelberg, Germany, 2010; Volume 139, pp. 89–112.

104. Caricato, A.P.; Martino, M.; Romano, F.; Mirchin, N.; Peled, A. Pulsed laser photodeposition of a-Se nanofilms by ArF laser. *Appl. Surf. Sci.* **2007**, *253*, 6517–6521. [CrossRef]

105. Hashimoto, Y.; Ueda, M.; Kohiga, Y.; Imura, K.; Hontsu, S. Application of fluoridated hydroxyapatite thin film coatings using KrF pulsed laser deposition. *Dent Mater J.* **2018**, *37*, 408–413. [CrossRef] [PubMed]

106. Zaki, A.M.; Blythe, H.J.; Heald, S.M.; Fox, A.M.; Gehring, G.A. Growth of high quality yttrium iron garnet films using standard pulsed laser deposition technique. *J. Magn. Magn. Mater.* **2018**, *453*, 254–257. [CrossRef]

107. Novotný, M.; Vondráček, M.; Marešová, E.; Fitl, P.; Bulíř, J.; Pokornýa, P.; Havlová, Š.; Abdellaoui, N.; Pereira, A.; Hubík, P.; et al. Optical and structural properties of ZnO:Eu thin films grown by pulsed laser deposition. *Appl. Surf. Sci.* **2019**, *476*, 271–275. [CrossRef]

108. Gyorgy, E.; Ristoscu, C.; Mihailescu, I.N. Role of laser pulse duration and gas pressure in deposition of AlN thin films. *J. Appl. Phys.* **2001**, *90*, 456. [CrossRef]

109. Boyd, I.W. Thin film growth by pulsed laser deposition. *Ceram. Int.* **1996**, *22*, 429–434. [CrossRef]

110. Popescu, A.C.; Duta, L.; Dorcioman, G.; Mihailescu, I.N.; Stan, G.E.; Pasuk, I.; Zgura, I.; Beica, T.; Enculescu, I.; Ianculescu, A.; et al. Radical modification of the wetting behavior of textiles coated with ZnO thin films and nanoparticles when changing the ambient pressure in the pulsed laser deposition process. *J. Appl. Phys.* **2011**, *110*, 064321. [CrossRef]

111. Murray, M.; Jose, G.; Richards, B.; Jha, A. Femtosecond pulsed laser deposition of silicon thin films. *Nanoscale Res. Lett.* **2013**, *8*, 272. [CrossRef] [PubMed]

112. Fehse, M.; Trócoli, R.; Hernández, E.; Ventosa, E.; Sepúlveda, A.; Morata, A.; Tarancón, A. An innovative multi-layer pulsed laser deposition approach for LiMn$_2$O$_4$ thin film cathodes. *Thin Solid Films* **2018**, *648*, 108–112. [CrossRef]

113. Stock, F.; Diebold, L.; Antoni, F.; Chowde Gowda, C.; Muller, D.; Haffner, T.; Pfeiffer, P.; Roques, S.; Mathiot, D. Silicon and silicon-germanium nanoparticles obtained by Pulsed Laser Deposition. *Appl. Surf. Sci.* **2019**, *466*, 375–380. [CrossRef]

114. Craciun, D.; Socol, G.; Stefan, N.; Miroiu, M.; Mihailescu, I.N.; Galca, A.C.; Craciun, V. Structural investigations of ITO-ZnO films grown by the combinatorial pulsed laser deposition technique. *Appl. Surf. Sci.* **2009**, *255*, 5288–5291. [CrossRef]

115. Rasoga, O.; Sima, L.; Chiritoiu, M.; Popescu-Pelin, G.; Fufa, O.; Grumezescu, V.; Socol, M.; Stanculescu, A.; Zgura, I.; Socol, G. Biocomposite coatings based on Poly(3-hydroxybutyrate-co-3-hydroxyvalerate)/calcium phosphates obtained by MAPLE for bone tissue engineering. *Appl. Surf. Sci.* **2017**, *417*, 204–212. [CrossRef]
116. Mihailescu, I.N.; Gyorgy, E. Pulsed Laser Deposition: An overview. In *International Trends in Optics and Photonics*; Asakura, T., Ed.; Part of the Springer Series in Optical Sciences book series; Springer: Heidelberg, Germany, 1999; Volume 74, pp. 201–214. [CrossRef]
117. Ozyegin, L.S.; Oktar, F.N.; Goller, G.; Kayali, S.; Yazici, T. Plasma-sprayed bovine hydroxyapatite coatings. *Mater. Lett.* **2004**, *58*, 2605–2609. [CrossRef]
118. Aguzzi, A. Prion diseases of humans and farm animals: Epidemiology, genetics, and pathogenesis. *J. Neurochem.* **2006**, *97*, 1726–1739. [CrossRef]
119. Paital, S.R.; Dahotre, N.B. Calcium phosphate coatings for bio-implant applications: Materials, performance factors, and methodologies. *Mater. Sci. Eng. R-Rep.* **2009**, *66*, 1–70. [CrossRef]
120. Park, J.-W.; Park, K.-B.; Suh, J.-Y. Effects of calcium ion incorporation on bone healing of Ti6Al4V alloy implants in rabbit tibiae. *Biomaterials* **2007**, *28*, 3306–3313. [CrossRef]
121. Bai, Y.; Park, I.S.; Lee, S.J.; Bae, T.S.; Duncan, W.; Swain, M.; Lee, M.H. One-step approach for hydroxyapatite-incorporated TiO_2 coating on titanium via a combined technique of micro-arc oxidation and electrophoretic deposition. *Appl. Surf. Sci.* **2011**, *257*, 7010–7018. [CrossRef]
122. Liu, L.-S.; Thompson, A.Y.; Heidaran, M.A.; Poser, J.W.; Spiro, R.C. An osteoconductive collagen/hyaluronate matrix for bone regeneration. *Biomaterials* **1999**, *20*, 1097–1108. [CrossRef]
123. Ratner, B.D.; Hoffman, A.S.; Schoen, F.J.; Lemons, J.E. *Biomaterials Science7—An Introduction to Materials in Medicine*, 2nd ed.; Elsevier Academic Press: San Diego, CA, USA, 2004; p. 864.
124. Mandracci, P.; Mussano, F.; Rivolo, P.; Carossa, S. Surface Treatments and Functional Coatings for Biocompatibility Improvement and Bacterial Adhesion Reduction in Dental Implantology. *Coatings* **2016**, *6*, 7. [CrossRef]
125. Traini, T.; Murmura, G.; Sinjari, B.; Perfetti, G.; Scarano, A.; D'Arcangelo, C.; Caputi, S. The Surface Anodization of Titanium Dental Implants Improves Blood Clot Formation Followed by Osseointegration. *Coatings* **2018**, *8*, 252. [CrossRef]
126. Han, A.; Tsoi, J.K.-H.; Matinlinna, J.P.; Chen, Z. Influence of Grit-Blasting and Hydrofluoric Acid Etching Treatment on Surface Characteristics and Biofilm Formation on Zirconia. *Coatings* **2017**, *7*, 130. [CrossRef]
127. Zhou, G.; Bi, Y.; Ma, Y.; Wang, L.; Wang, X.; Yu, Y.; Mutzk, A. Large current ion beam polishing and characterization of mechanically finished titanium alloy (Ti6Al4V) surface. *Appl. Surf. Sci.* **2019**, *476*, 905–913. [CrossRef]
128. Chen, R.; Li, S.; Wang, Z.; Lu, X. Mechanical model of single abrasive during chemical mechanical polishing: molecular dynamics simulation. *Tribol. Int.* **2019**, *133*, 40–46. [CrossRef]
129. Singh, H.; Sharma, V.S.; Singh, S.; Dogra, M. Nanofluids assisted environmental friendly lubricating strategies for the surface grinding of titanium alloy: Ti6Al4V-ELI. *J. Manuf. Process.* **2019**, *39*, 241–249. [CrossRef]
130. Eliaz, N.; Ritman-Hertz, O.; Aronov, D.; Weinberg, E.; Shenhar, Y.; Rosenman, G.; Weinreb, M.; Ron, E. The effect of surface treatments on the adhesion of electrochemically deposited hydroxyapatite coating to titanium and on its interaction with cells and bacteria. *J. Mater. Sci. Mater. Med.* **2011**, *22*, 1741–1752. [CrossRef]
131. Liu, C.-F.; Li, S.-J.; Hou, W.-T.; Hao, Y.-L.; Huang, H.-H. Enhancing Corrosion Resistance and Biocompatibility of Interconnected Porous β-type Ti-24Nb-4Zr-8Sn Alloy Scaffold through Alkaline Treatment and Type I Collagen Immobilization. *Appl. Surf. Sci.* **2019**, *476*, 325–334. [CrossRef]
132. Peng, F.; Shaw, M.T.; Olson, J.R.; Wei, M. Influence of surface treatment and biomimetic hydroxyapatite coating on the mechanical properties of hydroxyapatite/poly(L-lactic acid) fibers. *J. Biomater. Appl.* **2013**, *27*, 641–649. [CrossRef] [PubMed]
133. Dorozhkin, S.V. Calcium orthophosphate deposits: Preparation, properties and biomedical applications. *Mater. Sci. Eng. C* **2015**, *55*, 272–326. [CrossRef]
134. Koh, A.T.T.; Foong, Y.M.; Chua, D.H.C. Cooling rate and energy dependence of pulsed laser fabricated graphene on nickel at reduced temperature. *Appl. Phys. Lett.* **2010**, *97*, 114102. [CrossRef]
135. Yang, Z.; Hao, J. Progress in pulsed laser deposited two-dimensional layered materials for device applications. *J. Mater. Chem. C* **2016**, *4*, 8859–8878. [CrossRef]

136. Christen, H.M.; Eres, G. Recent advances in pulsed-laser deposition of complex oxides. *J. Phys. Condens. Matter.* **2008**, *20*, 264005. [CrossRef]

137. Saju, K.K.; Reshmi, R.; Jayadas, N.H.; James, J.; Jayaraj, M.K. Polycrystalline coating of hydroxyapatite on TiAl6V4 implant material grown at lower substrate temperatures by hydrothermal annealing after pulsed laser deposition. *Proc. Inst. Mech. Eng. Part H-J. Eng. Med.* **2009**, *223*, 1049–1057. [CrossRef]

138. Lynn, A.K.; DuQuesnay, D.L. Hydroxyapatite-coated Ti–6Al–4V Part 2: the effects of post-deposition heat treatment at low temperatures. *Biomaterials* **2002**, *23*, 1947–1953. [CrossRef]

139. Lu, Y.P.; Chen, Y.M.; Li, S.T.; Wang, J.H. Surface nanocrystallization of hydroxyapatite coating. *Acta Biomater.* **2008**, *4*, 1865–1872. [CrossRef]

140. Drevet, R.; Fauré, J.; Benhayoune, H. Thermal treatment optimization of electrodeposited hydroxyapatite coatings on Ti6Al4V substrate. *Adv. Eng. Mater.* **2012**, *14*, 377–382. [CrossRef]

141. Azis, S.A.A.; Kennedy, J.; Cao, P. Effect of annealing on microstructure of hydroxyapatite coatings and their behaviours in simulated body fluid. *Adv. Mater. Res.* **2014**, *922*, 657–662. [CrossRef]

142. Yang, Y.; Kim, K.H.; Agrawal, C.M.; Ong, J.L. Influence of post-deposition heating time and the presence of water vapor on sputter-coated calcium phosphate crystallinity. *J. Dent. Res.* **2003**, *82*, 833–837. [CrossRef]

143. Yang, Y.Z.; Kim, K.H.; Agrawal, C.M.; Ong, J.L. Effect of post-deposition heating temperature and the presence of water vapor during heat treatment on crystallinity of calcium phosphate coatings. *Biomaterials* **2003**, *24*, 5131–5137. [CrossRef]

144. Dinda, G.P.; Shin, J.; Mazumder, J. Pulsed laser deposition of hydroxyapatite thin films on Ti-6Al-4V: Effect of heat treatment on structure and properties. *Acta Biomaterialia* **2009**, *5*, 1821–1830. [CrossRef]

145. Zhou, J.; Zhang, X.; Chen, J.; Zeng, S.; De Groot, K. High temperature characteristics of synthetic hydroxyapatite. *J. Mater. Sci.: Mater. In Med.* **1993**, *4*, 83–85. [CrossRef]

146. Stan, G.E.; Pasuk, I.; Husanu, M.A.; Enculescu, I.; Pina, S.; Lemos, A.F.; Tulyaganov, D.U.; Mabrouk, K.E.L.; Ferreira, J.M.F. Highly adherent bioactive glass thin films synthesized by magnetron sputtering at low temperature. *J. Mater. Sci.-Mater. Med.* **2011**, *22*, 2693–2710. [CrossRef]

147. DFD®INSTRUMENTS. Available online: www.dfdinstruments.com (accessed on 29 April 2019).

148. Bao, Q.; Chen, C.; Wang, D.; Ji, Q.; Lei, T. Pulsed laser deposition and its current research status in preparing hydroxyapatite thin films. *Appl. Surf. Sci.* **2005**, *252*, 1538–1544. [CrossRef]

149. Mohseni, E.; Zalnezhad, E.; Bushroa, A.R. Comparative investigation on the adhesion of hydroxyapatite coating on Ti–6Al–4V implant: a review paper. *Int. J. Adhes. Adhes.* **2014**, *48*, 238–257. [CrossRef]

150. Stan, G.E. Adherent functional graded hydroxylapatite coatings produced by sputtering deposition techniques. *J. Optolelectron. Adv. Mater.* **2009**, *11*, 1132–1138.

151. Stan, G.E.; Morosanu, C.O.; Marcov, D.A.; Pasuk, I.; Miculescu, F.; Reumont, G. Effect of annealing upon the structure and adhesion properties of sputtered bio-glass/titanium coatings. *Appl. Surf. Sci.* **2009**, *255*, 9132–9138. [CrossRef]

152. Stan, G.E.; Popa, A.C.; Galca, A.C.; Aldica, G.; Ferreira, J.M.F. Strong bonding between sputtered bioglass–ceramic films and Ti-substrate implants induced by atomic inter-diffusion post-deposition heat-treatments. *Appl. Surf. Sci.* **2013**, *280*, 530–538. [CrossRef]

153. ISO 13779-2. *Implants for Surgery—Hydroxyapatite—Part 2: Coatings of Hydroxyapatite*; ISO: Geneva, Switzerland, 2008.

154. ISO 10993-5: 2009. *Biological Evaluation of Medical Devices—Part 5: Tests for in vitro cytotoxicity*; ISO: Geneva, Switzerland, 2009.

155. Cleare, A.; Pariante, C.M.; Young, A.H.; Anderson, I.M.; Christmas, D.; Cowen, P.J.; Dickens, C.; Ferrier, I.N.; Geddes, J.; Gilbody, S.; et al. Evidence-based guidelines for treating depressive disorders with antidepressants: A revision of the 2008 British Association for Psychopharmacology guidelines. *J. Psychopharmacol.* **2015**, *29*, 459–525. [CrossRef]

156. Cohen, O.; Rais, T.; Lepkifker, E.; Vered, I. Lithium carbonate therapy is not a risk factor for osteoporosis. *Horm. Metab. Res.* **1998**, *30*, 594–597. [CrossRef]

157. Nordenstrom, J.; Elvius, M.; Bagedahl-Strindlund, M.; Zhao, B.; Torring, O. Biochemical hyperparathyroidism and bone-mineral status in patients treated long-term with lithium. *Metabolism* **1994**, *43*, 1563–1567. [CrossRef]

158. Satija, N.K.; Sharma, D.; Afrin, F.; Tripathi, R.P.; Gangenahalli, G. High throughput transcriptome profiling of lithium stimulated human mesenchymal stem cells reveals priming towards osteoblastic lineage. *PLoS ONE* **2013**, *8*, e55769. [CrossRef] [PubMed]

159. Yang, M.L.; Li, J.J.; So, K.F.; Chen, J.Y.H.; Cheng, W.S.; Wu, J.; Wang, Z.M.; Gao, F.; Young, W. Efficacy and safety of lithium carbonate treatment of chronic spinal cord injuries: A double-blind, randomized, placebo-controlled clinical trial. *Spinal Cord* **2012**, *50*, 141–146. [CrossRef]

160. Gitlin, M. Lithium side effects and toxicity: Prevalence and management strategies. *Int. J. Bipolar Disord.* **2016**, *4*, 27. [CrossRef] [PubMed]

161. Ott, M.; Stegmayr, B.; Salander, R.E.; Werneke, U. Lithium intoxication: Incidence, clinical course and renal function—A population-based retrospective cohort study. *J. Psychopharmacol.* **2016**, *30*, 1008–1019. [CrossRef]

162. Rybakowski, J.K. Challenging the negative perception of lithium and optimizing its long-term administration. *Front. Mol. Neurosci.* **2018**, *11*, 349. [CrossRef]

163. Post, R.M. The new news about lithium: An underutilized treatment in the United States. *Neuropsychopharmacology* **2018**, *43*, 1174–1179. [CrossRef]

164. Kandori, K.; Kuroda, T.; Wakamura, M. Protein adsorption behaviors onto photocatalytic Ti(IV)-doped calcium hydroxyapatite particles. *Colloid Surf. B Biointerfaces* **2011**, *87*, 472–479. [CrossRef]

165. Wei, W.; Zhang, X.; Cui, J.; Wei, Z. Interaction between low molecular weight organic acids and hydroxyapatite with different degrees of crystallinity. *Colloid Surf. A Physicochem. Eng. Aspect* **2011**, *392*, 67–75. [CrossRef]

166. Pesakova, V.; Kubies, D.; Hulejova, H.; Himmlova, L. The influence of implant surface properties on cell adhesion and proliferation. *J. Mater. Sci.-Mater. Med.* **2007**, *18*, 465–473. [CrossRef]

167. Ng, C.H.; Rao, N.; Law, W.C.; Xu, G.; Cheung, T.L.; Cheng, F.T.; Wang, X.; Man, H.C. Enhancing the cell proliferation performance of NiTi substrate by laser diffusion nitriding. *Surf. Coat. Technol.* **2017**, *309*, 59–66. [CrossRef]

168. Webb, K.; Hlady, V.; Tresco, P.A. Relative importance of surface wettability and charged functional groups on NIH 3T3 fibroblast attachment, spreading, and cytoskeletal organization. *J. Biomed. Mater. Res.* **1998**, *41*, 422–430. [CrossRef]

169. Dos Santos, E.A.; Farina, M.; Soares, G.A.; Anselme, K. Surface energy of hydroxyapatite and beta-tricalcium phosphate ceramics driving serum protein adsorption and osteoblast adhesion. *J. Mater. Sci.: Mater. Med.* **2008**, *19*, 2307–2316. [CrossRef]

170. Redey, S.A.; Razzouk, S.; Rey, C.; Bernache-Assollant, D.; Leroy, G.; Nardin, M.; Cournot, G. Osteoclast adhesion and activity on synthetic hydroxyapatite, carbonated hydroxyapatite, and natural calcium carbonate: Relationship to surface energies. *J. Biomed. Mater. Res.* **1999**, *45*, 140–147. [CrossRef]

171. Zhao, G.; Schwartz, Z.; Wieland, M.; Rupp, F.; Geis-Gerstorfer, J.; Cochran, D.L.; Boyan, B.D. High surface energy enhances cell response to titanium substrate microstructure. *J. Biomed. Mater. Res.* **2005**, *74*, 49–58. [CrossRef]

172. Vidic, J.; Stankic, S.; Haque, F.; Ciric, D.; Goffic, R.L.; Vidy, A.; Jupille, J.; Delmas, B. Selective antibacterial effects of mixed ZnMgO nanoparticles. *J. Nanoparticle Res.* **2013**, *15*, 1595. [CrossRef] [PubMed]

173. Li, Y.; Liu, G.; Zhai, Z.; Liu, L.; Li, H.; Yang, K.; Tan, L.; Wan, P.; Liu, X.; Ouyang, Z.; et al. Antibacterial Properties of Magnesium In Vitro and in an In Vivo Model of Implant-Associated Methicillin-Resistant *Staphylococcus aureus* Infection. *Antimicrob Agents Chemother.* **2014**, *58*, 7586–7591. [CrossRef]

174. Sawai, J.; Kojima, H.; Igarashi, H.; Hashimoto, A.; Shoji, S.; Sawaki, T.; Hakoda, A.; Kawada, E.; Kokugan, T.; Shimizu, M. Antibacterial characteristics of magnesium oxide powder. *World J. Microbiol. Biotechnol.* **2000**, *16*, 187. [CrossRef]

175. Gayathri, B.; Muthukumarasamy, N.; Velauthapillai, D.; Santhosh, S.B.; asokan, V. Magnesium incorporated hydroxyapatite nanoparticles: Preparation, characterization, antibacterial and larvicidal activity. *Arab. J. Chem.* **2018**, *11*, 645–654. [CrossRef]

176. Applerot, G.; Lipovsky, A.; Dror, R.; Perkas, N.; Nitzan, Y.; Lubart, R.; Gedanken, A. Enhanced Antibacterial Activity of Nanocrystalline ZnO Due to Increased ROS-Mediated Cell Injury. *Adv. Funct. Matter.* **2009**, *19*, 842–852. [CrossRef]

177. Schwalfenberg, G.K.; Genuis, S.J. The Importance of Magnesium in Clinical Healthcare. *Scientifica (Cairo)* **2017**, *2017*, 4179326. [CrossRef] [PubMed]

178. Barrere, F.; van der Valk, C.M.; Meijer, G.; Dalmeijer, R.A.; Groot, K.; Layrolle, P. Osteointegration of biomimetic apatite coating applied onto dense and porous metal implants in femurs of goats. *J. Biomed. Mater. Res. B Appl. Biomater.* **2003**, *67*, 655–665. [CrossRef]

179. Surmenev, R.A.; Surmeneva, M.A. A critical review of decades of research on calcium phosphate–based coatings: How far are we from their widespread clinical application? *Curr. Opin. Biomed. Eng.* **2019**, *10*, 35–44. [CrossRef]

180. Chen, L.; Komasa, S.; Hashimoto, Y.; Hontsu, S.; Okazaki, J. *In Vitro* and *In Vivo* Osteogenic Activity of Titanium Implants Coated by Pulsed Laser Deposition with a Thin Film of Fluoridated Hydroxyapatite. *Int. J. Mol. Sci.* **2018**, *19*, 1127. [CrossRef]

181. Mróz, W.; Budner, B.; Syroka, R.; Niedzielski, K.; Golański, G.; Slósarczyk, A.; Schwarze, D.; Douglas, T.E. *In vivo* implantation of porous titanium alloy implants coated with magnesium-doped octacalcium phosphate and hydroxyapatite thin films using pulsed laser deposition. *J. Biomed. Mater. Res. Part B* **2015**, *103*, 151–158. [CrossRef]

182. Peraire, C.; Arias, J.L.; Bernal, D.; Pou, J.; Leon, B.; Arano, A.; Roth, W. Biological stability and osteoconductivity in rabbit tibia of pulsed laser deposited hydroxylapatite coatings. *J. Biomed. Mater. Res. Part A* **2006**, *77A*, 370–379. [CrossRef]

183. Mihailescu, I.N.; Lamolle, S.; Socol, G.; Miroiu, F.; Roenold, H.J.; Bigi, A.; Mayer, I.; Cuisinier, F.; Lyngstadaas, S.P. *In vivo* tensile tests of biomimetic titanium implants pulsed laser coated with nanostructured Calcium Phosphate thin films. *Optoelectron. Adv. Mater.-Rapid Commun.* **2008**, *2*, 337–341.

184. Dostalova, T.; Himmlova, L.; Jelinek, M.; Grivas, C. Osseointegration of loaded dental implant with KrF laser hydroxylapatite films on Ti6Al4V alloy by minipigs. *J. Biomed. Opt.* **2001**, *6*, 239–243. [CrossRef]

© 2019 by the authors. Licensee MDPI, Basel, Switzerland. This article is an open access article distributed under the terms and conditions of the Creative Commons Attribution (CC BY) license (http://creativecommons.org/licenses/by/4.0/).

Article

Shellac Thin Films Obtained by Matrix-Assisted Pulsed Laser Evaporation (MAPLE)

Simona Brajnicov [1,2], **Adrian Bercea** [1,3], **Valentina Marascu** [1], **Andreea Matei** [1,*] and **Bogdana Mitu** [1]

[1] National Institute for Lasers, Plasma, and Radiation Physics, 409 Atomistilor Street, P.O. Box MG-16, RO-077125 Magurele, Romania; brajnicov.simona@inflpr.ro (S.B.); bercea.adrian@inflpr.ro (A.B.); valentina.marascu@inflpr.ro (V.M.); mitub@infim.ro (B.M.)

[2] Faculty of Sciences, University of Craiova, RO-200585 Craiova, Romania

[3] CNRS, IRCER—UMR 7315, University of Limoges, F-87000 Limoges, France

* Correspondence: andreea.matei@inflpr.ro

Received: 14 June 2018; Accepted: 3 August 2018; Published: 7 August 2018

Abstract: We report on the fabrication of shellac thin films on silicon substrates by matrix-assisted pulsed laser evaporation (MAPLE) using methanol as matrix. Very adherent, dense, and smooth films were obtained by MAPLE with optimized deposition parameters, such as laser wavelength and laser fluence. Films with a root mean square (RMS) roughness of 11 nm measured on 40×40 μm^2 were obtained for a 2000-nm-thick shellac film deposited with 0.6 J/cm^2 fluence at a laser wavelength of 266 nm. The MAPLE films were tested in simulated gastric fluid in order to assess their capabilities as an enteric coating. The chemical, morphological, and optical properties of shellac samples were investigated by Fourier transform infrared spectroscopy (FTIR), X-ray photoelectron spectroscopy (XPS), scanning electron microscopy (SEM), and atomic force microscopy (AFM).

Keywords: thin films; matrix-assisted pulsed laser evaporation; shellac; enteric coatings

1. Introduction

Shellac is a biomaterial, a resin secreted by the female lac bug *Kerria lacca* on the trees in India and Thailand. It consists of esters and polyesters of polyhydroxy acids. The main components of shellac are aleuritic and shellolic acid [1]. Other components can be butolic and jalaric acid [2]. The color can vary between light yellow and dark red depending on the origin of the raw material.

Shellac possesses very interesting properties, such as being UV resistant, and it is a barrier for water and a very good dielectric [3]. As a thin film, shellac can be easily deposited by drop casting from a solution with ethanol or methanol. Shellac thin films have a dielectric constant between 3 and 5 [3] and very low surface roughness. They have been used as a substrate for organic field-effect transistor (OFET) and as a gate dielectric [3]. When used as a substrate, a concentrated solution is prepared. The solution has to evaporate for about half an hour at 50 °C and then it has to be heated for an hour at 70 °C in order to obtain roughness lower than 1 nm [3].

Shellac is used in the food industry as a coating for fruits as protection against dehydration and for the glossy and attractive aspect [4]. Shellac can be added to the composition of food packaging materials for protection against bacteria and as a water repellent element [5]. Nail polish [6], furniture and wood varnish, and the protection of art objects [7] are some other applications shellac can be used for.

Shellac biopolymer can be used as a new carbon source for growing bubble-decorated graphene oxide films with high magneto-resistance as reported by Singhbabu et al. in [8].

Improved properties for wood protection can be achieved for shellac films as a coating by dispersing inorganic nanoparticles (1–2%) in alcoholic shellac solution as demonstrated by Weththimuni et al. in [9]. Increased hardness and increased UV resistance can be obtained using ZrO$_2$

and ZnO nanoparticles. The low cost and low processing temperature make shellac a good candidate for microfluidics applications via hot embossing in the same way standard polydimethylsiloxane (PDMS) is used [10].

In pharmaceutics, shellac can be used for encapsulation and micro encapsulation [11] as an enteric coating due to its acidic character, delivering the drug to the colon for specific applications [12–14]. Kumpugdee-Vollrath et al. obtained controlled drug release using coatings of shellac and pectine in [15]. Coaxial electrospinning has been reported for the obtaining of shellac nanofibers for colon-targeted drug delivery by Wang et al. in [16].

In this work, we used a laser deposition technique to grow thin films of shellac for targeted drug release. The employed laser technique was Matrix-Assisted Pulsed Laser Evaporation (MAPLE), a variation of the pulsed laser deposition (PLD) procedure. Pulsed laser deposition is especially used in the deposition of inorganic thin films [17–19] and it cannot be used for all types of delicate materials. The PLD method can lead to their photochemical interaction when using laser emitting in UV. For polymers, organic materials, and materials with a complicated structure, the MAPLE system is preferred [20–22]. Important advantages arising from our approach using laser-based techniques are the good adherence, the high density of the films, and control over their thickness. Another important issue is the fact that MAPLE is a "dry technique". During the coating procedure with MAPLE, the drug is not in contact with the matrix (solvent) and drug–solvent chemical interactions are therefore avoided.

MAPLE-grown films have been tested in simulated gastric fluid, where the resin coating offers the needed resistance to the stomach and small intestine environment [23].

2. Materials and Methods

Shellac wax-free was purchased from Sigma-Aldrich (St. Louis, MO, USA) and used as received. The MAPLE technique was employed using a low concentration of shellac (1–2 wt % dissolved in methanol). The solutions were poured into a copper target holder and they were frozen with liquid nitrogen. During the experiments, the targets were kept in a solid state. The influence of various laser wavelengths (266, 355, 532, and 1064 nm) on the deposition process was investigated. The laser fluence was in the range 0.6–2 J/cm^2 for the experiments using the UV wavelength and up to 4 J/cm^2 for the visible and IR wavelengths. We will present the results just for UV laser beam irradiation because green and infrared light did not lead to continuous films, most probably due to the fact that the methanol matrix does not absorb light at these wavelengths [24] and the target evaporation took place mainly on defects. The laser spot was set in the range 0.9–1 mm^2 for all the experiments. The substrates of 10×10 mm^2, i.e., two side polished Si (100) wafer and quartz, were kept at room temperature and positioned at a distance of 4 cm from the target. The base pressure in the deposition chamber was $\sim 8 \times 10^{-6}$ mbar and it increased during laser irradiation to 10^{-4} mbar due to solvent evaporation. The number of pulses varied between 18,000 and 90,000 for a pulse repetition rate of 10 Hz. Before the deposition, the substrates were cleaned in an ultrasonic bath, using a standard procedure implying soap solution, water, acetone, and ethanol for 15 min each, and dried in a nitrogen gas flow.

For measuring the deposition rates of the films, contact profilometry studies were performed by means of a KLA-Tencor P-7 (Milpitas, CA, USA) step profiler equipped with a 2-μm radius and a 60° stylus working in contact with 2 mg applied force and scanning on a 3-mm length of the samples with a scan speed of 100 μm/s. Surface morphology was investigated by atomic force microscopy (AFM) and scanning electron microscopy (SEM) with an "XE-100" AFM produced by Park Systems (Suwon, South Korea) and a Quanta ESEM FEG 450 SEM (FEI, Hillsboro, OR, USA). Fourier transform infrared absorption spectra were acquired with a Jasco 6300 FTIR system (Oklahoma, OK, USA), in the range 4000–400 cm^{-1}, with 4 cm^{-1} resolution.

The wettability characteristics of the films were evaluated by contact angle measurements. The water contact angle was measured using an optical microscope KSV CAM101 (KSV Instruments Ltd., Helsinki, Finland) with water drops of 0.5–1 μL and by mediating the processed angles obtained upon five drops for each sample.

The simulated gastric fluid (SGF) dissolution media that is described in United States Pharmacopeia 33-28NF (2010) and European Pharmacopeia 7.0 (2010) was prepared as follows: 2.0 g NaCl, 3.2 g pepsin, 7.0 mL HCl, and 1000 mL milli q water. The solution used had a pH of 1.2. Shellac films deposited by MAPLE were immersed in SGF for 15, 30, 60, 120, and 240 min.

X-ray photoelectron spectroscopy (XPS) survey spectra and high-resolution XPS scan spectra were acquired for shellac dropcast, the initial film, and films immersed in SGF for different time durations using an Escalab Xi$^+$ system, Thermo Scientific (Waltham, MA, USA). The survey scans were acquired using an Al Kα gun with a spot size of 900 µm, a pass energy of 100.0 eV, and an energy step size of 1.00 eV. For the high-resolution XPS spectra, the pass energy was set to 10.0 eV, the energy step size was 0.10 eV, and 10 scans were accumulated.

3. Results and Discussion

The MAPLE-deposited films using a 266 nm irradiating wavelength are very smooth and the surface is droplet-free. The deposited films have low roughness: less than 1 nm on 5×5 µm^2 for films with a thickness in the range 400–2000 nm; however, at micrometric scale, the surface has a wave-like aspect (Figure 1).

The films present a slight hydrophobic tendency, with contact angles in the range 92–105° on samples prepared by MAPLE at a 266-nm wavelength.

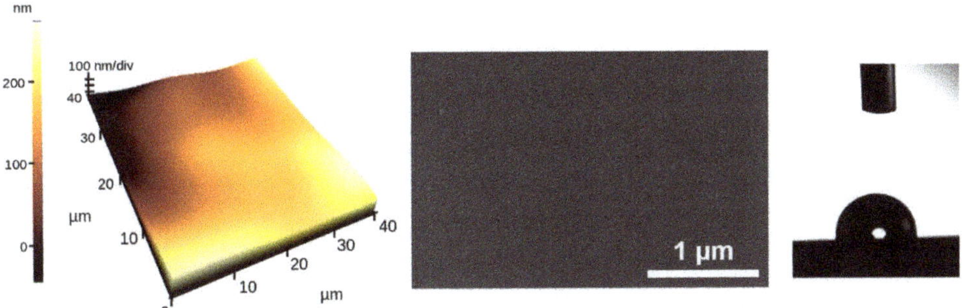

Figure 1. AFM and SEM images on a shellac film deposited from a 1 wt % shellac in a methanol frozen target on an Si substrate as a result of 72,000 pulses at a 266-nm wavelength with a fluence of 0.5 J/cm^2. RMS roughness is 23 nm on 40×40 µm^2. The water contact angle is 97.2°.

Profilometric studies on the shellac films deposited using UV-irradiated light revealed deposition rates between 0.01 and 0.3 Å/pulse for using fluences in the range 0.5–2 J/cm^2 (Figure 2). A significant increase was obtained when using a higher amount of shellac as a guest molecule in the target. The deposition rate as a function of fluence has very interesting behavior; there is no evident increase with fluence. More than that, a decrease of the deposition rate is encountered upon increasing the laser fluence when the target contained 2 wt % shellac. It has also been reported in Ref. [25] the case of depositing lysozyme and myoglobin proteins where the deposition rate increases up to a maximum, after which it has a clear decrease. This may be due to a low sticking coefficient on the already deposited film or to the ejection of the fragmented shellac scattered from the normal direction onto the target [25]. Additionally, at high fluences and higher shellac concentrations, the amount of the evaporated solvent molecules and fragments are higher, leading to a pressure increase in the deposition chamber, which blocks shellac to reach the substrate.

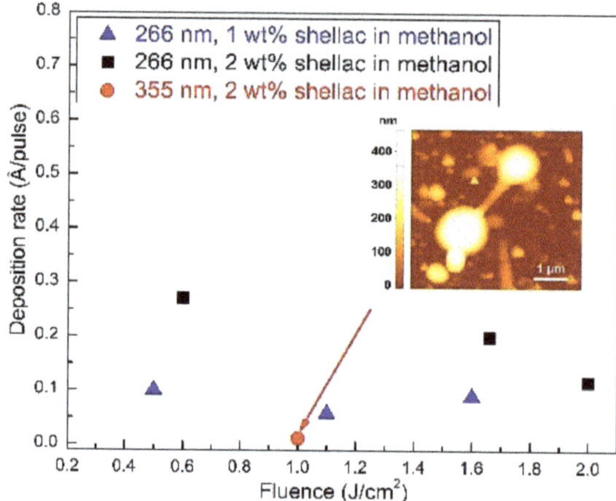

Figure 2. Dependence of shellac deposition rate on the laser fluence as resulting from profilometry on samples prepared by matrix-assisted pulsed laser evaporation (MAPLE). Inset AFM image on a shellac film deposited on an Si substrate using 355 nm laser wavelength as a result of 72,000 pulses at 1 J/cm^2 from a 2 wt % shellac in a methanol frozen target.

The very low deposition rate obtained for the experiments performed by using the 355 nm laser wavelength (the red circle on Figure 2) corroborated with the high roughness of the surface covered with droplets and elongated structures (inset Figure 2) preventing the utilization of this wavelength for shellac deposition in applications where a continuous and smooth film is needed. As already mentioned in the experimental part, the use of the 532 nm and 1064 nm irradiation wavelengths did not even lead to the deposition of continuous films. The few aggregates that could be seen on the substrate are related to the absorption of laser light by defects, seed electrons, and impurities in the frozen target, which lead to material transfer onto the substrate. Taking into consideration all these aspects, the following experiments involved the use of films deposited using the 266 nm laser wavelength.

The surface analysis revealed smooth films with low roughness grown at low fluences, unlike the case for higher fluences, where the surface is covered with wrinkle-like structures (Figure 3). Such features were previously explained upon molecular dynamics simulations of the MAPLE process [26,27]. They originate from large clusters of target material, including the solvent, which are ejected from the frozen target and deposited onto the substrate, followed in a second stage by solvent evaporation from the substrate and film formation.

As a conclusion, for experiments on the film's behavior upon immersion in SGF, the low range of fluences with values up to 0.6 J/cm^2 were chosen.

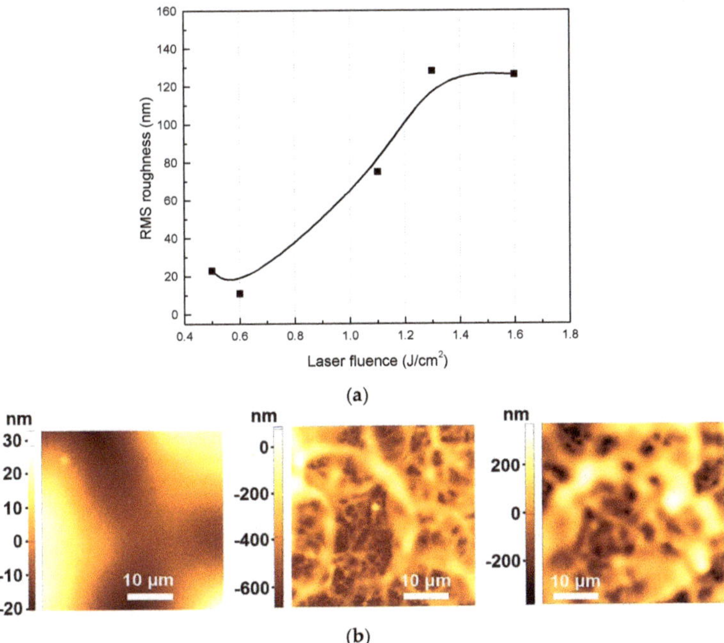

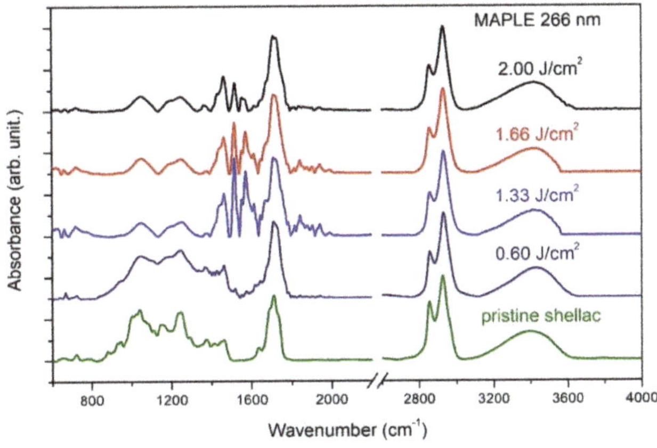

Figure 3. (**a**) The RMS roughness determined for areas of $40 \times 40\ \mu m^2$ for shellac films deposited by MAPLE. The line between the points is for eye-guidance; (**b**) From left to right: AFM images for 0.6, 1.3, and 1.6 J/cm^2.

Fourier transform infrared spectroscopy (FTIR) investigations were performed on films deposited by MAPLE and the results were compared to the pristine material. The laser fluence and the amount of shellac in the target solution were the investigated parameters (Figure 4).

Figure 4. FTIR spectra of shellac films.

The large band in the range 3100–3700 cm^{-1} with a maximum at about 3420 cm^{-1} is attributed to the O–H stretching vibration, while the O–H bending vibration is identified at 1250 cm^{-1}. The strong symmetric and asymmetric stretching vibrations of CH_2 can be easily identified at 2855 cm^{-1} and 2932 cm^{-1}, respectively. The other two strong vibrations peaked at 1712 cm^{-1} and 1733 cm^{-1} are attributed to the C=O stretching vibration of esters [23] and acids, respectively [28,29]. The other medium bands of the characteristic fingerprint are: 1636 cm^{-1} corresponding to the C=C stretching vibration of vinyl [30]; and 1463 cm^{-1} and 1376 cm^{-1} are attributed to CH_3 asymmetric and symmetric bending, respectively [31]. The weak signal at 1412 cm^{-1} was associated to the CH_2 group attached to the ester chain as reported in Zumbühl et al. in [32]. The absorption bands at 1175 and 1050 cm^{-1} may be due to stretching vibrations of C–O and C–C bonds [31]. Weak signals can be seen at 944 and 882 cm^{-1} for pristine shellac, dropcast, and the low fluence used of 0.6 J/cm^{-2}. They can be attributed to the O–H out-of-plane deformation in carboxylic acids, to the C–H out-of-plane deformation in aldehydes, or to their combination [33].

The FTIR spectra show that the general aspect of the spectra of shellac deposited using laser fluences up to 2 J/cm^2 is not changed, leading to the conclusion that a significant part of the film material remains unaffected by the laser transfer process. The best concordance with the pristine shellac of the intensities ratio was obtained for low fluences, where all shellac characteristic absorption bands can be found. However, at laser fluences of at least 1.33 J/cm^2, one can notice the appearance of two additional bands in the region 1400–1600 cm^{-1}, namely at 1515 and 1550 cm^{-1}, related to the symmetric and antisymmetric stretching modes of carboxylate moieties (COO–) formed upon deprotonation of the carboxylic acids of shellac [34]. This is accompanied by a decrease of C–O and C–C stretching bands at 1175 and 1050 cm^{-1}, suggesting a tendency for material crosslinking upon laser transfer, which was confirmed by the broadening of the bands and the loosening of the fine structure noticed at large fluences.

4. Acidic Solution Resistance on Samples Prepared by MAPLE Deposition

As mentioned in the experimental part, we have investigated the enteric capacity of shellac to be used as a coating for colon-targeted drugs. The chemical evolution of shellac in simulated gastric fluid (SGF) was analyzed at different time intervals. The time intervals were selected in accordance with [35]. Figure 5 presents the FTIR spectrum for a film immersed for 240 min in SGF. We have calculated and compared the intensities of the strong absorption bands from 3423 cm^{-1} (O–H) and 1711 cm^{-1} (C=O) normalized to the strongest band in the spectra appearing at 2933 cm^{-1} for the films immersed in SGF to the as-deposited films.

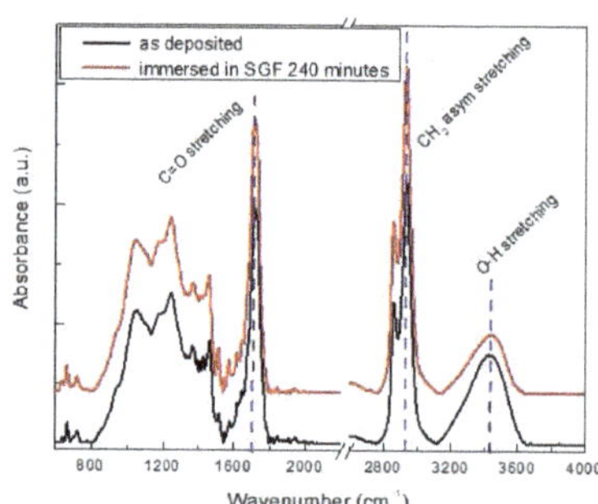

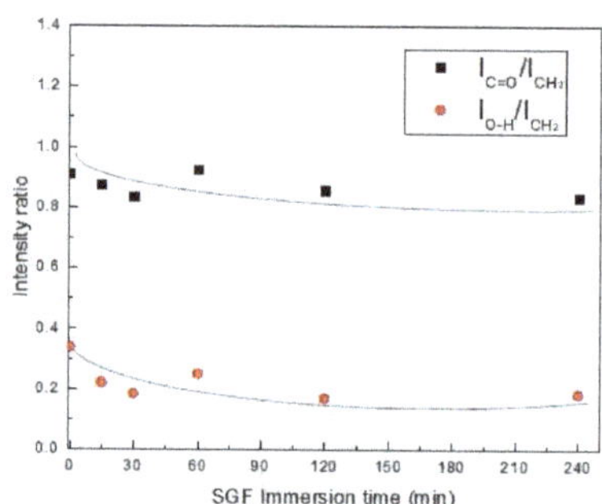

Figure 5. FTIR spectra of shellac film as-deposited and film immersed in simulated gastric fluid (SGF) for 240 min. Relative intensities of the absorption bands of O–H and C=O (3423 and 1711 cm^{-1}, respectively) normalized to the CH_2 band (the strongest band of at 2927 cm^{-1}).

It is clear that no significant modification in the material composition occurs even upon immersion of the shellac film into SGF for 240 min. Upon the first 30 min of exposure to SGF, a low decrease of the populations of the C=O stretching vibration of esters and the O–H stretching vibration relative to stretching vibrations of CH_2 identified at 2927 cm^{-1} was noticed. Nevertheless, upon a longer time of exposure in SGF, the shellac films were stable in the acidic gastric medium, no significant transformation could be identified, and the analyzed populations remained approximately constant up to the maximum investigated time of 240 min, which corresponds to the typical passage time through the digestive system.

The modification of surface chemical composition upon immersion in SGF, as revealed by the XPS survey spectra, is presented in Figure 6. It is shown that a slightly higher amount of oxygen is present in the MAPLE-deposited film as compared to the dropcasted shellac. Also, an incorporation of nitrogen onto the shellac surface appears and its concentration increases from 2.63% at 15 min of immersion up to 5.02% at 240 min of immersion. This is associated with pepsin interaction with shellac films, leading to the covalent attachment of nitrogen to the shellac structure.

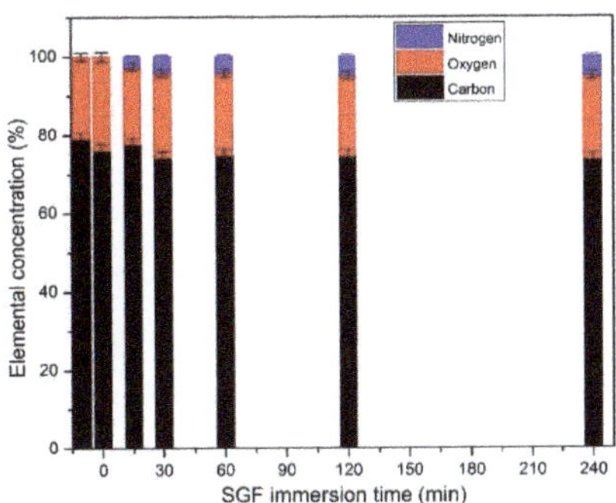

Figure 6. Dependence of the surface elemental concentration on the SGF immersion time.

The high-resolution XPS spectra in the carbon binding energy region were deconvoluted both for the shellac material and for those kept in SGF for various times; the results for the 240 min immersion are presented in Figure 7.

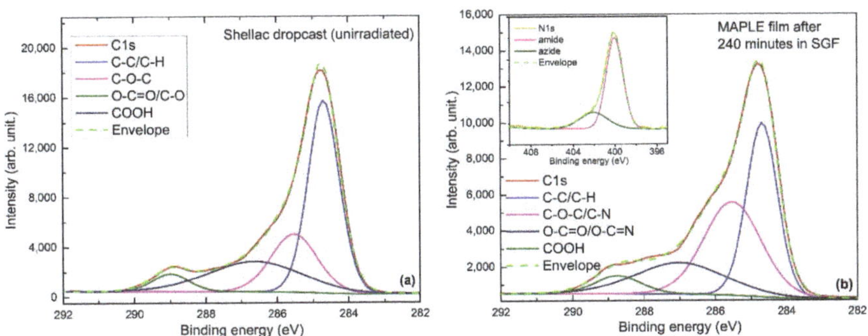

Figure 7. High resolution C 1s and fitting for non-irradiated shellac (**a**) and for a sample immersed in gastric fluid for 240 min (**b**).

The spectrum of the initial shellac film is dominated by the C–C/C–H peak at 284.8 eV, while larger peaks at 285.7 eV and 286.7 eV are assigned to *C–O–C* hydroxyl ether and O–C*=O/C*–O carbonyl peaks, respectively. At higher binding energies, around 289.1 eV the presence of a COOH carboxyl peak is noticed. Upon the interaction of the shellac film with the SGF, one can see the diminution of C–C/C–H-related bonds accompanied by a significant increase and widening of the peak at 285.7 eV.

This counts as C–N bonding, which superimposes on the *C–O–C* from the shellac, and it appears on the film surface upon the interaction with the pepsin. Looking into the N 1s high-resolution spectra, presented in the inset of Figure 7, one can see that the main nitrogen-related bonds are those related to the amide formation, peaked at 400.0 eV, as well as those associated with the protonated nitrogen obtained upon interaction with COOH functionalities [36]. The results are consistent with the increase of nitrogen atomic concentration with the time of immersion in SGF.

The thickness of the films remained intact for all the investigated samples. The interaction of the shellac film with the SGF was evaluated also by means of AFM measurements. We performed analysis of 2000-nm-thick films with an RMS roughness of 0.7–1 nm over a scanned area of $5 \times 5\ \mu m^2$. The step profile measurements show that the film's thickness does not modify significantly even upon 240 min immersion. Nevertheless, the roughness increases by 1 order of magnitude even for an immersion time as low as 15 min and reaches 12.3 nm upon 120 min of immersion. Small pores, both regarding their diameter and depth, had appeared on the shellac surface already upon 15 min of immersion (Figure 8). The pores' density and their depth increases when the exposure time to SGF increases; however, the largest pore identified after 4 h of immersion had a depth of 85 nm, which represents less than 5% of the film's thickness.

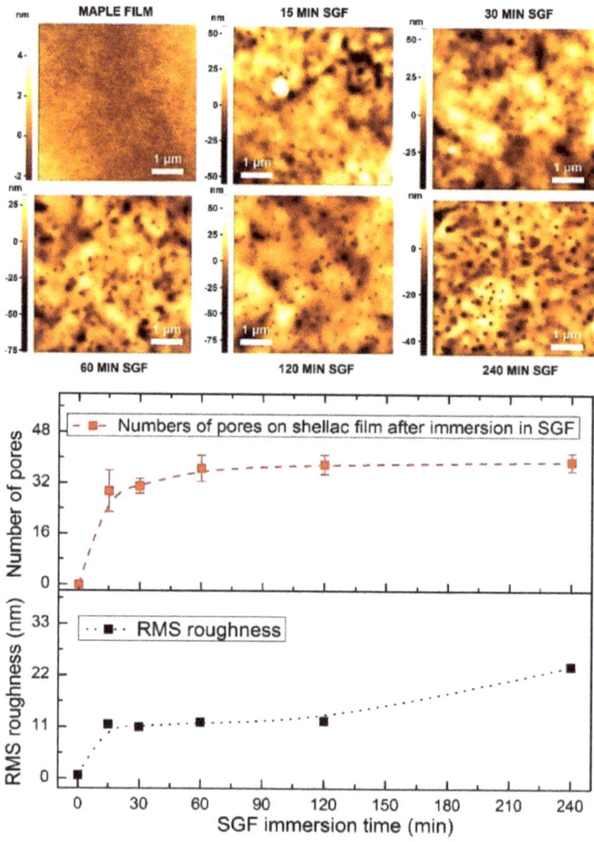

Figure 8. AFM images on $5 \times 5\ \mu m^2$ films of MAPLE shellac films immersed in SGF for different time durations. RMS Roughness on $5 \times 5\ \mu m^2$ and number of pores counted on $5\ \mu m^2$. The MAPLE films have a thickness of 2000 nm and were deposited using 2% shellac dissolved in methanol with a fluence of 0.6 J/cm^2 and 72,000 pulses.

The behavior of MAPLE-deposited shellac films upon immersion in SGF for 240 min was compared to that of films deposited by the dropcast method and submitted to the same environment. The AFM images are presented in Figure 9. The $5 \times 5\ \mu m^2$ AFM image for the dropcast film, before SGF immersion, presents a smooth film, with an RMS roughness of 0.6 nm. After SGF immersion, the surface of the film is strongly changed; the film is very rough and holes similar to those noticed in the case of MAPLE films can be identified. Nevertheless, the higher roughness of the dropcast films after immersion (88 nm measured on $5 \times 5\ \mu m^2$) compared to the MAPLE films also supports the hypothesis that the MAPLE technique produces more dense and compact films. This can be associated with the higher impinging energy of the material reaching the substrate in the case of MAPLE processing, which leads to a more compact deposit with respect to the softer dropcast process.

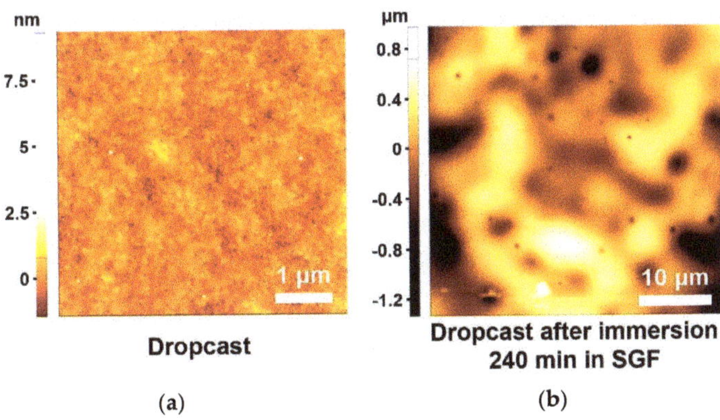

Figure 9. AFM image on $5 \times 5\ \mu m^2$ on a dropcast film (**a**); AFM image on $40 \times 40\ \mu m^2$ after 240 min immersion in SGF (**b**).

5. Conclusions and Outlook

The deposition of shellac thin films by MAPLE has proven to be successful by utilization of a UV laser. The obtained samples have low roughness and are droplet-free. We have obtained films with a thickness of 2000 nm and an RMS roughness less than 1% of the thickness for the optimal set of deposition parameters. It can be seen that laser fluence is an important factor in preserving the chemical structure of shellac. For low fluences, all the characteristic absorption bands were found and the best concordance to the pristine shellac was obtained. Upon immersion in Simulated Gastric Fluid, the thickness of the films remained intact even upon 240 min of immersion. The surface of the films changed even for an immersion time as low as 15 min, and the roughness increased as a result of the appearance of holes. The pores' density and size increased when the exposure time to SGF increased, but the biggest hole measured after 240 min in SGF had a depth of just 85 nm. This behavior, corroborated with the UV resistance [28] we have demonstrated previously, makes MAPLE shellac films promising for use as an enteric coating and opens the path for a variety of applications.

Author Contributions: Investigation, S.B.; Formal Analysis, A.B. and V.M.; Conceptualization, A.M.; Supervision, B.M.; Writing—Review & Editing, A.M., B.M., S.B., A.B. and V.M.

Funding: This research was funded by the Romanian National Authority for Scientific Research and Innovation, CNCS—UEFISCDI (PN-III-P2-2.1-PED-2016-0221 (contract PED 94/2017)—"IPOD").

Conflicts of Interest: The authors declare no conflict of interest.

References

1. Yates, P.; Field, G.F. Shellolic acid, a cedrenoid sesquiterpene from shellac. *J. Am. Chem. Soc.* **1960**, *82*, 5764–5765. [CrossRef]
2. Colombini, M.P.; Bonaduce, I.; Gautier, G. Molecular pattern recognition of fresh and aged shellac. *Chromatographia* **2003**, *58*, 357–364.
3. Irimia-Vladu, M.; Głowacki, E.D.; Schwabegger, G.; Leonat, L.; Zekiye Akpinar, H.; Sitter, H.; Bauer, S.; Sariciftci, N.S. Natural resin shellac as a substrate and a dielectric layer for organic field-effect transistors. *Green Chem.* **2013**, *15*, 1473–1476. [CrossRef]
4. Palou, L.; Valencia-Chamorro, S.A.; Pérez-Gago, M.B. Antifungal edible coatings for fresh citrus fruit: A review. *Coatings* **2015**, *5*, 962–986. [CrossRef]
5. Hult, E.L.; Iotti, M.; Lenes, M. Efficient approach to high barrier packaging using microfibrillar cellulose and shellac. *Cellulose* **2010**, *17*, 575–586. [CrossRef]
6. Loy, M.; Riddell, L. The effect of shellac nail polish on measurement of oxygen saturation by pulse oximetry. *Anaesthesia* **2014**, *69* (Suppl. S3), 42.
7. Sutherland, K.; del Río, J.C. Characterization and discrimination of various types of lac resin using gas chromatography mass spectrometry techniques with quaternary ammonium reagents. *J. Chromatogr. A* **2014**, *1338*, 149–163. [CrossRef] [PubMed]
8. Singhbabu, Y.N.; Choudhary, S.K.; Shukla, N.; Das, S.; Sahu, R.K. Observation of large positive magneto-resistance in bubble decorated graphene oxide films derived from shellac biopolymer: A new carbon source and facile method for morphology-controlled properties. *Nanoscale* **2015**, *7*, 6510–6519. [CrossRef] [PubMed]
9. Weththimuni, M.L.; Capsoni, D.; Malagodi, M.; Milanese, C.; Licchelli, M. Shellac/nanoparticles dispersions as protective materials for wood. *Appl. Phys. A* **2016**, *122*, 1058. [CrossRef]
10. Lausecker, R.; Badilita, V.; Gleißner, U.; Wallrabe, U. Introducing natural thermoplastic shellac to microfluidics: A green fabrication method for point-of-care devices. *Biomicrofluidics* **2016**, *10*, 044101. [CrossRef] [PubMed]
11. Campbell, A.L.; Stoyanov, S.D.; Paunov, V.N. Novel multifunctional micro-ampoules for structuring and encapsulation. *Chem. Phys. Chem.* **2009**, *10*, 2599–2602. [CrossRef] [PubMed]
12. Sinha, V.R.; Kumira, R. Coating polymers for colon specific drug delivery: A comparative in vitro evaluation. *Acta Pharm.* **2003**, *53*, 41–47. [PubMed]
13. Farag, Y.; Leopold, C.S. Development of shellac-coated sustained release pellet formulations. *Eur. J. Pharm. Sci.* **2011**, *42*, 400–405. [CrossRef] [PubMed]
14. Schad, B.; Smith, H.; Cheng, B.; Scholten, J.; VanNess, E.; Riley, T. Coating and taste masking with shellac. *Pharm. Technol.* **2013**, *2013* (Suppl. S5). Available online: http://www.pharmtech.com/coating-and-taste-masking-shellac?pageID=2 (accessed on 14 June 2018).
15. Kumpugdee-Vollrath, M.; Tabatabaeifar, M.; Helmis, M. New coating materials based on mixtures of shellac and pectin for pharmaceutical products. *Int. Sch. Sci. Res. Innov.* **2014**, *8*, 21–29.
16. Wang, X.; Yu, D.-G.; Li, X.-Y.; Bligh, S.W.A.; Williams, G.R. Electrospun medicated shellac nanofibers for colon-targeted drug delivery. *Int. J. Pharm.* **2015**, *490*, 384–390. [CrossRef] [PubMed]
17. Somacescu, S.; Scurtu, R.; Epurescu, G.; Pascu, R.; Mitu, B.; Osiceanu, P.; Dinescu, M. Thin films of SnO_2-CeO_2 binary oxides obtained by pulsed laser deposition for sensing application. *Appl. Surf. Sci.* **2013**, *278*, 146–152. [CrossRef]
18. Schöning, M.J.; Mourzina, Y.G.; Schubert, J.; Zander, W.; Legin, A.; Vlasov, Y.G.; Lüth, H. Pulsed laser deposition—An innovative technique for preparing inorganic thin films. *Electroanalysis* **2001**, *13*, 727–732. [CrossRef]
19. Brodoceanu, D.; Scarisoreanu, N.D.; Filipescu, M.M.; Epurescu, G.N.; Matei, D.G.; Verardi, P.; Craciun, F.; Dinescu, M. Pulsed laser deposition of oxide thin films. In *Plasma Production by Laser Ablation*; Gammino, S., Mezzasalma, A.M., Neri, F., Torrisi, L., Eds.; World Scientific Publishing Co. Pte. Ltd.: Singapore, 2004; pp. 41–46, ISBN 9789812702555.
20. Houser, E.J.; Chrisey, D.B.; Bercu, M.; Scarisoreanu, N.D.; Purice, A.; Colceag, D.; Constantinescu, C.; Moldovan, A.; Dinescu, M. Functionalized polysiloxane thin films deposited by matrix-assisted pulsed laser evaporation for advanced chemical sensor applications. *Appl. Surf. Sci.* **2006**, *252*, 4871–4876. [CrossRef]

21. Purice, A.; Schou, J.; Kingshott, P.; Dinescu, M. Production of active lysozyme films by matrix assisted pulsed laser evaporation at 355 nm. *Chem. Phys. Lett.* **2007**, *435*, 350–353. [CrossRef]

22. Birjega, R.; Matei, A.; Mitu, B.; Ionita, M.D.; Filipescu, M.; Stokker-Cheregi, F.; Luculescu, C.; Dinescu, M.; Zavoianu, R.; Pavel, O.D.; et al. Layered double hydroxides/polymer thin films grown by matrix assisted pulsed laser evaporation. *Thin Solid Films* **2013**, *543*, 63–68. [CrossRef]

23. Ravi, V.; Pramod Kumar, T.M.; Siddaramaiah. Novel colon targeted drug delivery system using natural polymers. *Indian J. Pharm. Sci.* **2008**, *70*, 111–113. [PubMed]

24. Warren, C.R. Rapid measurement of chlorophylls with a microplate reader. *J. Plant Nutr.* **2008**, *31*, 1321–1332. [CrossRef]

25. Matei, A.; Schou, J.; Constantinescu, C.; Kingshott, P.; Dinescu, M. Growth of thin films of low molecular weight proteins by matrix assisted pulsed laser evaporation (MAPLE). *Appl. Phys. A* **2011**, *105*, 629–633. [CrossRef]

26. Sellinger, A.; Leveugle, E.; Fitz-Gerald, J.-M.; Zhigilei, L.V. Generation of surface features in films deposited by matrix-assisted pulsed laser evaporation: The effects of the stress confinement and droplet landing velocity. *Appl. Phys. A* **2008**, *92*, 821–829. [CrossRef]

27. Leveugle, E.; Zhigilei, L.V. Molecular dynamics simulation study of the ejection and transport of polymer molecules in matrix-assisted pulsed laser evaporation. *J. Appl. Phys.* **2007**, *102*, 074914. [CrossRef]

28. Bercea, A.; Mitu, B.; Matei, A.; Marascu, V.; Brajnicov, S. Esterification process induced by UV irradiation of shellac thin films deposited by matrix assisted pulsed laser evaporation. Unpublished work, 2018.

29. Derrick, M.; Stulik, D.; Landry, J.M. *Infrared Spectroscopy in Conservation Science. Scientific Tools for Conservation*; Ball,, T., Tidwell, S., Eds.; Getty Publications: Los Angeles, CA, USA, 1999.

30. Derrick, M. Fourier transform infrared spectral analysis of natural resins used in furniture finishes. *JAIC* **1989**, *28*, 43–56.

31. Licchelli, M.; Malagodi, M.; Somaini, M.; Weththimuni, M.; Zanchi, C. Surface treatments of wood by chemically modified shellac. *Surf. Eng.* **2013**, *29*, 121–127. [CrossRef]

32. Zumbühl, S.; Hochuli, A.; Soulier, B.; Scherrer, N.C. Fluorination technique to identify the type of resin in aged vanishes and lacquers using infrared spectroscopy. *Microchem. J.* **2017**, *134*, 317–326. [CrossRef]

33. Shearer, G.L. An Evaluation of Fourier Transform Infrared Spectroscopy for the Characterization of Organic Compounds in Art and Archaeology. Ph.D. Thesis, University College London, London, UK, October 1989.

34. Oomens, J.; Steill, J.D. Free carboxylate stretching modes. *J. Phys. Chem. A Lett.* **2008**, *112*, 3281–3283. [CrossRef] [PubMed]

35. Camilleri, M.; Colemont, L.J.; Phillips, S.F.; Brown, M.L.; Thomforde, G.M.; Chapman, N.; Zinsmeister, A.R. Human gastric emptying and colonic filling of solids characterized by a new method. *Am. J. Physiol.* **1989**, *257*, 84–90. [CrossRef] [PubMed]

36. Alzate-Carvajal, N.; Basiuk, E.V.; Meza-Laguna, V.; Puente-Lee, I.; Farías, M.H.; Bogdanchikova, N.; Basiuk, V.A. Solvent-free one-step covalent functionalization of graphene oxide and nanodiamond with amines. *RSC Adv.* **2016**, *6*, 113596. [CrossRef]

© 2018 by the authors. Licensee MDPI, Basel, Switzerland. This article is an open access article distributed under the terms and conditions of the Creative Commons Attribution (CC BY) license (http://creativecommons.org/licenses/by/4.0/).

Review

In Vivo Assessment of Synthetic and Biological-Derived Calcium Phosphate-Based Coatings Fabricated by Pulsed Laser Deposition: A Review

Liviu Duta

Lasers Department, National Institute for Lasers, Plasma and Radiation Physics, 409 Atomistilor Street, 077125 Magurele, Romania; liviu.duta@inflpr.ro; Tel.: +40-21-457-4550 (ext. 2023)

Abstract: The aim of this review is to present the state-of-the art achievements reported in the last two decades in the field of pulsed laser deposition (PLD) of biocompatible calcium phosphate (CaP)-based coatings for medical implants, with an emphasis on their in vivo biological performances. There are studies in the dedicated literature on the in vivo testing of CaP-based coatings (especially hydroxyapatite, HA) synthesized by many physical vapor deposition methods, but only a few of them addressed the PLD technique. Therefore, a brief description of the PLD technique, along with some information on the currently used substrates for the synthesis of CaP-based structures, and a short presentation of the advantages of using various animal and human implant models will be provided. For an in-depth in vivo assessment of both synthetic and biological-derived CaP-based PLD coatings, a special attention will be dedicated to the results obtained by standardized and micro-radiographies, (micro) computed tomography and histomorphometry, tomodensitometry, histology, scanning and transmission electron microscopies, and mechanical testing. One main specific result of the in vivo analyzed studies is related to the demonstrated superior osseointegration characteristics of the metallic (generally Ti) implants functionalized with CaP-based coatings when compared to simple (control) Ti ones, which are considered as the "gold standard" for implantological applications. Thus, all such important in vivo outcomes were gathered, compiled and thoroughly discussed both to clearly understand the current status of this research domain, and to be able to advance perspectives of these synthetic and biological-derived CaP coatings for future clinical applications.

Keywords: calcium phosphate-based coatings; synthetic and natural hydroxyapatite; pulsed laser deposition; in vivo testing; biomedical applications

check for **updates**

Citation: Duta, L. In Vivo Assessment of Synthetic and Biological-Derived Calcium Phosphate-Based Coatings Fabricated by Pulsed Laser Deposition: A Review. *Coatings* **2021**, *11*, 99. https://doi.org/10.3390/coatings11010099

Received: 11 December 2020
Accepted: 12 January 2021
Published: 18 January 2021

Publisher's Note: MDPI stays neutral with regard to jurisdictional claims in published maps and institutional affiliations.

Copyright: © 2021 by the author. Licensee MDPI, Basel, Switzerland. This article is an open access article distributed under the terms and conditions of the Creative Commons Attribution (CC BY) license (https://creativecommons.org/licenses/by/4.0/).

1. Introduction

The biomedical domain has witnessed over the last decades a significant development due to an extensive demand for a wide variety of implants, grafts, and/or scaffolds. The bone tissue engineering field has therefore expanded to be able to address a wide spectrum of bone-related injuries and offer viable and efficient solutions. This is mainly achieved by combining the properties of bioactive materials and cells for an improved and faster bone tissue ingrowth. Implants' surface functionalization and modification with performant bioactive materials is of high interest as it provides various possibilities to modify the surface properties of biomaterials to make them suitable for specific medical applications. This technology is currently applied both for the prevention of failure and the prolongation of the bone implants' life [1]. The fabrication of resistant implants able to bypass the difficulties related to their rejection from the living bodies is therefore of huge research interest. It is important to note that the global market for implantable medical devices was valued at $72,265 million in 2015 and foreseen to attain $116,300 million in 2022 [2].

Calcium phosphates (CaP) represent the main inorganic component of bone tissues [3]. They are the most utilized bioceramics in the medical field (i.e., orthopedics and dentistry [4–7]), as coatings for a wide range of metallic implants [8]. In the last few decades, research emphasis was put on hydroxyapatite (HA), which is the most frequently used CaP due to some interesting characteristics, such as its role as scaffold for osteogenic differentiation [9] and its capacity to stimulate and accelerate new bone formation around implants [10–12]. One should note that there are currently two ways to obtain HA: the first one uses inorganic synthesis by different chemical routes (i.e., hydrothermal [13], co-precipitation [14], or sol–gel [15]). In the case of hydrothermal method, the synthesis temperature is relatively low, and the reaction conditions are moderate. The obtained products have high crystallinity, high purity, and controllable shape and size [16]. The chemical precipitation of nano-sized powders from salt solutions allows for the rapid synthesis of large amounts of material in a controlled manner [17]. The sol–gel process is a wet chemical method that involves atomic level molecular mixing, and provides good control over the composition and chemical homogeneity [18]. It should be emphasized that some of these approaches imply the use of complex processes, which could generate pollutant chemical wastes. Moreover, the composition of synthetic CaP is more complex than the one of biological-derived apatites due to both multiple lattice substitutions and the presence of ion vacancies [19]. As a consequence, a good alternative to classical chemical routes (i.e., extraction from renewable CaP resources) was introduced. One should note that the most important primary natural reservoirs of CaP are either bones (of mammalian [20,21] or fish [22,23] origin) or biogenic sources (egg-shells [24], mussel shells [25], or marine shells [26]). Unfortunately, these are generally treated as food industry wastes only [27]. Besides its low production cost, another great advantage of this fabrication route is the preservation of compositional and structural properties of the source material [28]. Furthermore, these as-fabricated biological-derived HA (BioHA) materials are well-suited to achieve a perfect synergy with the biological media due to their content in trace elements [4,8]. These elements have a determining role in the proper adjustment of biological properties (i.e., solubility, surface chemistry, and morphology), to keep compatible with the natural human bone [5,29]. As compared to chemically-synthesized HA, BioHA has different composition, stoichiometry, degree of crystallinity, degradation rate, and overall biological performance. From the biomimetic point of view, BioHA materials are more appropriate than synthetic HA to repair the skeletal system. It was reported that HA obtained from renewable, low-cost resources (i.e., animal or fish bones and egg-shells) can lead to physical-chemical characteristics and biological response comparable or even improved than those obtained in the case of synthetic ones, due to their resemblance with bone apatite [30,31]. The conversion of these by-products into HA is envisaged both as a strategy for wastes management and an economically feasible approach to reduce the overall costs for HA production. As compared to synthetic HA, the osseoconduction speed of BioHA takes place more rapid due to its higher solubility and content of Mg^{2+} and Na^+ ions. It is very important to note that these elements are generally implicated in the process of bone remodeling [32]. A detailed comparison between BioHA and synthetic HA materials can be found in Ref. [33].

In contrast to their excellent bone regeneration properties, CaP-based materials are brittle in bulk [34] (the brittle nature being in relation with their primary ionic bonds [3]), and characterized by poor mechanical properties. Their low impact resistance and tensile strength [35] represent critical drawbacks, which limit their wider clinical applications. However, the compressive strength value is rather good, exceeding that of the bone [35]. In general, an implant is fabricated from Ti or its medical-grade alloys. Despite their improved mechanical characteristics, Ti implants are characterized by low osseointegration rates. To overcome these shortcomings, HA can be used as a coating for Ti implants. This way, a combination between the ceramic's bioactivity and the metallic substrate's mechanical performances is attained [34,36]. To even accelerate coatings' osteoinduction and biomechanical fixation to the metallic substrate, some additions or HA doping with different ion concentrations were also reported [7,8]. Besides orthopedic prostheses and dental implants, HA coatings are currently being deposited also on (macro) porous scaffolds, thus expanding their applications towards bone regeneration therapies. Next to CaP, another important class of bioactive materials is represented by bioglasses (BG) [37,38], that demonstrated excellent osteointegrative characteristics in bulk form. Metallic implants' functionalization with BG coatings can be considered a good alternative for rapid bone reparation and regeneration [39].

In the field of thin film growth, the PLD method proved to be a simple, versatile, fast-processing, and cost-effective technique, which allows for a precise control over the growth rate and morphology to fabricate high-quality coatings [40,41]. In the case of CaP-based biomaterials, one of the most important and enabling characteristics of this technique is its capability to grow stoichiometric films [41]. This is mainly due to a high ablation rate that causes the evaporation of all elements at the same time [42]. The deposition temperature, substrate position, or background pressure can be independently controlled for the ease tailoring of the crystallinity, chemical composition, thickness and/or surface roughness of the fabricated structures [40]. One should emphasize upon that both film thickness and composition nonuniformity over high surface areas are two important drawbacks of the PLD technique, which can be fortunately overcome using special experimental set-ups (i.e., laser beam rastering over large diameter targets) [43]. In this respect, excimer lasers represent one economical choice for large-area commercial scale-up of the PLD process, advancing this technique as a viable alternative for future industrial applications [41].

There are studies in the dedicated literature on the in vivo testing of CaP-based coatings (especially HA) synthesized by many physical vapor deposition methods as alternatives to plasma spraying (PS), but only a few of them addressed the PLD technique. Moreover, to the best of my knowledge, there are no any review papers to summarize all the efforts dedicated in this direction, and the main objective of this review is to fill in this gap of knowledge. Therefore, the current review is focused on the gathering, compilation, and thoroughly discussion of the in vivo results pertaining to the studies reported in the last two decades on either synthetic or biological-derived CaP coatings only, synthesized by PLD technique, for various medical applications.

2. Search Strategy

2.1. Study Selection

A comprehensive literature search in the Web of Science (https://apps.webofknowledge.com/) database up to 11 December 2020, was carried out. The search included both animal- and human-related studies, and only the publications written in English were considered. The applied search strategy was: ("in vivo" [Topic], "Ti implants" [Topic], AND "Pulsed laser deposition" [Topic]), OR ("hydroxyapatite coatings" [Topic], "in vivo" [Topic] AND "Pulsed laser deposition" [Topic]), OR ("hydroxyapatite" [Topic], "Pulsed laser deposition" [Topic], AND "patient" [Topic]), OR ("hydroxyapatite" [Topic], "in vivo" [Topic] AND "Laser" [Topic]), OR ("hydroxylapatite" [Topic], "osseointegration" [Topic], AND "KrF laser" [Topic]), OR ("hydroxyapatite" [Topic], "osteoconduction" [Topic], AND "laser" [Topic]), OR ("calcium phosphate" [Topic], "in vivo" [Topic], AND "Pulsed laser deposition" [Topic]), OR ("calcium phosphate" [Topic], "in vivo" [Topic], AND "osseointegration" [Topic]).

2.2. Inclusion of Studies

These searches resulted in the identification of more than 700 studies. These titles were initially screened for possible inclusion, resulting in further consideration of approximatively 100 publications. A careful screening of the abstracts led to 40 full-text articles. Considering the aim of this review, less than half of these articles (i.e., 15) met the inclusion criteria and were chosen and assessed in detail, and parts of the reported results were included and discussed in the present review.

3. Pulsed Laser Deposition Technique

Bioceramic coatings are currently used for various biomedical applications [9] to modify the implant surface (by increasing its surface roughness), to promote osseointegration. Nowadays, PS is the only industrial technology used for coating dental and orthopedic implants with CaP materials. In this respect, in the dedicated literature there are numerous and very interesting research works that point out and critically evaluate the physical-chemical and biological characteristics of the structures fabricated by this technique, along with their clinical performances [44,45]. The CaP-based coatings fabricated by conventional thermal PS onto medical implants function as an intermediate layer between the tissues and the metallic implants [46]. Despite its wide commercial availability, this technique still has some drawbacks, such as: (i) the synthesized coatings generally consist of several phases (i.e., β-TCP forms at 1200 °C, and it transforms into TTCP at T > 1400 °C); (ii) at higher synthesis temperatures, the negative influence of the mismatch between the thermal expansion coefficients of HA and TCP and the ones corresponding to Ti-based alloys (11–15 × 10^{-6} cm/(cm·K) vs. 8–10 × 10^{-6} cm/(cm·K), respectively), limit the obtaining of good CaP coatings onto metallic substrates [47]; (iii) it supplies very thick structures with low adherence to the substrate (because the coatings' tensile stresses have a greater tendency to initiate cracks and cause film delamination [47]); (iv) surface morphology, phase composition, or uniformity of crystallization [32,48] are difficult to be controlled; and (v) it is a line-of-sight method [3]. Therefore, no coating technique can be considered perfect and all these drawbacks gradually supported research efforts focused on the introduction of various, alternative coating techniques to PS (i.e., radio frequency magnetron sputtering, PLD, electrochemical deposition, etc.) [1,3,7,8,11]. Among these, PLD technique is worth mentioning due to some important advantages over PS method, such as: (i) a much faster surface deposition process; (ii) a stoichiometric transfer of the material's composition from the target in the synthesized coating; (iii) a better morphological and compositional uniformity; (iv) a lower porosity; (v) precise thickness control; (vi) effectiveness for coating small implants, with complex shapes; (vii) a decreased tendency of the synthesized structures to crack or delaminate; and, very important, (viii) a high adherence to the metallic substrate [1,49]. In the biomedical field, for the fabrication of CaP-based coatings for bone implant applications, one of the most applied plasma-assisted methods is PLD [7]. This technique is able to fabricate dense and extremely adherent films. In this respect, synthetic HA and BioHA coatings obtained by PLD previously demonstrated high adherence to metallic substrates [50]. Moreover, the composition of these coatings is consistent with the one corresponding to the raw (base) materials, along with an improved crystallinity [50]. One should note that in PLD, after the ablation of the target by laser pulses, a plasma plume is generated. When the species existing in the plume reach the surface of the substrate, they may deposit onto the surface and form a film [41]. The number of the deposited species depends both on their density in the plume and their energy. In the case of low energies, the species may not deposit on the substrate surface even if they arrive at the surface [51]. If the substrate temperature is high, the energy of the species can be compensated. Consequently, the number of deposited species onto the substrate surface will increase, along with the density of the droplets. High substrate temperatures also contribute to the atomic diffusion, which, in turn, can determine the appearance of two phenomena: the first one is the atomic rearrangements and crystallization of the film and the second one, the improvement of the film−substrate bonding state [52].

The laser sources appropriate for the ablation of a wide range of materials use wavelengths in the UV domain due to some important advantages over IR and/or visible laser sources, such as (i) a higher penetration depth of the laser beam in the target material and (ii) a higher energy of the photons that allows for a more efficient vaporization of the target [53]. In this respect, the laser sources used for PLD experiments are either excimer lasers (i.e., ArF [54], KrF [55], or XeCl [56], emitting at wavelengths of 193, 248, or 308 nm, respectively) or solid-state lasers (i.e., Nd:YAG [57], emitting at 266 nm). It should be also emphasized that to increase the amount of evaporated material from the ablated target to the detriment of expulsed liquid or solid phases, lasers emitting in a pulsed regime with pulse durations in either the nanoseconds or picoseconds range are generally used [58]. In these regimes, the absorption process takes place more quickly than in the case of thermal diffusion processes. More insights on the PLD and PS techniques are well described elsewhere [33,41,49].

It is important to mention that post-deposition treatments are generally applied to transform CaP phases with lower Ca/P ratio to crystalline HA. Thus, there are two commercially used post-treatments: sintering [33,59,60] and soaking in alkaline solutions [61,62]. These treatments are generally applied for several hours, in the range of 600–800 °C. Their aim is to transform the water trapped in the film during the synthesis process in OH^- ions, to stabilize the crystalline structure [34].

There are reports in the dedicated literature on the in vivo testing of CaP-based coatings (especially HA) fabricated by different physical vapor deposition methods, but, to the best of my knowledge, only a few of them addressed the PLD technique. Therefore, for an easy access to information, Table 1 introduces the PLD experimental details given in the papers considered in this review.

Table 1. Pulsed laser deposition (PLD) experimental details given in the papers considered in this review, in the chronological order in which they were reported in the literature.

Laser	Repetition Rate (Hz)	Substrate Temperature (°C)	Fluence (J/cm^2)	Atmosphere/ Pressure (Pa)	Coating Thickness (µm)	Post-Treatment	Ref
KrF	-	500	3	Ar–water vapor	1	-	[63]
KrF, CO$_2$	2–10	RT	3–10	-	1–4	550 °C, 1 h, 10^{-2} Pa	[64]
KrF	-	490	3	Ar–water mixture	1–1.2	-	[65]
KrF	30	650–730	4–7	Ar/H$_2$O/40–80	1–3	-	[66]
ArF	20	490	0.9	Water vapor/45	2	-	[67]
KrF	2	150–400	2	O$_2$/13–50	0.4	150–400 °C, 6 h	[68]
ArF	-	460	4.2	Water vapor/45	0.05–0.1	-	[69]
-	-	-	-	-	0.05/0.3	800 °C, 8 h	[32]
ArF	10	RT	1	Water vapors/0.1	-	380 °C, 1 h	[70]
KrF	5	200,600	5	Ar/45	-	-	[52]
KrF	10	400	5	Water vapors/50	-	400 °C, 6 h	[71]
-	3	186,410	4.6	Water vapors	-	-	[72]
KrF	-	-	-	-	-	-	[73]
KrF	5	600	5	Ar/45	-	-	[74]
KrF	10	500	3.5	Water vapor/50	-	600 °C, 6 h	[75]

4. In Vivo Assessment of PLD CaP-Based Coatings

In vitro tests play key-roles in the overall evaluation of a biomaterial by delivering important information on its potential behavior inside a living system. To better understand the complex processes occurring in a living environment and to deliver accurate data for the validation of the biomaterials' performances that target clinical trials, in vivo tests (using either animal or human models) are of key-importance and should, therefore, follow thorough in vitro acceptance [75].

Biocompatibility is an essential and required characteristic of the biomedical materials introduced inside the living systems, presenting a beneficial interaction with the surrounding bone tissues. Immediately after CaP-based functionalized implants are surgically inserted into the living bodies, there are tissue responses that take place at the implanted materials—soft/hard tissues interface. The formation of connective tissue fibers (2–3 weeks), which make a fiber mesh containing carbonated apatite, is one such response that occurs initially [76,77]. This tissue-free layer develops onto the ceramic surface and actively contributes both to a strong fixation of the implant to the surrounding bone tissues and to the acceleration of bone integration and healing at early implantation times [78–81]. These aspects are of key-importance for any implant system [82]. The amount and nature of osseointegration in metallic implants are determined by several factors such as surface topography, which includes surface roughness [83], wettability [84,85], and surface morphology [86,87]. It was demonstrated that an initial stability and a further stronger interface with the bone tissues are more likely to be achieved in the case of implants with rough surfaces as compared to smooth ones [88], in both animal and human models, with bone ongrowth interface [89,90]. Furthermore, the bone–implant interfacial shear strength is directly related to the degree of surface roughness. However, for an optimal clinical performance of metallic implants in bone, both an ideal type/shape of the implant and the surface roughness degree still remain unknown. It is also important to mention that initial stability is also achieved by drilling the bone socket to a diameter slightly inferior to the implant's dimensions, but to the proper matching length of the implants. If the opening hole is drilled through the bone and enlarged progressively, and the implant is screwed in with low momentum, the surrounding bone is going to be compressed. This provides a firmer fit to the implant itself and it also increases the calcium concentration in the interface. This method decreases also the blood clot thickness around the implants. If growth factors are used, a boost in the healing process will be observed, while in their absence, the osteoconduction process only will occur.

A nowadays challenge in implantology is the capability of a surface to assist cells' colonization and differentiation. Cell migration, adhesion, and proliferation onto implant surfaces are key-processes for the initiation of tissue regeneration [91]. Osteoconduction and osseointegration represent two mechanisms that involve the adhesion or proliferation of cells and integration in the CaP [35,92]. The adhesion of cells is directly related to the CaP ability to adsorb the proteins from the extracellular matrix. For CaPs, this is strongly determined by important parameters, such as the surface roughness, crystallinity degree, Ca/P ratio, solubility, phase content, grain and particle sizes, and surface energy [92].

Osteoinduction is the property of a material to induce the differentiation of progenitor cells to osteoblasts [35,92]. It was demonstrated that CaPs in the absence of supplements are osteoinductive materials [92]. In turn, the osteoinduction ability depends on several CaP properties, i.e., surface charge, morphology, and chemistry, which can influence the adsorption of proteins [92].

The success rate of a medical implant is in relation to several factors such as the implant's design, the structure, and properties of the used material, the applied loads' magnitude, the employed surgical technique, health conditions of either animals, or human patients [82], and the existence of an inflammation-free environment. In the latter case, there exist two major problems: (i) the area intended for the implant insertion might be infected and (ii) the implant's surface might get contaminated during surgery. Both situations can be easily by-passed, by appropriate dental work and by a proper isolation during the surgical act, respectively. All aspects related to the biological interactions to improve the long-time stability and reliability of medical devices inside living systems have pushed forward the research of a wide range of surface modification techniques. The aim was to achieve rapid healing and an improved bone-implant interaction, for an early osseointegration. Therefore, in the last two decades, the surface functionalization of implants by increasing the bioactivity using various biological and chemical processes or by surface macro- and microtexturing was one of the major research topics in the biomedical domain [82].

4.1. Used Substrates

Currently, in comparison to polymers and ceramics, metallic biomaterials are extensively used in orthopaedics, dentistry, and oral and maxillofacial surgeries. This is due to several important advantages, such as the high mechanical strength, superior biocompatibility, high resistance to corrosion, and improved chemical stability under biological conditions [93–95]. Bioinert metallic implants (i.e., stainless steel (316L) and Ti, Ti-based, and cobalt-chromium alloys [96]) are generally utilized for a wide range of medical applications, but there is currently a new generation of biodegradable metals (i.e., Mg, Zn or Fe) that has been intensively employed for temporary applications [97,98]. Because they do not chemically bond to bone tissue unless their surfaces are modified, the use of the latter ones is generally limited.

4.2. Animal and Human Implant Models

After the in vitro validation of biomaterials' surfaces, the use of laboratory in vivo models is a step forward biocompatibility assessment [99] and future clinical outcome of metallic implants introduced in bones. In this respect, in vivo testing on animal models is a key-parameter for both understanding and evaluation of the biological processes that occur in a living system.

The general animal spectrum used for in vivo testing of CaP-based coatings synthesized by the PLD technique only is limited to rats, rabbits, (mini) pigs, dogs, goats, and sheep [34], and each animal model having its own advantages and limitations [100]. Small animal models demonstrated some important advantages over larger ones, i.e., overall lower costs, osteogenic ability, the possibility to both shorten the implantation time-periods [101,102], and monitor bone formation by imaging methods (e.g., μCT). Thus, the possibility to carry out longer experimental times is worth considering. In this respect, rat or mice models are generally used for subcutaneous examination of implants [103,104], while rabbits represent the easiest way to investigate the interaction between the coating and the femoral bones [105–107]. It should be noted that the rabbit bone model was previously indicated [108,109] as a valuable screening tool able to select favorable implant surface characteristics/technologies before moving to the next step (i.e., human trials). It is currently used in various medical tests [110], due to both its size and ease of handling. In addition, it was demonstrated that its skeleton reaches maturity very fast (~24 weeks) [111]. In comparison, large animal models (i.e., dogs, sheep, or goats) are generally indicated for verifying the practicability of implants closer to real clinical situations. In such large animal models, the investigation of the osteoinduction process is easier to be attained than in the case of small ones. Large animals have the advantage of an immune system more similar to humans than in the case of small animals and can offer the possibility to host different types of test materials [112]. Despite this, such large animal models are not so intensively utilized for the evaluation of osteoinduction because of their lower metabolic rates, besides higher costs for management and maintenance [113].

Usually, implants are surgically inserted in either the femur, tibia, or mandible bones [114]. The rabbit's bones manifest faster changes as compared to the case of larger animal models [115]. Considering the difficulties met when extrapolating the results obtained on rabbit bones to the human case, various screenings for implant design and validation of the tested materials are still needed to be performed [75], before their testing on larger animal models. Dog models are generally used for testing dental implants [116]. One should emphasize upon that the focus of the most studies is directed to the biological response of the living bone to CaP-based materials.

For an easier follow-up of the text, Table 2 introduces information on the sample codes, which will be further used in the paper.

For an ease access to information presented in the papers considered in this review, Tables 3–6 introduce data on the investigated coating materials (along with the type and dimension of substrates, bone implantation sites, length of studies, and all performed analyses). All the information was grouped according to the applied small and large animal models, respectively (Table 3—rats/mice, Table 4—rabbits, Table 5—minipigs, and Table 6—dogs/sheep/humans).

It should be emphasized that because of the reduced number of in vivo studies on the PLD synthesis of CaP-based coatings, the use of a certain animal model as the most suitable one for the optimal assessment of metallic implants' osteoconduction has not been established yet. The reports on clinical trials (using human patients) are therefore scarce [71].

Table 2. Sample acronyms related to different materials used in the review and their explanation.

Samples Code	Description of Samples
Ti/Ti6Al4V	Titanium/titanium alloy (control specimen or deposition substrate)
HA	Synthetic hydroxyapatite
HA/TTCP	Synthetic hydroxyapatite/tetracalcium phosphate
OCP	Octacalcium phosphate
Mn-CHA	Manganese-doped carbonated hydroxyapatite
BHA/HA	Bilayer bovine hydroxyapatite/hydroxyapatite
HATW	Titanium web scaffold functionalized with hydroxyapatite
HA/45S5; HA/BG	Hydroxyapatite/bioglass composite
Mg:OCP	Octacalcium phosphate doped with magnesium
Mg:HA	Synthetic hydroxyapatite doped with magnesium
FHA	Fluoridated hydroxyapatite
Li-C	Biological hydroxyapatite doped with lithium carbonate
Li-P	Biological hydroxyapatite doped with lithium phosphate

Table 3. Information related to the investigated coating materials, according to the applied small animal models (i.e., rats/mice).

Investigated Material	Animal Model/Number	Used Substrate/Dimensions	Bone Implantation Site/Length of Study (Weeks)	Performed Analyses	Ref
HA	Adult rats/36	Teflon, polyethylene, Ti6Al4V/2.5 × 2 × 1 mm^3	Femurs/2–8	SEM, histology	[64]
HA	KSN nude mice/12	Ti web discs/5 mm diameter, 1.5 mm thickness	Backs/2–12	Histology	[70]
Fluoridated HA	Rats/16	Ti disc implants and screws/15 mm diameter, 1 mm thickness and 1.2 mm external diameter, 12 mm length	Distal femur/4–8	Microcomputed tomography, histology	[73]

Table 4. Information related to the investigated coating materials, according to the applied small animal models (i.e., rabbits).

Investigated Material	Animal Model/Number	Used Substrate/Dimensions	Bone Implantation Site/Length of Study (Weeks)	Performed Analyses	Ref
HA, HA/TTCP	New Zealand White Rabbits	Ti6Al4V	Proximal tibia, distal femurs/8	Histomorphometry	[66]
HA	New Zealand White Rabbits/12	Grit-blasted commercial titanium rods/7 mm length, 2 mm diameter	Proximal tibia/24	Histology, histomorphometry	[67]
HA, Mn-CHA, OCP	New Zealand White rabbits/12	Coins of Ti implants/1.95 mm thick, 6.25 mm diameter	Tibia/8	Tensile (pullout) tests	[68]
HA/45S5	New Zealand White rabbits	Ti6Al4V/20 mm diameter, 1 mm thickness	Shin bones/4–12	Histology	[52]
Mg-doped octacalcium phosphate and HA	Rabbits/23	Porous Ti6Al4V scaffolds/(10.3 × 2.5 × 2.5) mm^3	Femoral condyles/24	Histopathology, planimetric analysis, μCT	[72]
HA/45S5	New Zeeland White rabbits	Ti6Al4V plates/Φ20 mm × 1 mm	Shin bones/4–8	Histology	[74]
Lithium-doped biological HA	New Zealand White rabbits/26	3D Ti implants/3.5 mm internal and 5.5 mm external diameters, 9 mm length	Femoral condyles/4–9	Computed tomography, mechanical testing, scanning electron microscopy	[75]

Table 5. Information related to the investigated coating materials, according to the applied small animal models (i.e., mini-pigs).

Investigated Material	Animal Model/Number	Used Substrate/Dimensions	Bone Implantation Site/Length of Study (Weeks)	Performed Analyses	Ref
HA	Mini-pigs/3	Dental cylindrical implants/13 mm length, 3.3 mm diameter	Lower jaws/16	(Micro)radiography, transmission and fluorescent microscopies	[63]
HA	Mini-pigs/4	Ti6Al4V dental implants/12 mm length, 3.3 mm diameter	Lower jaws/16	Polarized and fluorescent light, percentage of osseointegration	[65]

Table 6. Information related to the investigated coating materials, according to the applied large animal models (i.e., dogs/sheep/human patients).

Investigated Material	Animal Model/Number	Used Substrate/Dimensions	Bone Implantation Site/Length of Study (Weeks)	Performed Analyses	Ref
BHA/HA	Beagle dogs/8	Commercial Ti screw implants/8 mm length, 3.7 mm diameter	Bilateral femoral Shafts/4–24	Histology	[32]
HA	Sheep/20	Cylindrical Ti implants/10 mm long, 5 mm diameter	Tibia/8–12	Histomorphometry	[69]
HA	Human patients/12	3D Ti mesh implants/(10 × 5) cm² size, 0.2 cm in holes diameter, thickness of 0.25 cm	Skulls/12–24	Tomodensitometry	[71]

4.3. Characterization Methods

For the in vivo assessment of the CaP-based coatings considered in this review, the performed investigations, on which we will focus our attention, are standardized radiography, microradiography, (micro) computed tomography (μCT), tomodensitometry, histology (under polarized and fluorescent light), histomorphometry, planimetric analysis, fluorescent microscopy, scanning and transmission electron microscopies (SEM, TEM), and mechanical testing. While mechanical testing aims to determine the bonding strength between the newly formed bone tissue and the implant, histological investigations are used for a wide range of purposes, such as the measurement of the new bone area, bone apposition ratio, etc. [105–107,117]. It is important to note that light microscopy or optical microscopy is the most common laboratory technique used for biological investigations. It is a cheap, robust, and typically noninvasive method that uses visible light to detect and magnify small objects. The resolution limit is an intrinsic property due to the wavelength of visible light radiation [118]. Polarized light microscopy is a contrast-enhancing technique with a high degree of sensitivity that can be used for both qualitative and quantitative investigations. It can provide information on both absorption color and optical path boundaries and the structure and composition of materials that are invaluable for identification and diagnostic purposes. The technique of fluorescence microscopy has become an essential tool in biology and biomedical sciences, as well as in materials science, being capable of revealing the existence of single molecules. SEM and TEM microscopies enable the characterization of microstructures at many different length scales, from micro- to nanoscale, within an imaging session. They all generate a highly focused beam of electrons, which impacts the specimens inside a vacuum chamber. SEM microscopy is used to examine material surfaces, in comparison to TEM microscopy, which primarily focuses on investigation of the internal structure of the specimens [118].

4.3.1. Standardized Radiography and Microradiography

Dostalova et al. [63] evaluated quantitatively and qualitatively the Ti implants by radiographical measurements. After 16 weeks of implantation, the osseointegration process was confirmed by the presence of newly formed bone around all implants, along with a strong bone–implant connection. When measured in the long-axis cross-section, the implant area was in the range of ~25–28 mm for Ti implants and ~21–37 mm in the case of the HA-coated ones. This corresponded to an osseointegration area of ~18–22 mm for Ti implants and of ~15–30 mm for the HA-functionalized ones. The inferred percentage of osseointegration varied from 73% to 79% for Ti implants and from 70% to 86% for the HA-coated ones. No differences between various implants were found to be statistically important.

4.3.2. Computed Tomography

Using μCT, Mroz et al. [72] performed qualitative and quantitative investigations on Mg:HA samples and Ti implants, and the obtained results revealed significantly higher differences between these groups ($p = 0.0311$).

In another study, Chen et al. [73] demonstrated by μCT that for both 4 and 8 weeks, the ratio of bone volume to total volume, mean trabecular number, and mean trabecular thickness were significantly higher ($p < 0.05$) in the case of FHA-coated implants, which was indicative for an accelerated osteogenesis process in the region of interest. Moreover, for both time periods, the mean trabecular separation was found to be significantly lower ($p < 0.05$).

Duta et al. [75] used CT scans, at 4 weeks after the surgical procedure, to observe the correct placement and the good integration of all implants into the surrounding bone. Thus, the presence of the peripheral osteosclerosis and no inflammatory process of the soft tissues were indicated. The obtained results pointed out to an increase in the osseous density, for both investigated time periods (4 and 9 weeks, respectively). Thus, the bone density values corresponding to the Li-C and Li-P coatings, measured at 9 weeks, were ~1.2 times higher than those inferred at 4 weeks after implantation. At 4 weeks, both Li-C and Li-P structures indicated bone density values ~1.3 times higher than those obtained for Ti implants. Moreover, at 9 weeks, the density values inferred in the case of functionalized 3D Ti implants were ~1.4 times higher as compared to Ti ones.

4.3.3. Tomodensitometry

In a pioneering contribution, Duta et al. [71] used the tomodensitometry analysis to evidence the osteogenesis process of simple and functionalized Ti meshes implanted in human patients, at 3 and 6 months after surgery. After 3 months, no changes were detected for any of the investigated patients in terms of tissue density on Hounsfield (HU) scale. After 6 months, in the case of two patients with Ti mesh and for all with Ti meshes functionalized with bioactive surfaces, changes in measurements of osteoinductive and osseointegration phenomena were evidenced. In the first case, these changes were assessed only on the edge area, which was in direct contact with the bone. In the second case, four patients from the total of six, showed changes of tissue density, both on the edge and interjacent areas of the mesh. Up to 6 months, no patient evidenced those changes in the center region of the implants. Thus, the inferred values for the functionalized Ti meshes were of 583 and 412 HU (for the edge and interjacent region, respectively), in comparison to 78 HU, obtained in the case of Ti meshes.

4.3.4. Histology

Antonov et al. [64] performed histological evaluations of their synthesized structures for three different time periods of 15, 30, and 60 days, respectively. It is important to note that no inflammation was observed around the sites of implantation for any of the investigated structures. However, in the case of Ti samples, a significant fibrous tissue formation was pointed out. For both types of structures (synthesized by either KrF or CO_2 lasers), a thin fibrous layer could be observed after 15 days of implantation. At 30 days, a partially direct bone–implant contact was indicated, while very little fibrous tissue and the formation of new bone, which was in direct contact with the implant surface, were shown at 60 days. A very interesting observation was that the osteointegration rate was slightly superior in the case of annealed samples, rather than for the non-annealed ones. Between the two types of synthesized coatings (using either KrF or CO_2 lasers), no significant statistical differences in the osteogenesis process were inferred.

The histological evaluation under polarized and fluorescent light was used by Dostalova et al. [65] to indicate the presence of newly formed bone tissue around the investigated implants. No osteoclasts, macrophages, or any inflammatory reaction cells could be observed in the ground sections. In the case of synthesized coatings, a fibrous connective tissue occurred in ~23% of the implant surface, but without the formation of a continuous layer. In contrast, the fibrous connective tissue between the implant and the newly formed bone tissue occupied ~35% of the surface, especially in the middle part of the implant. However, it was indicated that these differences were not statistically significant ($p = 0.05$). Under fluorescent light, a uniform distribution of the fluorescent label in the whole bone, most probably due to a remodeling process of early formed bone, was observed. These results were supported by the higher magnification investigations, in which active bone cells were observed.

Using histological analyses, Peraire et al. [67] succeeded to demonstrate bone up-growth in the endosteal areas, with a great amount of lacunae area in contact with them, while in the center, the bone apposition was indicated to be totally absent. The fiber mesh either disappeared in some implants or was very thin. In the case of the Ti rods functionalized with HA coatings by plasma spraying (HA–PS), a good bone regeneration at the ends of the implants (endosteal zones) was observed, but the newly formed bone tissue in contact with the implant surface evidenced a significant large amount of lacunae area. This active remodeling process might occur because of the coating degradation. HA–PS coatings partially disappeared in some areas of the implants, were delaminated or even detached, and HA particles could be therefore observed. In the central zone, bone apposition areas could be hardly seen, with a slightly thicker fiber mesh. Inflammatory cells (i.e., lymphocytes, macrophages, or neutrophils) were also present. In the case of HA coatings synthesized by PLD (HA–PLD), a good bone regeneration at both endosteal ends of the implant was indicated. All samples presented superior (grade 4) responses in bone reaction and interface analysis parameters. The newly formed bone tissue was similar to the cortical one, and the presence of mature osteocytes was detected. Under optical microscopy, HA–PLD coatings could not be evidenced because of their low thickness, while under polarized microscopy, the mineralized matrix in apposition to the implants could be observed for all investigated areas (cortical insertion area, opposite endosteal area, and bone marrow). The central area in contact with the bone marrow indicated bone apposition areas interspersed in a thin fiber mesh, with large trabeculae growing from the coating surface.

Hayami et al. [32] performed histological investigations of the interface between bone and the tip of each implant. After 4 weeks, large gaps at the Ti implant–bone interface could be observed, bone and connective tissues being intermingled. In the case of the sprayed implants, a thick adhering coating could be clearly seen, along the observable length of the interface. For the BHA/HA bilayered implants, no detectable gap was visible, and the bone and the implant closely adhering to each other along the full length of the interface. Thus, the process of biointegration was assumed to be complete after only 4 weeks of implantation. This strong bone–bilayered implants connection was also observed at 8 and 24 weeks post-operation. At 24 weeks after surgery, normal bony structures with osteocytes were indicated surrounding the implants.

Using histological analysis, Hontsu et al. [70] could observe that the HA-functionalized implants were surrounded by a mesh of thin fibrous tissue. Inside the implants, the mesh porosity was filled with fibrous tissue containing capillaries. Inside the HATW structure, a slight amount of new immature bone formation was indicated. This amount clearly increased from the second to the fourth week of implantation, for all types of scaffolds. At 4 and 8 weeks, more sites containing osteoblast-like cells and osteoid tissue expressing active bone formation could be observed, while at 12 weeks, the structure of the bone was further matured and contained larger areas of remodeled lamellar bone.

After observing the stained tissue sections, Wang et al. [52] indicated an obvious crack between the newly formed bone tissue and the 200 °C-synthesized film. This was mainly because of the preparation process of the tissue sections.

Although SEM results demonstrated good connections between the newly formed bone tissues and the surfaces of the two type of films (Figure 1a,b), the histological micrographs indicated a superior bone growth of the structures synthesized at 600 °C, in comparison to those prepared at 200 °C. The osteoblasts in the newly formed bone tissue were shown clear and homogenous, with a growth oriented along certain directions.

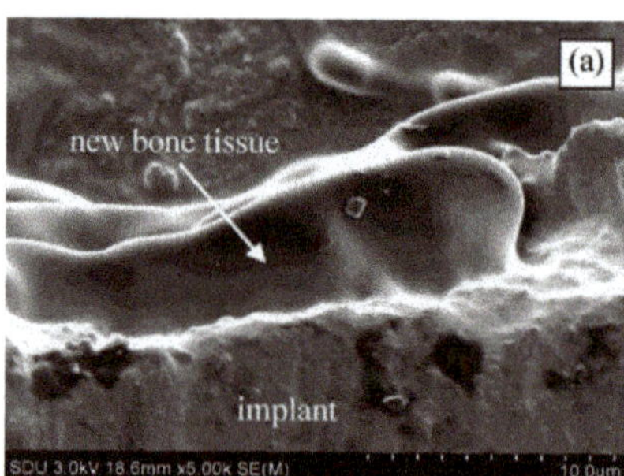

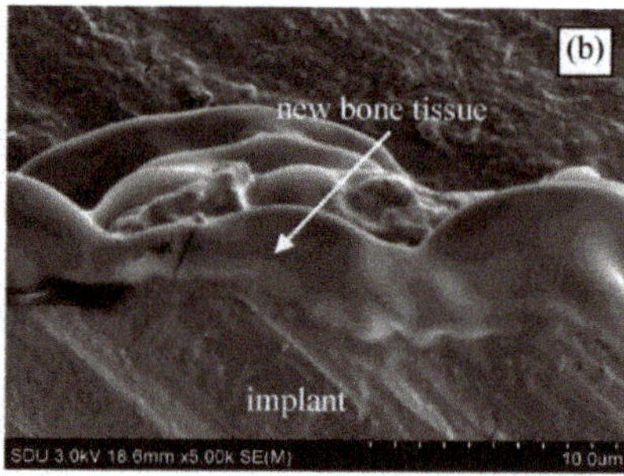

Figure 1. Morphologies of the implants after embedding for 1 month. (**a**,**b**) are the SEM morphologies of 200 and 600 °C films. (Reprinted with permission from [52]. Copyright 2013 Elsevier.)

It was therefore concluded that all tissue sections from the implants deposited at 600 °C were good, with a satisfactory connection between the film and newly formed bone tissue. In contrast, for the structures synthesized at 200 °C, a lower adherence between the newly formed bone tissue and the films was demonstrated. This was mainly influenced by the presence of a dark area between the film and the new bone tissue, which seemed to appear because of the high shearing stress during the cutting process.

Using the histopathological evaluation, Mroz et al. [72] showed that Ti implants had a tight adherence to the mineralized bone tissue, being in many cases within the fatty marrow. Small fields of connective tissue were also observed directly adjacent to the implant, along with a direct penetration of regular bone into the implants' pores. In the case of Mg:OCP samples, a direct integration of the bone and penetration into the pores and the structure of the implant were indicated. For the Mg:HA group, evidence of a direct bone–implant integration was shown, with no sign of osteoclastic or osteoblastic reaction around the implants.

Longitudinal sections were collected by Chen et al. [73] to assess the newly bone tissue formation around the implants. No adverse inflammatory reactions or gaps at the bone–implant interface were observed. At 8 weeks after surgery, the bone area ratio and bone–implant contact were significantly higher (~6 and ~1.5 times, respectively) around FHA-coated implants in comparison to Ti ones.

Wang et al. [74] performed histological investigations to assess the osteoconduction capacity of HA/BG composite coatings, with three different BG concentrations. Thus, the 90%HA + 10%BG structure was shown to connect well with the bone tissue. However, the bone matrix filled the implant–bone tissue interspace, and osteoblast cells (OB) were not observed in this region before 2 months after implantation. In contrast, the 80%HA + 20%BG film exhibited better osteoconduction behavior as new OB could be observed near the implant surface in the first month after surgery. It is important to note that the improved biological activity of the film was indicated to be mainly due to their *c*-axis orientation. Macroscopically, a good connection with the bone tissue was also observed in the case of the 20%HA + 80%BG structure. In this case, a crack was indicated between the film and the bone tissue, which corresponded to a poor bond state (mainly because of the shearing force coming from the inner circle saw, used to cut the specimen into slices).

4.3.5. Histomorphometry

Dostalova et al. [65] inferred the area of the bone–implant interface and found the values to vary from ~65%, for Ti implants, to ~78%, for HA-functionalized ones. It is important to note that between both types of surfaces, no significant statistical differences were inferred (Student's *t*-test, $p = 0.05$). As a consequence, a similar osseointegration trend for both Ti implants and HA-functionalized ones was indicated.

In the case of the implants inserted by Kim et al. [66] in the femur, a value of ~48% of the mineralized bone at the bone–coating interface for the synthesized HA coatings and of ~63% for the HA/TTCP biphasic ones, was inferred. For the case of tibia-implanted samples, values of ~51% and ~56% for the mineralized bone were indicated in the case of HA and HA/TTCP biphasic coatings, respectively. Very important, all samples were shown to exhibit a good integration, with no significant foreign body response.

Among the three materials evaluated by Peraire et al. [67], there were no statistically significant differences in bone in-growth around the drilling hole. However, the HA–PLD implants presented the highest value of the percentage of bone contact (86%), and the lowest value of percentage of lacunae contact (14%). This difference was statistically significant (Scheffe Test, $p < 0.05$,). When referring to the ratio between the total bone surface and length of the evaluated area, the HA–PLD implants showed a significant increase in comparison to the HA–PS group (Scheffe test, $p < 0.05$). This did not apply as well to the Ti group. In the case of the HA–PS group, even though the bone response was slightly different than the one corresponding to the Ti, no statistically significant differences were found, in any of the quantified variables.

Paz et al. [69] used in their study two methods to evaluate the percentage of bone–implant contact (% BIC): conventional light transmission microscopy and environmental scanning electron microscopy (ESEM). The obtained results demonstrated no statistically significant differences between these two methods, although ESEM is believed to offer greater precision when characterizing bone areas in close contact with implant surfaces. ESEM analysis showed a considerable improvement of the bioactivity in the case of HA-coated samples, the bone–implant interface being more homogeneous and continuous in comparison to Ti implants. Moreover, the bone was observed to enter deeper into the craters of the macrostructure for the HA-coated samples than for the Ti implants. In areas with low bone density, the HA-coated structures presented a superior behavior in contrast to Ti. ESEM analyses showed an improvement of the percent of total BIC for both HA-synthesized coatings (50 and 100 nm-thick samples, respectively) with regard to Ti implants. There were no significant differences between the two coated samples, although the percent apical BIC indicated better characteristics of the thicker coating over the thinner one. The conventional light transmission microscopy images showed similar results as ESEM, but with a modification in the percent total BIC; for this investigation, the thinner HA coating showed superior behavior in comparison to the thicker one, while for the percent apical BIC, the thicker coating presented a better behavior, but with a great increase in the standard deviation for the thinner one.

4.3.6. Planimetric Analysis

The results of planimetric investigations performed by Mroz et al. [72] revealed the best bone integration in the case of Mg:HA samples, this group having the highest average length of the bone–implant interface and also the best reproducibility of the results. No significant differences were inferred between the Mg-doped (OCP and HA) implants and the Ti ones.

4.3.7. Fluorescent Microscopy

When fluorescent microscopy was used by Dostalova et al. [63], the active bone formation could be observed, both in the neck and in the bottom of the implants, surrounding osteons and braiding the fibrous connective tissue on the implant.

Chen et al. [73] investigated by fluorescence microscopy the bone formation around the implanted structures. As determined by confocal laser scanning microscopy, the bone area between the implant surface and the boundaries observed at 1, 4, and 8 weeks, respectively, was significantly higher for FHA-coated implants as compared to Ti ones ($p < 0.05$). These implants indicated 1.5 and more than 4 times more new bone tissue formation at 1 week and 4 and 8 weeks, respectively, as compared with the case of Ti implants.

4.3.8. Scanning Electron Microscopy

Antonov et al. [64] used SEM to investigate the in vivo behavior of the synthesized coatings. New bone formation was therefore observed, which surrounded implantation site of the HA-coated alloy samples.

SEM observations performed by Peraire et al. [67] allowed for the detection of the HA–PLD coatings on the grit-blasted Ti surfaces, at 24 weeks after surgery. The images obtained using backscattered electrons showed different contrasts for the bone and the coating, which allowed for their clear differentiation.

Hayami et al. [32] investigated by SEM the holes from the removed implants. Because the threads of the screw implants had a long pitch, the troughs left in the bone were trapezoidal in shape. For Ti implants, bone growth was slow and insufficient because of the imperfect trapezium formation, in comparison to the BHA/HA-bilayered implants, where the growth rate and quantity of newly formed bone tissue were greater due to the formation of regular trapezia. In the *V*-shaped groove of the thermal-sprayed HA coatings, a flaky piece was indicated, which appeared to be a fragment that peeled away from the substrate.

Wang et al. [52] investigated the morphologies of the implants at 1 month after implantation. These are shown in Figure 1a,b.

A layer of newly formed bone tissue was clearly seen, developing along both types of structures (synthesized at 200 and 600 °C, respectively). This layer intimately adhered to the implants' surface and had a thickness of ~5 μm. After 3 months of implantation, the newly formed bone tissue occupied the bone−implant interspace (Figure 2a,b).

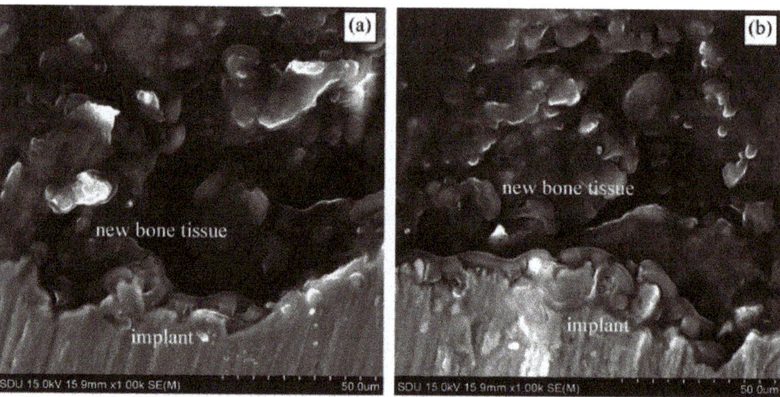

Figure 2. Morphologies of the implants after embedding for 3 months. (**a,b**) are the SEM morphologies of 200 and 600 °C films. (Reprinted with permission from [52]. Copyright 2013 Elsevier.)

A good bioactivity of the films was therefore indicated, the two types of implants being capable of inducing new bone growth on their surfaces.

Using SEM (Figure 3), coupled with the analysis of backscattering electrons, Duta et al. [75] managed to infer the adherence ratio of the remaining bone fragments onto the surface of the extracted implants.

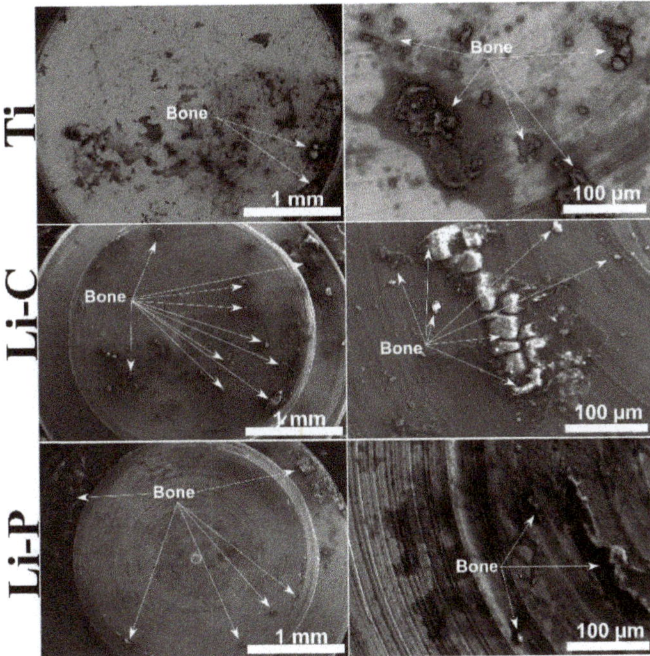

Figure 3. SEM micrographs indicating bone detachment on the surface of a control and functionalized (with Li-C and Li-P coatings) Ti implant, at 4 weeks after surgery.

Thus, adherence ratios up to ~38% higher in the case of functionalized 3D Ti implants (with Li-C and Li-P coatings) as compared to the Ti ones were indicated. This result, corroborated with the higher values of the detachment force obtained in the case of functionalized 3D Ti implants, in comparison to Ti ones, were indicative for an enhanced osseointegration process. Moreover, the presence of such osseous structures onto the surface of the implants suggested, besides the beginning of the implant integration process into the bone, the absence of any adverse reactions at the implantation site.

4.3.9. Transmission Electron Microscopy

Using TEM, Dostalova et al. [63] have shown no marks of irritation and inflammation in the surrounded bone. OB could be seen both in the border of the implants and newly formed bone tissue. A low number of foreign body cells were indicated in the close vicinity of the implant cover. In the case of Ti implants, a fibrous connective tissue between the implants and the newly formed bone was seen, while in the case of the synthesized HA coatings, this layer could be observed only seldom.

4.3.10. Mechanical Testing

The pullout (tensile) test was used by Mihailescu et al. [68] to compare the strength of the biological–chemical bonding between the bone and Ti implant surfaces functionalized with HA, Mn-CHA, and OCP coatings, respectively. The inferred values for all three tested groups demonstrated a significantly improved bone attachment strength value ($p \leq 0.05$), which was about twice as high as the one associated to the Ti implants (~5 N). In comparison to the strength value corresponding to the synthetic HA (~8 N), up to 10% (in the case of OCP, ~9 N) and 25% (in the case of Mn-CHA, ~11 N) higher values were obtained.

Duta et al. [75] evaluated the quality of the implants' osseointegration by mechanical (tensile) tests. In none of the cases, alteration or disruption of the implants were present.

The detachment force (F_{max}) of implants under tensile pull-out testing, inferred for Ti and functionalized (with Li-C and Li-P coatings) 3D Ti implants, at 4 and 9 weeks after surgery, are represented in Figures 4 and 5, respectively.

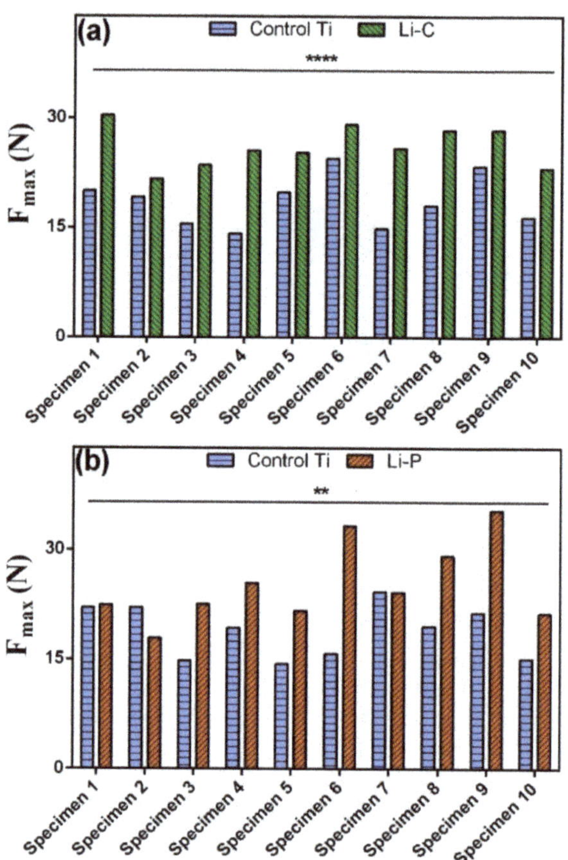

Figure 4. Detachment force, F_{max}, of implants (n = 10) under tensile pull-out testing, inferred in the case of control 3D Ti implants (marked in blue color) and of those functionalized with (**a**) Li-C (marked in green color) and (**b**) Li-P (marked in orange color) coatings, at 4 weeks after surgery. **** Represents highly significant differences ($p \leq 0.0001$). ** Represents significant differences ($p \leq 0.01$).

At 4 weeks after surgery (Figure 4), the obtained mean detachment force values demonstrated significant and highly significant differences between the Ti group and the Li-P one, and between the Ti group and the Li-C one, respectively. When referring to the extractions performed at longer periods of time, i.e., 9 weeks from implantation (Figure 5), the inferred mean detachment force values indicated highly significant differences for both investigated groups.

It is important to mention that the failure loads of the implants functionalized with both Li-C and Li-P coatings measured at 9 weeks were ~5 times higher in comparison to those inferred at 4 weeks after surgery, respectively. Moreover, for both time periods, the Li-C and Li-P functionalized implants demonstrated a bone attachment strength of ~2 times stronger than the one corresponding to Ti implants. One could therefore indicate that both the PLD surface functionalization of the implants and a longer implantation time period could induce a positive influence on the overall bone bonding strength characteristics of the investigated medical devices. It should be emphasized that the fabrication of novel BioHA implant coatings derived from sustainable and inexpensive CaP resources, with improved mechanical properties, correlated with an increased bone fixation in vivo, could stand for a pioneering contribution to the progress of advanced medical devices.

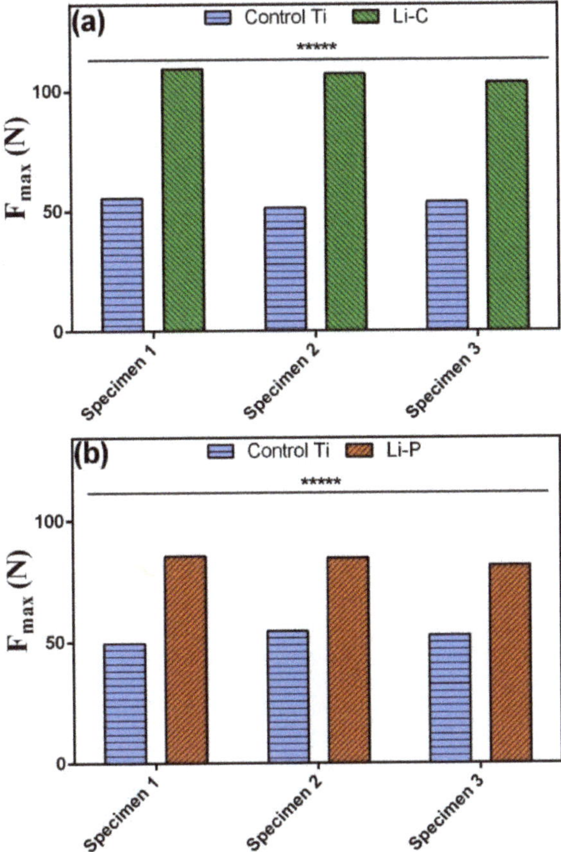

Figure 5. Detachment force, F_{max}, of implants (n = 3) under tensile pull-out testing, inferred in the case of control 3D Ti implants (marked in blue color) and of those functionalized with (**a**) Li-C (marked in green color) and (**b**) Li-P (marked in orange color) coatings, at 9 weeks after surgery. ***** Represents highly significant differences ($p \leq 0.00001$).

5. Discussion

Because the biomaterials' osteoinduction mechanism is not yet entirely understood, one could not precisely answer the question whether if the sole biomaterial or an interaction between the biomaterial and the relevant proteins present in the living system are responsible for the osteoinduction process. Because most of the implants do not possess the capability to induce bone growth, specific material properties are required to activate the osteoinduction process. To begin the differentiation of the undifferentiated inducible osteoprogenitor cells into bone-forming cells, it was suggested that both the chemistry and the geometry of the biomaterial in contact with these cells represent critical factors to be considered [119].

Metallic implants (including Ti) are generally used for various biomedical applications, mainly due to their resistance to corrosion and favorable mechanical characteristics [120]. Because of its bioinert nature, bulk Ti is not capable to form a biochemical bond with the bone, and this biological inactivity often generates a fibrous tissue that surrounds the implanted device [121]. To improve both osseointegration rates and longevity of Ti implants, the deposition of CaP-based coatings onto their surfaces is envisaged. It was therefore demonstrated that implants' surface functionalization with CaP-based coatings could promote the formation of real bonds with the surrounding bone, due to their proved chemical similarity with natural bone tissue and their high biocompatibility [122]. This process occurs rapidly along the entire surface of the coating, in comparison to the case of simple Ti implants (used as controls in the experiments) [47].

A nowadays growing research interest in the field of biomaterials is related to the use of biological-derived CaP materials as viable, safe, and low-cost alternatives to synthetic CaP-ones [33]. It should be emphasized that, unfortunately, the Earth's available mineral resources are threatened to become limited in the near future because of the rapid demographic increase and economic growth. The access to sustainable resources is therefore critical. Consequently, this will generate a beneficial economic and environmental impact over the society, allowing for an intelligent use of these renewable resources.

The mechanical properties of CaP-based coatings are responsible with the overall success rate of an implant [123]. Thus, the optimal functioning of an endosseous implant is directly influenced by the biomaterial's mechanical stability, which can be easily evaluated by extraction tests. To obtain information on the force that occurs between the bone tissue and implanted materials, various experimental study models have been developed, each of them with their own particularities [115,124,125]. In this respect, the investigation of the coating bond strength is typically performed by scratch [126–128], pull-off [129], tensile adhesion [130–132], or shear strength tests [133], respectively. It is important to emphasize upon that the ISO 13779-2:2008 standard requirement for tensile adhesion strength of CaP-based coatings, used for load-bearing applications, is of 15 MPa [134]. It was reported that, in general, CaP-based coatings synthesized by the PLD technique easily surpass this imposed value [33,59,135]. There are some studies in the literature concentrated on tensile strength measurements [136,137], which, very importantly, can provide a direct measurement of the attachment between the bone and the implant surface, being therefore influenced only by the chemical bonding between those two [138–140]. This way, the effects of friction and of mechanical forces introduced by surface roughness can be minimized [141]. In the case of animal trials, the implant's increased bone retention is considered a clinically relevant indicator of improved stability and capacity of the implants to carry loads without detaching. Unfortunately, this type of information cannot be acquired by histological or SEM investigations, which provide only limited information on the functional performance of an implant. One should note that in none of the studies included in this review, related to mechanical testing of CaP-based coatings, alteration or disruption of the implants were present. In general, the inferred values for all functionalized Ti implants demonstrated significantly improved bone attachment in comparison to Ti ones. Next to the PLD surface functionalization of metallic implants, a longer implantation time period was demonstrated to induce a positive influence on the overall bone bonding strength characteristics of the investigated medical devices. One should note that the fabrication by PLD of novel BioHA implant coatings derived from sustainable and inexpensive CaP-based resources, with improved mechanical properties, correlated with an increased bone fixation in vivo, could stand for a pioneering contribution to the progress of advanced medical devices.

In general, the results of the studies included in this review, obtained using standardized radiography and microradiography [63], computed tomography [72,73,75], histomorphometry [65–67,69], and tomodensitometry [71], have confirmed the osseointegration process (pointing to an increase in the osseous density), along with a strong bone–implant connection and no inflammatory process of the soft tissues. The histological investigations, performed with various microscopy techniques [32,52,63–65,67,70,72–75], have indicated also new active bone formation and demonstrated no adverse inflammatory reactions or gaps around the sites of implantation or at the bone–implant interface, for any of the investigated structures. In addition, the bone and the implant were shown to tightly adhere to each other along the full length of the interface. One interesting observation was that the temperature applied during the deposition process seemed to play an important role in the bone growth of the synthesized structures, the osteoblasts in the newly formed bone tissue being shown clear and homogenous [52]. Moreover, the osteointegration rate was demonstrated to be slightly superior in the case of annealed samples, rather than for the non-annealed ones [64]. It seemed also that no matter what the laser source used (i.e., KrF or CO_2), between the synthesized coatings, no significant statistical differences in the osteogenesis process were inferred [64]. Significantly higher values of the bone area between the implant surface and the boundaries and bone adherence ratios were inferred at various implantation time periods in the case of functionalized implants in comparison to control ones [73,75]. In this respect, in the case of control samples, a fibrous connective tissue between the metallic implants and the newly formed bone was shown, while in the case of the synthesized coatings, this layer could be observed only seldom [63].

Even though it is generally accepted that CaP-based coatings deposited by PLD improve bone strength and the initial osseointegration rate, the coatings' properties necessary to achieve an optimum bone response are yet to be determined. This is mainly because of the limited number of in vivo studies available in the dedicated literature. It should be emphasized upon that the in vivo testing should demonstrate stability in biological environment for up to 1 month, which corresponds to the initial healing phase of the implants [142]. The limitations on such experiments can be related to (i) the difficulty to select a suitable animal model in order to properly simulate the actual mechanical loading and unloading conditions in which an implant should function inside a living system; (ii) the need to sacrifice a large number of animals to reach a significant statistical relevance, able to validate the obtained results; (iii) the demands for high costs and long time frame in the case of clinical trials; (iv) the lack of coordination among material scientists and biologists and thus an insufficient understanding of this interdisciplinary subject; and (v) the serious ethical concerns related to the used animals (including also the choice of their correct number), as they might be sometimes subjected to painful procedures or toxic exposures during the experimental trials [143].

Taking into consideration all these aspects, a future important progress of CaP-based materials might be linked to a shift of the focus from osteoconduction to osteoinduction, e.g., by additive manufacturing of scaffolds with complex, controlled three-dimensional porous structures and development of novel ion-substituted CaPs with increased biological activity. Moreover, new strategies, possibly based on self-assembling and/or nanofabrication might be developed for the successful fabrication of load-bearing bone graft substitutes. In the future, the composition, microstructure, and molecular surface chemistry of various types of CaPs might be tailored in such a way to match the specific biological and metabolic requirements of tissues or disease states. The multilayer composite coating systems, fabricated by the PLD technique, should represent also a future trend, able to provide multifunctional properties for the biomedical implants.

6. Conclusions

This review summarizes a two decades achievements reported in the field of in vivo assessment of calcium phosphate (CaP)-based coatings deposited onto metallic implants by one of the most frequently used plasma-assisted techniques, i.e., pulsed laser deposition (PLD). Due to their proven biocompatibility, mechanical (high adherence), and osseointegration and osteoconduction properties, CaP-based bioceramics are widely used in the field of bone regeneration, both in orthopedics and dentistry. For an in-depth in vivo assessment of various CaP-based coatings synthesized by the PLD method only, the results of the studies included in this review were obtained using a wide range of investigation techniques, among which a special focus was put on standardized and (micro) radiographies, (micro) computed tomography and histomorphometry, tomodensitometry, histology, scanning and transmission electron microscopies, and mechanical testing. It is important to note that all these results indicate superior osseointegration characteristics of the metallic (generally Ti) implants functionalized with CaP-based coatings when compared to simple (uncoated) Ti ones, which are considered as the "gold standard" for implantological applications.

In the last two decades, research studies performed on CaP-coated metallic implants by PLD resulted in an interesting progress in vitro and in vivo, while not enough comparable clinical results were delivered so far for an easier assessment. This was mainly because of the lack of standardization of the coating properties and in vivo models. Therefore, additional testing is still needed in this direction, both to be able to advance a certain "recipe" to obtain optimum in vivo results, and to further reveal the relative influence of implant design, surgical procedure, and coating characteristics (thickness, structure, porosity, and surface morphology, which includes the wettability behavior), on either short-term or long-term clinical beneficial effects of the CaP-based coatings. In addition, one should emphasize upon the growing interest on the biological-derived CaP-materials as viable, safe, and cheap alternatives to the CaP synthetic ones, along with their improved biological properties and a greater resemblance to the mineral part of the human bones. The domain of PLD synthesis of natural-CaP sustainable coatings is in its first stages of development, and therefore, various possibilities to expand in the near future in terms of natural-CaP new resources, different concentrations of doping agents, or morphology and structural control of the obtained coatings are envisaged.

Funding: L.D. acknowledges the financial support of the Romanian Ministry of Education and Research under Romanian National Nucleu Program LAPLAS VI—contract No. 16N/2019.

Institutional Review Board Statement: The animal surgical protocols complied with the regulations and precautions of ISO-10993-Part 2 and Part 6, National Animal Care Guidelines and EU Council Directive of 22 September 2010, regarding the care and use of laboratory animals for scientific purposes (2010/63/EU), and were approved by the Institutional Animal-Care Committee, the Local Ethical Committee for the Affairs of Experiments on Animals in Lodz (22.12.2008, decree No. 56/ŁB 440/2008), the Osaka Dental University Ethics Committee, Japan (approval No. 16-08002, 2 August 2016), and the "Committee of Ethics and Academic and Scientific Deontology" at the UMF in Craiova, Romania (document No. 135/20.12.2019).

Informed Consent Statement: Informed consent was obtained from all subjects involved in the study.

Data Availability Statement: The data presented in this review are available in Refs. [32,52,63–75].

Acknowledgments: L.D. acknowledges F.N. Oktar (Marmara University, Istanbul) for providing animal-origin powders that were used for a part of the studies reported in this review. L.D. thanks George E. Stan for his useful comments and kind help.

Conflicts of Interest: The author declares no conflict of interest.

References

1. Surmenev, R.A.; Surmeneva, M.A.; Ivanova, A.A. Significance of calcium phosphate coatings for the enhancement of new bone osteogenesis—A review. *Acta Biomater.* **2014**, *2*, 557–579. [CrossRef] [PubMed]
2. Allied Market Research. Available online: https://www.alliedmarketresearch.com/implantable-medical-devices-market (accessed on 10 December 2020).
3. Eliaz, N.; Metoki, N. Calcium Phosphate Bioceramics: A Review of Their History, Structure, Properties, Coating Technologies and Biomedical Applications. *Materials* **2017**, *10*, 334. [CrossRef] [PubMed]
4. Oladele, I.O.; Agbabiaka, O.; Olasunkanmi, O.G.; Balogun, A.O.; Popoola, M.O. Non-synthetic sources for the development of hydroxyapatite. *J. Appl. Biotechnol. Bioeng.* **2018**, *5*, 88–95. [CrossRef]
5. Akram, M.; Ahmed, R.; Shakir, I.; Ibrahim, W.A.W.; Hussain, R. Extracting hydroxyapatite and its precursors from natural resources. *J. Mater. Sci.* **2014**, *49*, 1461–1475. [CrossRef]
6. Šupová, M. Substituted hydroxyapatites for biomedical applications: A review. *Ceram. Int.* **2015**, *41*, 9203–9231. [CrossRef]
7. Graziani, G.; Boi, M.; Bianchi, M. A Review on Ionic Substitutions in Hydroxyapatite Thin Films: Towards Complete Biomimetism. *Coatings* **2018**, *8*, 269. [CrossRef]

8. Tite, T.; Popa, A.C.; Balescu, L.M.; Bogdan, I.M.; Pasuk, I.; Ferreira, J.M.F.; Stan, G.E. Cationic substitutions in hydroxyapatite: Current status of the derived biofunctional effects and their in vitro interrogation methods. *Materials* **2018**, *11*, 2081. [CrossRef]

9. Ballini, A.; Mastrangelo, F.; Gastaldi, G.; Tettamanti, L.; Bukvic, N.; Cantore, S.; Cocco, T.; Saini, R.; Desiate, A.; Gherlone, E.; et al. Osteogenic differentiation and gene expression of dental pulp stem cells under low-level laser irradiation: A good promise for tissue engineering. *J. Biol. Regul. Homeost. Agents* **2015**, *29*, 813–822.

10. Barrere, F.; van der Valk, C.M.; Meijer, G.; Dalmeijer, R.A.; Groot, K.; Layrolle, P. Osteointegration of biomimetic apatite coating applied onto dense and porous metal implants in femurs of goats. *J. Biomed. Mater. Res. B Appl. Biomater.* **2003**, *67*, 655–665. [CrossRef]

11. Surmenev, R.A.; Surmeneva, M.A. A critical review of decades of research on calcium phosphate–based coatings: How far are we from their widespread clinical application? *Curr. Opin. Biomed. Eng.* **2019**, *10*, 35–44. [CrossRef]

12. Dorozhkin, S.V. Biphasic, triphasic, and multiphasic calcium orthophosphates. In *Advanced Ceramic Materials*; Tiwari, A., Gerhardt, R.A., Szutkowska, M., Eds.; Wiley, Scrivener Publishing: Austin, TX, USA, 2016; pp. 33–95.

13. Kang, Z.; Zhang, J.; Niu, L. A one-step hydrothermal process to fabricate superhydrophobic hydroxyapatite coatings and determination of their properties. *Surf. Coat. Technol.* **2018**, *334*, 84–89. [CrossRef]

14. Ciobanu, C.S.; Popa, C.L.; Predoi, D. Cerium-doped hydroxyapatite nanoparticles synthesized by the co-precipitation method. *J. Serb. Chem. Soc.* **2016**, *81*, 433–446. [CrossRef]

15. Raucci, M.G.; Demitri, C.; Soriente, A.; Fasolino, I.; Sannino, A.; Ambrosio, L. Gelatin/nano-hydroxyapatite hydrogel scaffold prepared by sol-gel technology as filler to repair bone defects. *J. Biomed. Mater. Res. Part A* **2018**, *106*, 2007–2019. [CrossRef] [PubMed]

16. Guoqing, M. Three common preparation methods of hydroxyapatite. *IOP Conf. Ser. Mater. Sci. Eng.* **2019**, *688*, 033057. [CrossRef]

17. Hare'l, G.; Ravi, B.G.; Chaim, R. Effects of solvent and agitation on microstructural characteristics of sol–gel derived nanocrystalline Y-TZP powders. *Mater. Lett.* **1999**, *39*, 63–68. [CrossRef]

18. Eshtiagh-Hosseini, H.; Housaindokht, M.R.; Chahkandi, M. Effects of parameters of sol–gel process on the phase evolution of sol–gel-derived hydroxyapatite. *Mater. Chem. Phys.* **2007**, *106*, 310–316.

19. Rey, C.; Combes, C.; Drouet, C.; Glimcher, M.J. Bone mineral: Update on chemical composition and structure. *Osteoporos. Int.* **2009**, *20*, 1013–1021. [CrossRef]

20. Niakan, A.; Ramesh, S.; Ganesan, P.; Tan, C.Y.; Purbolaksono, J.; Chandran, H.; Teng, W.D. Sintering behaviour of natural porous hydroxyapatite derived from bovine bone. *Ceram. Int.* **2015**, *41*, 3024–3029. [CrossRef]

21. Ofudje, E.A.; Rajendran, A.; Adeogun, A.I.; Idowu, M.A.; Kareem, S.O.; Pattanayak, D.K. Synthesis of organic derived hydroxyapatite scaffold from pig bone waste for tissue engineering applications. *Adv. Powder Technol.* **2018**, *29*, 1–8. [CrossRef]

22. Panda, N.N.; Pramanik, K.; Sukla, L.B. Extraction and characterization of biocompatible hydroxyapatite from fresh water fish scales for tissue engineering scaffold. *Bioprocess Biosyst. Eng.* **2014**, *37*, 433–440. [CrossRef]

23. Boutinguiza, M.; Pou, J.; Comesaña, R.; Lusquiños, F.; de Carlos, A.; León, B. Biological hydroxyapatite obtained from fish bones. *Mater. Sci. Eng. C Mater. Biol. Appl.* **2012**, *32*, 478–486. [CrossRef]

24. Iriarte-Velasco, U.; Sierra, I.; Zudaire, L.; Ayastuy, J.L. Conversion of waste animal bones into porous hydroxyapatite by alkaline treatment: Effect of the impregnation ratio and investigation of the activation mechanism. *J. Mater. Sci.* **2015**, *50*, 7568–7582. [CrossRef]

25. Shavandi, A.; El-Din, A.; Bekhit, A.; Ali, A.; Sun, Z. Synthesis of nano-hydroxyapatite (nHA) from waste mussel shells using a rapid microwave method. *Mater. Chem. Phys.* **2015**, *149–150*, 607–616. [CrossRef]

26. Tâmâsan, M.; Ozyegin, L.S.; Oktar, F.N.; Simon, V. Characterization of calcium phosphate powders originating from Phyllacanthus imperialis and Trochidae Infundibulum concavus marine shells. *Mater. Sci. Eng. C* **2013**, *33*, 2569–2577. [CrossRef] [PubMed]

27. Terzioğlu, P.; Öğütc, H.; Kalemtaş, A. Natural calcium phosphates from fish bones and their potential biomedical applications. *Mater. Sci. Eng. C* **2018**, *91*, 899–911. [CrossRef] [PubMed]

28. Herliansyah, M.K.; Hamdi, M.; Ide-Ektessabi, A.; Wildan, M.W.; Toque, J.A. The influence of sintering temperature on the properties of compacted bovine hydroxyapatite. *Mater. Sci. Eng. C* **2009**, *29*, 1674–1680. [CrossRef]

29. Pietak, A.M.; Reid, J.W.; Stott, M.J.; Sayer, M. Silicon substitution in the calcium phosphate bioceramics. *Biomaterials* **2007**, *28*, 4023–4032. [CrossRef]

30. Hoyer, B.; Bernhardt, A.; Heinemann, S.; Stachel, I.; Meyer, M.; Gelinsky, M. Biomimetically Mineralized Salmon Collagen Scaffolds for Application in Bone Tissue Engineering. *Biomacromolecules* **2012**, *13*, 1059–1066. [CrossRef]

31. Rincón-López, J.A.; Hermann-Muñoz, J.A.; Giraldo-Betancur, A.L.; De Vizcaya-Ruiz, A.; Alvarado-Orozco, J.M.; Muñoz-Saldaña, J. Synthesis, characterization and in vitro study of synthetic and bovine-derived hydroxyapatite ceramics: A comparison. *Materials* **2018**, *11*, 333. [CrossRef]

32. Hayami, T.; Hontsu, S.; Higuchi, Y.; Nishikawa, H.; Kusunoki, M. Osteoconduction of a stoichiometric and bovine hydroxyapatite bilayer-coated implant. *Clin. Oral Implants Res.* **2011**, *22*, 774–776. [CrossRef]

33. Duta, L.; Popescu, A.C. Current Status on Pulsed Laser Deposition of Coatings from Animal-Origin Calcium Phosphate Sources. *Coatings* **2019**, *9*, 335. [CrossRef]

34. León, B.; John, J. *Thin Calcium Phosphate Coatings for Medical Implants*; Springer: New York, NY, USA, 2009; pp. 1–328.

35. Ambard, A.J.; Mueninghoff, L. Calcium phosphate cement: Review of mechanical and biological properties. *J. Prosthodont.* **2006**, *15*, 321–328. [CrossRef] [PubMed]

36. Sima, L.E.; Stan, G.E.; Morosanu, C.O.; Melinescu, A.; Ianculescu, A.; Melinte, R.; Neamtu, J.; Petrescu, S.M. Differentiation of mesenchymal stem cells onto highly adherent radio frequency-sputtered carbonated hydroxylapatite thin films. *J. Biomed. Mater. Res.* **2010**, *95*, 1203–1214. [CrossRef] [PubMed]

37. Popa, A.C.; Stan, G.E.; Besleaga, C.; Ion, L.; Maraloiu, V.A.; Tulyaganov, D.U.; Ferreira, J.M.F. Submicrometer Hollow Bioglass Cones Deposited by Radio Frequency Magnetron Sputtering: Formation Mechanism, Properties, and Prospective Biomedical Applications. *ACS Appl. Mater. Interfaces* **2016**, *8*, 4357–4367. [CrossRef] [PubMed]

38. Stan, G.E.; Popescu, A.C.; Mihailescu, I.N.; Marcov, D.A.; Mustata, R.C.; Sima, L.E.; Petrescu, S.M.; Ianculescu, A.; Trusca, R.; Morosanu, C.O. On the bioactivity of adherent bioglass thin films synthesized by magnetron sputtering techniques. *Thin Solid Films* **2010**, *518*, 5955–5964. [CrossRef]

39. Stan, G.E.; Marcov, D.A.; Pasuk, I.; Miculescu, F.; Pina, S.; Tulyaganov, D.U.; Ferreira, J.M.F. Bioactive glass thin films deposited by magnetron sputtering technique: The role of working pressure. *Appl. Surf. Sci.* **2010**, *256*, 7102–7110. [CrossRef]

40. Chrisey, D.B.; Hubler, G.K. (Eds.) *Pulsed Laser Deposition of Thin Films*, 1st ed.; John Wiley & Sons: Hoboken, NJ, USA, 1994; pp. 1–649.

41. Eason, R. *Pulsed Laser Deposition of Thin Films-Applications-Led Growth of Functional Materials*; Wiley-Interscience: Hoboken, NJ, USA, 2006; pp. 1–682.

42. Bao, Q.; Chen, C.; Wang, D.; Ji, Q.; Lei, T. Pulsed laser deposition and its current research status in preparing hydroxyapatite thin films. *Appl. Surf. Sci.* **2005**, *252*, 1538–1544. [CrossRef]

43. Rath, M.; Varadarajan, E.; Natarajan, V.; Rao, M.S.R. A comparative study on macroscopic and nanoscale polarization mapping on large area PLD grown PZT thin films. *Ceram. Int.* **2018**, *44*, 8749–8755. [CrossRef]

44. Cheang, P.; Khor, K.A. Addressing processing problems associated with plasma spraying of hydroxyapatite coatings. *Biomaterials* **1996**, *17*, 537–544. [CrossRef]

45. Epinette, J.A.; Manley, M.T. *Fifteen Years of Clinical Experience with Hydroxyapatite Coatings in Joint Arthroplasty*, 1st ed.; Springer: Paris, France, 2004; p. 452. [CrossRef]

46. Bose, S.; Tarafder, S.; Bandyopadhyay, A. Hydroxyapatite coatings for metallic implants. In *Hydroxyapatite (Hap) for Biomedical Applications*; Mucalo, M., Ed.; Woodhead Publishing Series in Biomaterials; Elsevier: Amsterdam, The Netherlands, 2015; pp. 143–157.

47. Yingchao, S.; Irsalan, C.; Yufeng, Z.; Liping, T.; Yi-Xian, Q.; Donghui, Z. Biofunctionalization of metallic implants by calcium phosphate coatings. *Bioact. Mat.* **2019**, *4*, 196–206. [CrossRef]

48. Yoshinari, M.; Hayakawa, T.; Wolke, J.G.; Nemoto, K.; Jansen, J.A. Influence of rapid heating with infrared radiation on RF magnetron-sputtered calcium phosphate coatings. *J. Biomed. Mater. Res.* **1997**, *37*, 60–67. [CrossRef]

49. Surmenev, R.A. A review of plasma-assisted methods for calcium phosphate-based coatings fabrication. *Surf. Coat. Technol.* **2012**, *206*, 2035–2056. [CrossRef]

50. Duta, L.; Oktar, F.N.; Stan, G.E.; Popescu-Pelin, G.; Serban, N.; Luculescu, C.; Mihailescu, I.N. Novel doped hydroxyapatite thin films obtained by pulsed laser deposition. *Appl. Surf. Sci.* **2013**, *265*, 41–49. [CrossRef]

51. Zeng, H. *Evaluation of Bioceramic Coatings Produced by Pulsed Laser Deposition and Ion Beam Sputtering*; University of Alabama at Birmingham: Birmingham, AL, USA, 1997.

52. Wang, D.G.; Chena, C.Z.; Ma, Q.S.; Jin, Q.P.; Li, H.C. A study on in vitro and in vivo bioactivity of HA/45S5 composite films by pulsed laser deposition. *Appl. Surf. Sci.* **2013**, *270*, 667–674. [CrossRef]

53. Schneider, C.W.; Lippert, T. Laser Ablation and Thin Film Deposition. In *Laser Processing of Materials*; Schaaf, P., Ed.; Springer Series in Materials Science; Springer: Berlin/Heidelberg, Germany, 2010; Volume 139, pp. 89–112.

54. Caricato, A.P.; Martino, M.; Romano, F.; Mirchin, N.; Peled, A. Pulsed laser photodeposition of a-Se nanofilms by ArF laser. *Appl. Surf. Sci.* **2007**, *253*, 6517–6521. [CrossRef]

55. Hashimoto, Y.; Ueda, M.; Kohiga, Y.; Imura, K.; Hontsu, S. Application of fluoridated hydroxyapatite thin film coatings using KrF pulsed laser deposition. *Dent. Mater. J.* **2018**, *37*, 408–413. [CrossRef]

56. Zaki, A.M.; Blythe, H.J.; Heald, S.M.; Fox, A.M.; Gehring, G.A. Growth of high quality yttrium iron garnet films using standard pulsed laser deposition technique. *J. Magn. Magn. Mater.* **2018**, *453*, 254–257. [CrossRef]

57. Novotný, M.; Vondráček, M.; Marešová, E.; Fitl, P.; Bulíř, J.; Pokorná, P.; Havlová, Š.; Abdellaoui, N.; Pereira, A.; Hubík, P.; et al. Optical and structural properties of ZnO:Eu thin films grown by pulsed laser deposition. *Appl. Surf. Sci.* **2019**, *476*, 271–275. [CrossRef]

58. Gyorgy, E.; Ristoscu, C.; Mihailescu, I.N. Role of laser pulse duration and gas pressure in deposition of AlN thin films. *J. Appl. Phys.* **2001**, *90*, 456. [CrossRef]

59. Zhang, Z.; Dunn, M.F.; Xiao, T.; Tomsia, A.P.; Saiz, E. Nanostructured Hydroxyapatite Coatings for Improved Adhesion and Corrosion Resistance for Medical Implants. *Mat. Res. Soc. Symp. Proc.* **2002**, *703*, 291–296. [CrossRef]

60. Jain, P.; Mandal, T.; Prakash, P.; Garg, A.; Balani, K. Electrophoretic deposition of nanocrystalline hydroxyapatite on Ti6Al4V/TiO2 substrate. *J. Coat. Technol. Res.* **2013**, *10*, 263–275. [CrossRef]

61. Wen, C.; Guan, S.; Peng, L.; Ren, C.; Wang, X.; Hu, Z. Characterization and degradation behavior of AZ31 alloy surface modified by bone-like hydroxyapatite for implant applications. *Appl. Surf. Sci.* **2009**, *255*, 6433–6438. [CrossRef]

62. Bakhsheshi-Rad, H.; Idris, M.; Abdul-Kadir, M. Synthesis and in vitro degradation evaluation of the nano-HA/MgF2 and DCPD/MgF2 composite coating on biodegradable Mg–Ca–Zn alloy. *Surf. Coat. Technol.* **2013**, *222*, 79–89. [CrossRef]

63. Dostálová, T.; Jelínek, M.; Himmlová, L.; Grivas, C. Laser–Deposited Hydroxyapatite Films on Dental Implants—Biological Evaluation in vivo. *Laser Phys.* **1998**, *8*, 182–186.
64. Antonov, E.N.; Bagratashvili, V.N.; Popov, V.K.; Sobol, E.N.; Howdle, S.M.; Joiner, C.; Parker, K.G.; Parker, T.L.; Doktorov, A.A.; Likhanov, V.B.; et al. Biocompatibility of laser-deposited hydroxyapatite coatings on titanium and polymer implant materials. *J. Biomed. Opt.* **1998**, *3*, 423–428. [CrossRef]
65. Dostálová, T.; Himmlová, L.; Jelínek, M.; Grivas, C. Osseointegration of loaded dental implant with KrF laser hydroxylapatite films on Ti6Al4V alloy by minipigs. *J. Biomed. Opt.* **2001**, *6*, 239–243. [CrossRef]
66. Kim, H.; Vohra, Y.K.; Louis, P.J.; Lacefield, W.R.; Lemons, J.E.; Camata, R.P. Biphasic and Preferentially Oriented Microcrystalline Calcium Phosphate Coatings: In-vitro and In-vivo Studies. *Key Eng. Mat.* **2005**, *284–286*, 207–210. [CrossRef]
67. Peraire, C.; Arias, J.L.; Bernal, D.; Pou, J.; Leon, B.; Arano, A.; Roth, W. Biological stability and osteoconductivity in rabbit tibia of pulsed laser deposited hydroxylapatite coatings. *J. Biomed. Mater. Res. Part A* **2006**, *77*, 370–379. [CrossRef]
68. Mihailescu, I.N.; Lamolle, S.; Socol, G.; Miroiu, F.; Roenold, H.J.; Bigi, A.; Mayer, I.; Cuisinier, F.; Lyngstadaas, S.P. In vivo tensile tests of biomimetic titanium implants pulsed laser coated with nanostructured Calcium Phosphate thin films. *Adv. Mater.-Rapid Commun.* **2008**, *2*, 337–341.
69. Paz, M.D.; Chiussi, S.; González, P.; Serra, J.; León, B.; Alava, J.I.; Güemes, I.; Sánchez-Margallo, F.M. Osseointegration of Calcium Phosphate Nanofilms on Titanium Alloy Implants. *Key Eng. Mater.* **2008**, *361–363*, 645–648. [CrossRef]
70. Hontsu, S.; Hashimoto, Y.; Yoshikawa, Y.; Kusunoki, M.; Nishikawa, H.; Ametani, A. Fabrication of Hydroxyl Apatite Coating Titanium Web Scaffold Using Pulsed Laser Deposition Method. *J. Hard. Tissue Biol.* **2012**, *21*, 181–188. [CrossRef]
71. Duta, L.; Stan, G.E.; Popescu, A.C.; Socol, G.; Miroiu, F.M.; Mihailescu, I.N.; Ianculescu, A.; Poeata, I.; Chiriac, A. Hydroxyapatite thin films synthesized by Pulsed Laser Deposition onto titanium mesh implants for cranioplasty applications. In Proceedings of the ROMOPTO International Conference on Micro- to Nano-Photonics III, Bucharest, Romania, 3–6 September 2012; Volume 8882, p. 888208. [CrossRef]
72. Mroz, W.; Budner, B.; Syroka, R.; Niedzielski, K.; Golanski, G.; Slosarczyk, A.; Schwarze, D.; Douglas, T.E.L. In vivo implantation of porous titanium alloy implants coated with magnesium-doped octacalcium phosphate and hydroxyapatite thin films using pulsed laser deposition. *J. Biomed. Mater. Res. B Appl. Biomater.* **2015**, *103*, 151–158. [CrossRef]
73. Chen, L.; Komasa, S.; Hashimoto, Y.; Hontsu, S.; Okazaki, J. In Vitro and In Vivo Osteogenic Activity of Titanium Implants Coated by Pulsed Laser Deposition with a Thin Film of Fluoridated Hydroxyapatite. *Int. J. Mol. Sci.* **2018**, *19*, 1127. [CrossRef]
74. Wang, D.G.; Chen, C.Z.; Yang, X.X.; Ming, X.C.; Zhang, W.L. Effect of bioglass addition on the properties of HA/BG composite films fabricated by pulsed laser deposition. *Ceram. Int.* **2018**, *44*, 14528–14533. [CrossRef]
75. Duta, L.; Neamtu, J.; Melinte, R.P.; Zureigat, O.A.; Popescu-Pelin, G.; Chioibasu, D.; Oktar, F.N.; Popescu, A.C. In Vivo Assessment of Bone Enhancement in the Case of 3D-Printed Implants Functionalized with Lithium-Doped Biological-Derived Hydroxyapatite Coatings: A Preliminary Study on Rabbits. *Coatings* **2020**, *10*, 992. [CrossRef]
76. Rosengren, A.; Danielsen, N.; Bjursten, L.M. Inflammatory reaction dependence on implant localization in rat soft tissue models. *Biomaterials* **1997**, *18*, 979–987. [CrossRef]
77. Black, J.; Hastings, G. *Handbook of Biomaterial Properties*; Springer: New York, NY, USA, 2013; p. 676.
78. Habibovic, P.; Kruyt, M.C.; Juhl, M.V.; Clyens, S.; Martinetti, R.; Dolcini, L.; Theilgaard, N.; van Blitterswijk, C.A. Comparative in vivo study of six hydroxyapatite-based bone graft substitutes. *J. Orthop. Res.* **2008**, *26*, 1363–1370. [CrossRef]
79. Marini, E.; Ballanti, P.; Silvestrini, G.; Valdinucci, F.; Bonucci, E. The presence of different growth factors does not influence bone response to hydroxyapatite: Preliminary results. *J. Orthopaed. Traumatol.* **2004**, *5*, 34–43. [CrossRef]
80. Rabiee, S.; Moztarzadeh, F.; Solati-Hashjin, M. Synthesis and characterization of hydroxyapatite cement. *J. Mol. Struct.* **2010**, *969*, 172–175. [CrossRef]
81. Brånemark, P.I.; Hansson, B.O.; Adell, R.; Breine, U.; Lindström, J.; Hallén, O.; Ohman, A. Osseointegrated implants in the treatment of the edentulous jaw. Experience from a 10-year period. *Scand. J. Plast. Reconstr. Surg. Suppl.* **1977**, *16*, 1–132. [PubMed]
82. Choi, A.H.; Karacan, I.; Ben-Nissan, B. Surface modifications of titanium alloy using nanobioceramic-based coatings to improve osseointegration: A review. *Mater. Technol.* **2018**, 742–751. [CrossRef]
83. Martin, J.Y.; Schwartz, Z.; Hummert, T.W.; Schraub, D.M.; Simpson, J.; Lankford, J., Jr.; Dean, D.D.; Cochran, D.L.; Boyan, B.D. Effect of titanium surface roughness on proliferation, differentiation, and protein synthesis of human osteoblast-like cells (MG63). *J. Biomed. Mater. Res. A* **1995**, *29*, 389–401. [CrossRef]
84. Rupp, F.; Gittens, R.A.; Scheideler, L.; Marmur, A.; Boyan, B.D.; Schwartz, Z.; Geis-Gerstorfera, J. A review on the wettability of dental implant surfaces I: Theoretical and experimental aspects. *Acta Biomater.* **2014**, *10*, 2894–2906. [CrossRef]
85. Gittens, R.A.; Scheideler, L.; Rupp, F.; Hyzy, S.L.; Geis-Gerstorfer, J.; Schwartz, Z.; Boyanc, B.D. A review on the wettability of dental implant surfaces II: Biological and clinical aspects. *Acta Biomater.* **2014**, *10*, 2907–2918. [CrossRef]
86. Elias, C.N.; Oshida, Y.; Lima, J.H.; Muller, C.A. Relationship between surface properties (roughness, wettability and morphology) of titanium and dental implant removal torque. *J. Mech. Behav. Biomed. Mater.* **2008**, *1*, 234–242. [CrossRef]
87. Massaro, C.; Rotolo, P.; De Riccardis, F.; Milella, E.; Napoli, A.; Wieland, M.; Textor, M.; Spencer, N.D.; Brunette, D.M. Comparative investigation of the surface properties of commercial titanium dental implants. Part I: Chemical composition. *J. Mater. Sci. Mater. Med.* **2002**, *13*, 535–548. [CrossRef]
88. Lee, T.Q.; Danto, M.I.; Kim, W.C. Initial stability comparison of modular hip implants in synthetic femurs. *Orthopedics* **1998**, *21*, 885–888.

89. Wong, M.; Eulenberger, J.; Schenk, R.; Hunziker, E. Effect of surface topology on the osseointegration of implant materials in trabecular bone. *J. Biomed. Mater. Res.* **1995**, *29*, 1567–1575. [CrossRef]

90. Svehla, M.; Morberg, P.; Zicat, B.; Bruce, W.; Sonnabend, D.; Walsh, W.R. Morphometric and mechanical evaluation of titanium implant integration: Comparison of five surface structures. *J. Biomed. Mater. Res.* **2000**, *51*, 15–22. [CrossRef]

91. Lavenus, S.; Louarn, G.; Layrolle, P. Nanotechnology and Dental Implants. *Int. J. Biomater.* **2010**, *915327*, 9. [CrossRef]

92. Samavedi, S.; Whittington, A.R.; Goldstein, A.S. Calcium phosphate ceramics in bone tissue engineering: A review of properties and their influence on cell behavior. *Acta Biomater.* **2013**, *9*, 8037–8045. [CrossRef]

93. Szcześ, A.; Hołysz, L.; Chibowski, E. Synthesis of hydroxyapatite for biomedical applications. *Adv. Colloid Interface Sci.* **2017**, *249*, 321–330. [CrossRef]

94. Harun, W.S.W.; Asri, R.I.M.; Alias, J.; Zulkifli, F.H.; Kadirgama, K.; Ghani, S.A.C.; Shariffuddin, J.H.M. A comprehensive review of hydroxyapatite-based coatings adhesion on metallic biomaterials. *Ceram. Int.* **2018**, *44*, 1250–1268. [CrossRef]

95. Park, J.B.; Lakes, R.S. Metallic implant materials, Ch 5. In *Biomaterials. An introduction*; Park, J.B., Lakes, R.S., Eds.; Springer: New York, NY, USA, 2007; pp. 99–137. [CrossRef]

96. Niinomi, M.; Nakai, M.; Hieda, J. Development of new metallic alloys for biomedical applications. *Acta Biomater.* **2012**, *11*, 3888–3903. [CrossRef]

97. Zheng, Y.F.; Gu, X.N.; Witte, F. Biodegradable Metals. *Mater. Sci. Eng. R Rep.* **2014**, *77*, 1–34. [CrossRef]

98. Zhu, D.; Su, Y.; Young, M.L.; Ma, J.; Zheng, Y.; Tang, L. Biological responses and mechanisms of human bone marrow mesenchymal stem cells to Zn and Mg biomaterials. *ACS Appl. Mater. Interfaces* **2017**, *33*, 27453–27461. [CrossRef]

99. Coelho, P.G.; Granjeiro, J.M.; Romanos, G.E.; Suzuki, M.; Silva, N.R.F.; Cardaropoli, G.; Thompson, V.P.; Lemons, J.E. Basic research methods and current trends of dental implant surfaces. *J. Biomed. Mater. Res. Appl. Biomater.* **2009**, *88*, 579–596. [CrossRef]

100. Lu, Y.P.; Chen, Y.M.; Li, S.T.; Wang, J.H. Surface nanocrystallization of hydroxyapatite coating. *Acta Biomater.* **2008**, *4*, 1865–1872. [CrossRef]

101. Barradas, A.M.C.; Yuan, H.; van Blitterswijk, C.A.; Habibovic, P. Osteoinductive biomaterials: Current knowledge of properties, experimental models and biological mechanisms. *Eur. Cell Mater.* **2011**, *21*, 407–429. [CrossRef]

102. Cui, Q.; Dighe, A.S.; Irvine, J.N., Jr. Combined angiogenic and osteogenic factor delivery for bone regenerative engineering. *Curr. Pharm. Des.* **2013**, *19*, 3374–3383. [CrossRef] [PubMed]

103. Sarikaya, B.; Aydin, H.M. Collagen/beta-tricalcium phosphate based synthetic bone grafts via dehydrothermal processing. *BioMed Res. Int.* **2015**, *2015*, 576532. [CrossRef] [PubMed]

104. Ghanaati, S.; Barbeck, M.; Orth, C.; Willershausen, I.; Thimm, B.W.; Hoffmann, C.; Rasic, A.; Sader, R.A.; Unger, R.E.; Peters, F.; et al. Influence of β-tricalcium phosphate granule size and morphology on tissue reaction in vivo. *Acta Biomater.* **2010**, *6*, 4476–4487. [CrossRef]

105. Lakstein, D.; Kopelovitch, W.; Barkay, Z.; Bahaa, M.; Hendel, D.; Eliaz, N. Enhanced osseointegration of grit-blasted, NaOH-treated and electrochemically hydroxyapatite-coated Ti–6Al–4V implants in rabbits. *Acta Biomater.* **2009**, *5*, 2258–2269. [CrossRef]

106. He, F.; Ren, W.; Tian, X.; Liu, W.; Wu, S.; Chen, X. Comparative study on in vivo response of porous calcium carbonate composite ceramic and biphasic calcium phosphate ceramic. *Mater. Sci. Eng. C* **2016**, *64*, 117–123. [CrossRef]

107. Okuda, T.; Ioku, K.; Yonezawa, I.; Minagi, H.; Kawachi, G.; Gonda, Y.; Murayama, H.; Shibata, Y.; Minami, S.; Kamihira, S.; et al. The effect of the microstructure of beta-tricalcium phosphate on the metabolism of subsequently formed bone tissue. *Biomaterials* **2007**, *28*, 2612–2621. [CrossRef]

108. Sennerby, L.; Dasmah, A.; Larsson, B.; Iverhed, M. Bone tissue responses to surface-modified zirconia implants: A histomorphometric and removal torque study in the rabbit. *Clin. Implant Dent. Relat. Res.* **2005**, *7*, S13–S20. [CrossRef]

109. Susin, C.; Qahash, M.; Hall, J.; Sennerby, L.; Wikesjo¨, U.M. Histological and biomechanical evaluation of phosphorylcholine-coated titanium implants. *J. Clin. Periodontol.* **2008**, *35*, 270–275. [CrossRef]

110. Neyt, J.G.; Buckwalter, J.A.; Carroll, N.C. Use of animal models in musculoskeletal research. *Iowa Orthop. J.* **1998**, *18*, 118–123.

111. Gilsanz, V.; Roe, T.F.; Gibbens, D.T.; Schulz, E.E.; Carlson, M.E.; Gonzalez, O.; Boechat, M.I. Effect of sex steroids on peak bone density of growing rabbits. *Am. J. Physiol.* **1988**, *255*, E416–E421. [CrossRef]

112. Ferguson, J.C.; Tangl, S.; Barnewitz, D.; Genzel, A.; Heimel, P.; Hruschka, V.; Redl, H.; Nau, T. A large animal model for standardized testing of bone regeneration strategies. *BMC Vet. Res.* **2018**, *14*, 330. [CrossRef]

113. Kruyt, M.C.; de Bruijn, J.D.; Wilson, C.E.; Oner, F.C.; van Blitterswijk, C.A.; Verbout, A.J.; Dhert, W.J.A. Viable osteogenic cells are obligatory for tissue-engineered ectopic bone formation in goats. *Tissue Eng.* **2003**, *9*, 327–336. [CrossRef]

114. Cancedda, R.; Giannoni, P.; Mastrogiacomo, M. A tissue engineering approach to bone repair in large animal models and in clinical practice. *Biomaterials* **2007**, *28*, 4240–4250. [CrossRef]

115. Pearce, A.I.; Richards, R.G.; Milz, S.; Schneider, E.; Pearce, S.G. Animal models for implant biomaterial research in bone: A review. *Eur. Cells Mater.* **2007**, *13*, 1–10. [CrossRef]

116. Mostafa, A.A.; Zaazou, M.H.; Chow, L.C.; Mahmoud, A.A.; Zaki, D.Y.; Basha, M.; Abdel Hamid, M.A.; Khallaf, M.E.; Sharaf, N.F.; Hamdy, T.M. Injectable nano amorphous calcium phosphate based in situ gel systems for the treatment of periapical lesions. *Biomed. Mater.* **2015**, *10*, 065006. [CrossRef]

117. Habibovic, P.; Li, J.; van der Valk, C.M.; Meijer, G.; Layrolle, P.; van Blitterswijk, C.A.; de Groot, K. Biological performance of uncoated and octacalcium phosphate-coated Ti6Al4V. *Biomaterials* **2005**, *26*, 23–36. [CrossRef]

118. Inkson, B.J. Scanning electron microscopy (SEM) and transmission electron microscopy (TEM) for materials characterization, Chapter 2. In *Materials Characterization Using Nondestructive Evaluation (NDE) Methods*, 1st ed.; Hübschen, G., Altpeter, I., Tschuncky, R., Herrmann, H.G., Eds.; Elsevier: Amsterdam, The Netherlands, 2016; Volume 43, pp. 17–43. [CrossRef]

119. Fujibayashi, S.; Neo, M.; Kim, H.M.; Kokubo, T.; Nakamura, T. Osteoinduction of porous bioactive titanium metal. *Biomaterials* **2004**, *25*, 443–450. [CrossRef]

120. Yeo, I.-S.L. Modifications of dental implant surfaces at the microand nano-level for enhanced osseointegration. *Materials* **2020**, *13*, 89. [CrossRef]

121. Katta, P.P.K.; Nalliyan, R. Corrosion resistance with self-healing behavior and biocompatibility of Ce incorporated niobium oxide coated 316L SS for orthopedic applications. *Surf. Coat. Technol.* **2019**, *375*, 715–726. [CrossRef]

122. Dorozhkin, S.V. Multiphasic calcium orthophosphate (CaPO4) bioceramics and their biomedical applications. *Ceram. Int.* **2016**, *42*, 6529–6554. [CrossRef]

123. Pichugin, V.F.; Surmenev, R.A.; Shesterikov, E.V.; Ryabtseva, M.A.; Eshenko, E.V.; Tverdokhlebov, S.I.; Prymak, O.; Epple, M. The preparation of calcium phosphate coatings on titanium and nickel-titanium by rf-magnetron-sputtered deposition: Composition, structure and micromechanical properties. *Surf. Coat. Technol.* **2008**, *202*, 3913–3920. [CrossRef]

124. Salou, L.; Hoornaert, A.; Louarn, G.; Layrolle, P. Enhanced osseointegration of titanium implants with nanostructured surfaces: An experimental study in rabbits. *Acta Biomater.* **2015**, *11*, 494–502. [CrossRef]

125. Sul, Y.-T.; Johansson, C.; Albrektsson, T. A novel in vivo method for quantifying the interfacial biochemical bond strength of bone implants. *J. R. Soc. Interface* **2009**, *7*, 81–90. [CrossRef]

126. Dinda, G.P.; Shin, J.; Mazumder, J. Pulsed laser deposition of hydroxyapatite thin films on Ti-6Al-4V: Effect of heat treatment on structure and properties. *Acta Biomater.* **2009**, *5*, 1821–1830. [CrossRef]

127. Łatka, L.; Pawlowski, L.; Chicot, D.; Pierlot, C.; Petit, F. Mechanical properties of suspension plasma sprayed hydroxyapatite coatings submitted to simulated body fluid. *Surf. Coat. Technol.* **2010**, *205*, 954–960. [CrossRef]

128. ISO 20502:2005(E). *Fine Ceramics (Advanced Ceramics, Advanced Technical Ceramics)—Determination of Adhesion of Ceramic Coatings by Scratch Testing*; ISO: Geneva, Switzerland, 2005; Available online: www.iso.org (accessed on 11 December 2020).

129. Gadow, R.; Killinger, A.; Stiegler, N. Hydroxyapatite coatings for biomedical applications deposited by different thermal spray techniques. *Surf. Coat. Technol.* **2010**, *205*, 1157–1164. [CrossRef]

130. Yi, J.; Song, L.; Liu, X.; Xiao, Y.; Wu, Y.; Chen, J.; Wu, F.; Gu, Z. Hydroxyapatite Coatings Deposited by Liquid Precursor Plasma Spraying: Controlled Dense and Porous Microstructures and Osteoblastic Cell Responses. *Biofabrication* **2010**, *2*, 045003. [CrossRef]

131. Dey, A.; Mukhopadhyay, A.K.; Gangadharan, S.; Sinha, M.K.; Basu, D.; Bandyopadhyay, N.R. Nanoindentation study of microplasma sprayed hydroxyapatite coating. *Ceram. Int.* **2009**, *35*, 2295–2304. [CrossRef]

132. Singh, G.; Singh, S.; Prakash, S. Surface characterization of plasma sprayed pure and reinforced hydroxyapatite coating on Ti6Al4V alloy. *Surf. Coat. Technol.* **2011**, *205*, 4814–4820. [CrossRef]

133. Gomes, P.S.; Botelho, C.; Lopes, M.A.; Santos, J.D.; Fernandes, M.H. Evaluation of human osteoblastic cell response to plasma-sprayed silicon-substituted hydroxyapatite coatings over titanium substrates. *J. Biomed. Mater. Res. Appl. Biomater.* **2010**, *94B*, 337–346. [CrossRef]

134. ISO 13779-2:2008. Implants for Surgery—Hydroxyapatite—Part 2: Coatings of Hydroxyapatite. Available online: http://www.iso.org/ (accessed on 11 December 2020).

135. Duta, L.; Mihailescu, N.; Popescu, A.C.; Luculescu, C.R.; Mihailescu, I.N.; Çetin, G.; Gunduz, O.; Oktar, F.N.; Popa, A.C.; Kuncser, A.; et al. Comparative physical, chemical and biological assessment of simple and titanium-doped ovine dentine-derived hydroxyapatite coatings fabricated by pulsed laser deposition. *Appl. Surf. Sci.* **2017**, *413*, 129–139. [CrossRef]

136. Edwards, J.T.; Brunski, J.B.; Higuchi, H.W. Mechanical and morphologic investigation of the tensile strength of a bone hydroxyapatite interface. *J. Biomed. Mater. Res.* **1997**, *36*, 454–468. [CrossRef]

137. Skripitz, R.; Aspenberg, P. Tensile bond between bone and titanium: A reappraisal of osseointegration. *Acta Orthop. Scand.* **1998**, *69*, 315–319. [CrossRef]

138. Shannon, F.J.; Cottrell, J.M.; Deng, X.-H.; Crowder, K.N.; Doty, S.B.; Avaltroni, M.J.; Warren, R.F.; Wright, T.M.; Schwartz, J. A novel surface treatment for porous metallic implants that improves the rate of bony ongrowth. *J. Biomed. Mater. Res. A* **2008**, *86*, 857–864. [CrossRef]

139. Schumacher, T.C.; Tushtev, K.; Wagner, U.; Becker, C.; Große Holthaus, M.; Hein, S.B.; Haack, J.; Engelhardt, C.H.M.; Khassawna, T.E.; Rezwan, K. A novel, hydroxyapatite-based screw-like device for anterior cruciate ligament (ACL) reconstructions. *Knee* **2017**, *24*, 933–939. [CrossRef]

140. Aparicioa, C.; Padrósb, A.; Gil, F.-J. In vivo evaluation of micro-rough and bioactive titanium dental implants using histometry and pull-out tests. *J. Mech. Behav. Biomed.* **2011**, *4*, 1672–1682. [CrossRef]

141. Nakamura, T.; Yamamuro, T.; Higashi, S.; Kokubo, T.; Itoo, S. A new glass-ceramic for bone replacement: Evaluation of its bonding to bone tissue. *J. Biomed. Mater. Res.* **1985**, *19*, 685–698. [CrossRef]

142. Pezeshki, P.; Lugowski, S.; Davies, J.E. Dissolution behavior of calcium phosphate nanocrystals deposited on titanium alloy surfaces. *J. Biomed. Mater. Res.* **2010**, *94A*, 660–666. [CrossRef]

143. Paital, S.R.; Dahotre, N.B. Calcium phosphate coatings for bio-implant applications: Materials, performancefactors, and methodologies. *Mater. Sci. Eng.* **2009**, *R66*, 1–70. [CrossRef]

MDPI
St. Alban-Anlage 66
4052 Basel
Switzerland
Tel. +41 61 683 77 34
Fax +41 61 302 89 18
www.mdpi.com

Coatings Editorial Office
E-mail: coatings@mdpi.com
www.mdpi.com/journal/coatings

www.ingramcontent.com/pod-product-compliance
Lightning Source LLC
LaVergne TN
LVHW071249170726
843515LV00006BA/1363